OTHER BOOKS BY DR. WINNIFRED B. CUTLER

Menopause: A Guide for Women and the Men Who Love Them, 1983, coauthored by C. R. Garcia, M.D., and D. A. Edwards, Ph.D.

The Medical Management of Menopause and Premenopause: Their Endocrinologic Basis, 1984, coauthored by C. R. Garcia, M.D.

Hysterectomy Before and After, 1988

Menopause: A Guide for Women and the Men Who Love Them, 1991, complete revised and expanded edition, coauthored by C. R. Garcia, M.D.

LOVE CYCLES

VILLARD BOOKS *New York 1991*

LOVE CYCLES

The Science of Intimacy

WINNIFRED B. CUTLER, Ph.D.

Published in the United States by Villard Books,
a division of Random House, Inc., New York, and simultaneously
in Canada by Random House of Canada Limited, Toronto.

Villard Books is a registered trademark of Random House, Inc.

All inquiries regarding motion picture, dramatic, translation, and other
related rights should be addressed to the author's representative:
Loretta Barrett Books, Inc.
121 West 27th Street
New York, NY 10001

Grateful acknowledgment is made to Harcourt Brace Jovanovich, Inc., and Faber and Faber Ltd. for permission to use excerpts from "Burnt Norton" and "Dry Salvages" from *Four Quartets* by T. S. Eliot. Published by Harcourt Brace Jovanovich, Inc., in the U.S.A. Published by Faber and Faber Ltd. in the UK under the title *Complete Poems 1909–1962*. Copyright 1943 by T. S. Eliot. Copyright renewed 1971 by Esme Valerie Eliot. Reprinted by permission of Harcourt Brace Jovanovich, Inc., and Faber and Faber Ltd.

Pamela Thompson Sinkler generated the art that was newly drawn for *Love Cycles*.

Grateful acknowledgment is expressed to HarperCollins for their permission to reproduce graphs that Pamela Sinkler drew in the book *Hysterectomy Before and After*. These include the graphs numbered 3-2, 3-6, 4-4, 7-1, 7-4, and 7-5.

We thank the editors of *Physiology and Behavior* for permission to reproduce Tables 1, 2, and 3 in Chapter 5, and Figure 1-3.

We thank the editor of *American Journal of Obstetrics and Gynecology* for permission to reproduce graphs 8-3 and 8-4.

We thank the editors of *Psychoneuroendocrinology* for permission to reproduce Figures 1-1 and 1-2.

We thank the editors of the *Journal of Human Biology* for permission to reproduce Figures 8-6, 8-7, 8-8, and 8-9.

Library of Congress Cataloging-in-Publication Data

Cutler, Winnifred Berg.
Love cycles: the science of intimacy/Winnifred B. Cutler.
p. cm.
Includes bibliographical references and index.
ISBN 0-679-40048-6
1. Sex. 2. Sex customs—United States. 3. Intimacy (Psychology).
I. Title.
HQ21.C86 1991
306.7′0943—dc20 91-50058

9 8 7 6 5 4 3 2
First Edition

Book design by *Debbie Glasserman*

This book is dedicated to
my children
Jodie Elizabeth Cutler
and
Evan Karl Cutler
and children everywhere
in hopes that
their love cycles
might be superior to ours

Man's curiosity searches past and future
And clings to that dimension. But to apprehend
The point of intersection of the timeless
With time, is an occupation for the saint— . . .

For most of us, there is only the unattended
Moment, the moment in and out of time.
The distraction fit, lost in a shaft of sunlight,

The wild thyme unseen, or the winter lightning
Or the waterfall, or music heard so deeply
That it is not heard at all, but you are the music
While the music lasts.

These are only hints and guesses
Hints followed by guesses; and the rest
Is prayer, observance, discipline, thought and action.
The hint half guessed, the gift half understood. . . .

—T. S. ELIOT, "THE DRY SALVAGES" *(The Four Quartets)*

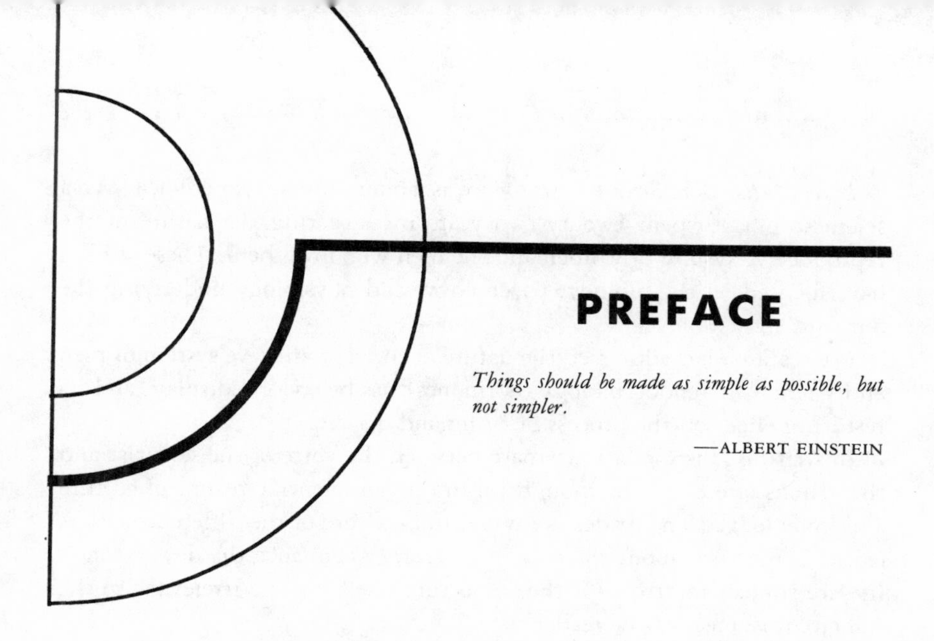

PREFACE

Things should be made as simple as possible, but not simpler.

—ALBERT EINSTEIN

Intimate relations between men and women are complex and constantly shifting. Love cycles are subject to internal physiological forces and are inevitably connected to cosmic forces beyond the awareness or control of the individual.

The primal reproductive rhythm of men and women is the menstrual cycle of women. Both women and men are subject to its beat. It helps to understand what the natural rhythm of a woman's body produces because the monthly ebb and flow of her sex hormones profoundly affects her own physical, mental, and spiritual rhythms and those of the people with whom she is intimate. The woman's hormonal cycle inevitably forms the subtle wave motion that carries her and those around her.

Imagine falling off a sailboat in rough waters and floating on the waves until they bring you to shore. Compare this action of harmonious behavior in the context of rough waves with the behavior of one who fights the current and panics.

Wise people understand what knowledgeable sailors know: They use rough waves to enhance rather than to consume their power. In the love cycles of the fertile years the waves are the hormonal tides that form a woman's menstrual cycle. The timing of the tides is paced by the cosmos in what I call the cosmic dance.

Love Cycles: The Science of Intimacy is about human experience. As a scientist I have spent twenty-two years investigating the nature of the reproductive system of women and the men who love them. These studies have focused on the biology, psychology, and physiology underlying the outward show of love.

My research has addressed the nature of the reproductive system of men and women, its endocrinology (hormones), its behavioral display, and its resulting effect on the process of aging and disease.

In writing *Love Cycles* I alternate between the voice of the scientist and the visions offered to the thoughtful in the nonscientific realms of culture and knowledge. The reader is invited to be aware of the difference. Facts exist. Conjecture about the way to use facts to enhance the days of one's life are subject to error. Or the conjecture itself may be irrelevant to the life circumstances of the reader.

My collaborative studies led to a series of discoveries published in the scientific journals between 1979 and 1987. These are described in the following chapters: 1, "A Time to Embrace"; 2, "The Life Cycle in the Love Cycle"; 5, "Fertility Cycles of Women and Men"; 6, "Pheromones: Male and Female Essences"; and 8, "Cosmic Cycles—Lunar Events." Just as those discoveries were built upon the work of scientists who went before me, investigators continue to build knowledge through their own experiments, and some of their work can use parts of the foundation laid by my own.

The study of the scholarly work of others led me to write four books before this one. Those books—on menopause, hysterectomy, and what women could do to promote their own wellness—were the product of my systematic analysis of more than thirty-five hundred individual scientific experiments published in the biomedical literature of the 1970s and '80s. My studies convinced me that I could make a contribution to the health care of women by writing books for them. Lisa Biello, a medical-book acquisitions editor, asked me to write a medical textbook for physicians who minister to women. Several were written in close collaboration with Celso R. Garcia, M.D., professor of gynecology at the University of Pennsylvania School of Medicine.

Collaborative work is wonderful. It can expand one's own perceptions. At times I found myself in easy agreement with my colleagues. At other times my own conclusions about the meaning and value of the published studies was at variance with theirs.

The scientific process allows for diversity of conclusion. In *Love Cycles: The Science of Intimacy* I set out my own thoughts, based on science, about the relevance of love cycles as a critical and controlling force in the lives of men and women. Stephanie Young, health and fitness editor at *Glamour* magazine, provided critical feedback on five of the chapters.

Learning to ride the waves of biological cycles brings us into harmony with the cosmic dance. And this harmony promotes our survival as a species. In order to learn to dance, one first needs to know the steps. Only then can a great deal of practice enhance the gracefulness of the motion. Particularly in the reproductive dance between men and women, graceful motion refines the experience. For many of us love is lovelier after we have had some practice, because we have begun to learn some of the rhythms. We understand what it takes and we can be more graceful about it.

Once they begin to penetrate the available knowledge, women and the men who love them will have the tools to enhance their own cosmic dance.

As we learn the natural rhythms of our bodies as well as those of our intimate partners, we gain the power that comes from knowledge. Finding the delicate balance between stability and stimulation can become a kind of "high art" in the living of one's life. Our own biological rhythms are affected by the people with whom we share space and intimate connections. And these dynamic combinations between partners form the basis of harmony or disharmony in our human love cycles.

With perceptions made sophisticated by our learning the biological realities, we can know when to "go with the flow" and when to apply energy to overcome distress. We know when to push forward and when to pull back.

The people with whom we share space, from the closest intimate partner to the people standing around us as we move through the events of the day, are all similarly bound into their own inevitable ebbs and flows. If you imagine yourself floating in a canoe that is easily toppled over, you will know the need for gentleness as you move through space. Gentleness and balance then provide the protection against capsizing. The principle that works in the water when we are inside a vulnerable canoe also applies in our human relationships. Since intimate relationships expose our vulnerabilities, those who practice gentility are rewarded by a greater internal stability.

Acknowledging and understanding the tremendous power of various love cycles on our hormones and the reactions that follow allow us to ride the waves instead of fighting them.

Those who understand love cycles have a renewed opportunity to enhance their intimate relationships. By understanding the nature of the science of intimacy, women and men are better able to get more of what we all want—a more loving and loved existence. Ultimately the science of intimacy points to the poetry and magic of human experience.

CONTENTS

LIST OF ILLUSTRATIONS

LOVE CYCLES

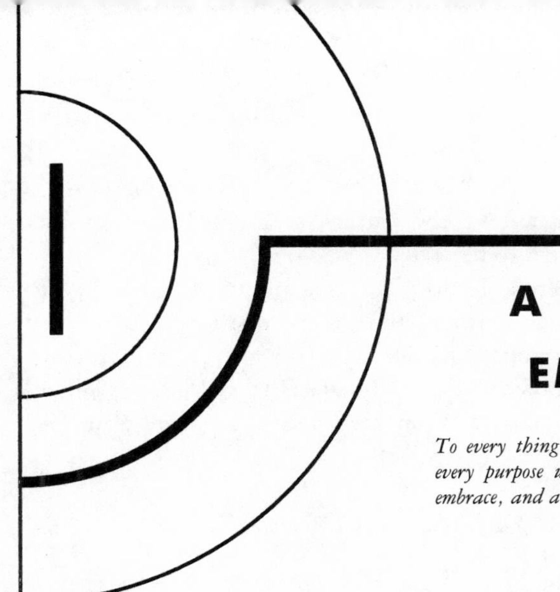

A TIME TO EMBRACE

To every thing there is a season, and a time to every purpose under the heaven: . . . a time to embrace, and a time to refrain from embracing.

—ECCLESIASTES *3: 1, 5*

In 1974, when I began graduate school, I chose to center my focus on the biology of human reproduction, in large part because I wanted to understand myself better. I was thirty, had been married for twelve years, and had two school-age children. The oral contraceptive had liberated women from the "contraceptive failure" so common to the other methods. With that early freedom women's liberationists had begun to proclaim the right to sexual pleasures that were unheard of while pregnancy loomed as a possibility. By 1974 talking about sex, fertility, and "reproductive behavior" was acceptable. Even so, while talk was common, substantive knowledge was not. I wanted to know more about what comprises female sexuality and reproduction. I sensed that there were mysteries to be explored, that there were realms of knowledge to discover, and that these would alter my own experience of love and life. The textbooks did not address these mysteries, but I believed they existed and could be understood. As a graduate student that was my job: to discover, and to make a contribution to knowledge.

For the next fifteen years, as I studied anatomy and biology, I gathered the evidence and conducted the experiments that produced the first proof: proof that the frequency of sexual behavior influences a woman's hormonal pattern, her general health, and her fertility. The body of this research hints at the enormous power of the pattern of biological forces that provoke a man and a woman to be sexual.

As my research uncovered, healthy sex is a matter of timing. The frequency of sexual behavior either "tunes" or "disrupts" the natural rhythm of a woman's hormonal functioning. This chapter explores just what that timing is and how it specifically affects the bodies of women. The repercussions of this discovery on men and women are examined throughout this book.

Rx for Women: Regular Sex—at least once a week; and until this is available, remain celibate

While you'll probably never get a prescription like this from a doctor, you should know that it is actually sound advice. Regular sex within a monogamous, loving relationship is good for you. Your sexual behavior can have a significant impact on your hormones, and in consequence on the way your body and mind function. In the broadest sense, your social life profoundly affects your biological life; your sexual choices influence the quality of your health.

These discoveries suggest a new understanding of the relationship between the sexes. In the pages that follow, I bring the clinical and scientific data into the realm of everyday life by showing the role of sexual intimacy in human biology. It is my purpose to unravel the mysterious design that underlies human sexuality—what I call the love cycle.

SEXUAL HEALTH—A MATTER OF TIMING

My scientific discoveries may be new, but their poetic description was recorded by the ancients. In the Bible, even before the wisdom of Solomon taught that there is "a time to embrace," the Ten Commandments offered some very specific rules. The commandment "Remember the Sabbath Day, to keep it holy. Six days shalt thou labour and do all thy work. But the seventh day is the Sabbath . . . in it thou shalt not do any work." is reported in the Book of Exodus (20). There, we find the basic human rhythm of the seven-day week. As we will see, the seven-day pattern forms the basis for human sexual cycles. The same point appears again, in the Book of Leviticus, a composite body of rules for sexual and social behavior. Here it advises that a man should tend to his wife sexually at the beginning of the Sabbath *each week* without fail. More frequent tending was fine—weekly was a minimum. According to Leviticus 15 the only exceptions were those days when a woman was menstruating or "with issue." With this advice men and women of those times could follow the dictum to "Be fruitful and multiply." Nothing has changed in the last three thousand years. My research with more than seven hundred women confirms the value of weekly sexual contact, a weekly love cycle. Regular weekly sex is still good for women, helping to ensure hormonal levels that promote good health, retard aging, and increase fertility. With a period of abstention when the menstrual blood is flowing, regular sex is ideal.

AVOIDING SEX DURING MENSTRUAL BLOOD FLOW

My studies indicate that abstention from sex when the blood is flowing is optimal. During menstruation, sex is not necessary for the beneficial effects described above and, according to several preliminary indications, may actually be dangerous to a woman's health. I believe sex during menses may exacerbate the vulnerability to endometriosis, and I explain why in Chapter 3 (see page 70). In Chapter 9 (page 248) I describe how sex during menses may increase the tendency for excessively heavy menstrual bleeding during the premenopausal stage and predispose women to a medical prescription for hysterectomy.

CYCLE LENGTH AND FERTILITY

There is an optimal menstrual-cycle length for assuring fertility. The discovery came out of Switzerland.

In the mid-1960s Dr. Rudi F. Vollman, a Swiss gynecologist, sat down with his wife, Emmie, to enlist her help with the menstrual records of hundreds of his patients. He had been getting his patients to record their menstrual onsets and their daily temperatures in hopes of learning more about the menstrual-cycle patterns and fertility. Mrs. Vollman spent months, under his direction, recording and plotting the data. She looked to see if successful pregnancy might be related to particular kinds of menstrual patterns. She pored through several thousand menstrual records, working by hand, long before the days of home computers. The results became the basis of Dr. Vollman's highly respected textbook on the menstrual cycle of women, which brought stature to his medical career. From analyzing the menstrual records of thousands of women, Dr. Vollman and Mrs. Vollman (whose name never appeared as coauthor) discovered that women whose menstrual-cycle lengths approach 29.5 days are virtually guaranteed to be fertile that cycle. Women with shorter or longer cycles are less likely to be as fertile. Three days shorter or longer on either side of 29.5 "perfection" (26.5 and 32.5, respectively) still qualify as a fertile cycle. Beyond these boundaries fertility drops off. The farther from "perfect" cycle length, the less fertile you are likely to be.

MENSTRUAL AND MOON CYCLES

The menstrual cycle is the time between the onset of one menstrual flow, and the last moment before the onset of the next. Among fertile women the average cycle length is 29.5 days. During these 29.5 days a predictable and repeating series of physical events occur in the ovaries, the uterus, the blood circulation, and the brain. An egg proceeds (ovulates) from the ovary into the fallopian tube. The uterus builds a nest to house the egg should the egg be fertilized that month. Sex hormones rise. And sex hormones fall. This exquisite hormonal symphony of woman is detailed in Chapter 3. For now what is particularly relevant is the stability of the pattern—once in every 29.5 (± 3) days.

The moon also forms a cycle—from one full moon to the next. Its cycle length of 29.5 days coincides with the optimum cycle length of fertile women. As we'll see in Chapter 8, the moon cycle and the menstrual cycle of fertile women are closely correlated.

Although seemingly the end event of the cycle, the onset of menstruation is counted as the first day (Day 1) of the cycle. At the onset of menstruation the lining of the uterus begins a new cycle of nest development, eliminating the old nest and setting the stage to build a new one.

Many women use calendars to keep track of their menstrual cycles. Month after month a calendar can reflect the predictable pattern of a fertile woman, with the onset of menstruation approximately every 29.5 days. While this kind of recording is certainly practical and captures the basic cyclical pattern, marking the day does not fully reflect on the finely wrought timing of the cycle. With a 29.5-day pattern, if the onset of menstruation occurs in the morning one month, it will occur a half day later the next, therefore in the evening. Knowing your timing precisely lets you see how close to "perfection" your own cycle is. According to the Vollmans' work, if your cycle is like clockwork, occurring every 29.5 days, you are 98 percent likely to conceive if a healthy sperm arrives in time to await the ovulation of your egg.

FERTILITY IS A REFLECTION OF GOOD HEALTH

Even if you do not want to get pregnant, being fertile is beneficial in itself. A fertile-type cycle is a sign of hormonal health and well-being. It is a biological statement that the body is functioning well. Although the physical details of reproductive-system anatomy and physiology are covered in Chapter 3, the philosophy can be addressed here.

From the perspective of evolution, think of the continuity of life: from an individual to her offspring. First the woman must live and be well. Then, maybe, she will be strong enough to bear the biological cost of reproduction. Survival of the individual is necessary before there can be survival of the species. When survival is at stake, automatic mechanisms tend to shut down all nonessential systems. The process is comparable to an austerity program that goes into effect when a company is in danger of "going under": The frills are temporarily eliminated. In the case of the human body, the shutdown conserves the energy needed to heal, to survive, and to maintain the individual.

Reproduction is costly. It consumes vast amounts of a woman's energy. In the short run the process is not necessary. For these reasons a woman's reproduction is usually the most vulnerable to temporary system shutdown when she is under stress. Her heart and her lungs do not stop functioning, but the part that is temporarily expendable does. Women under stress may even stop menstruating. Olympic-caliber athletes, college freshmen, and the newly married, for example, often develop temporary menstrual disturbances. The most stunning example of this mechanism came out of Hitler's Germany. Reports have consistently stated that women placed in the horror of concentration camps immediately stopped menstruating. They became infertile. The complex defense mechanism of the body helped to conserve their chances for survival.

Reproductive biology teaches the lesson that shutting down a nonessential system is the body's automatic way to save energy. Similarly, when all systems are functioning optimally, they reflect a condition of vibrant health. The more fertile the cycle, the more precisely timed the internal synchrony of hormonal ebbs and flows, the healthier the physical system. The bones will be stronger, the cardiovascular health better, and the general joie de vivre higher. Even for those women who want to avoid pregnancy, a fertile hormonal system enhances the quality of life.

SETTING THE CYCLE

You might think that there is not a whole lot you can do to adjust your cycle. But there is. Beginning in 1974, when I was a graduate student and teaching fellow, I sent teams of undergraduate biology students armed with questionnaires, clipboards, and stethoscopes to solicit participation in a series of studies that varied as they progressed. Each study was designed to uncover mysteries of the menstrual cycle and the sexual-behavior patterns that might relate to it. Somewhat like saleswomen, my students were assigned exclusive territories of undergraduate dormitories.

In my first pilot study of sixty women, age eighteen to twenty-two, I found that those who had sex at least once a week (the "weekly" women) all had menstrual cycles that approached the classic fertile length (see Figure 1-1). Women who had sex less frequently, in no discernible pattern (the "sporadic" women), had more variable cycles. And among women who did not have any sex (the "never" women) during the fourteen weeks of my study, cycle lengths varied from short to long. These celibate women were better off than their sporadic counterparts because proportionately more had cycles within the 29.5-day, fertile-type range. A stable, regular pattern of "weekly" sexual behavior seemed to promote the optimal menstrual-cycle length.

This pattern (weekly sex = fertile-length menstrual cycles) was replicated by a larger study of 248 women the next year. Figure 1-2 shows a composite analysis of the 248 menstrual cycle lengths, categorized in terms of sexual frequency—weekly, sporadic, and never.[1]

Regular weekly sex with a man was associated with a high likelihood of menstrual cycles of close to 29.5 days. A few of those women had longer cycles; none had shorter ones. Statistically we had demonstrated that both sporadic and celibate sexual behavior were associated with abnormal cycles. The same patterns for sporadic and celibate women turned up here as in the pilot study. This replication began to convince me. A second replication study, seven years later, proved it. The same results emerged.

---

[1] The privacy of these very sensitive sexual data of research subjects was carefully protected through a complex system of code names and other measures.

The women who were active on a weekly basis are fewer in number in this study because sexually bonded women may have been more likely to use the birth control pill or IUDs and were not qualified to participate in the study. These contraceptive methods artificially alter menstrual cycles and hormones, and I was interested in discovering elements of the natural menstrual-cycle pattern. Despite their relatively smaller group size, the weekly-active women showed the same statistical relationship as in the pilot study.

FIGURE 1-1

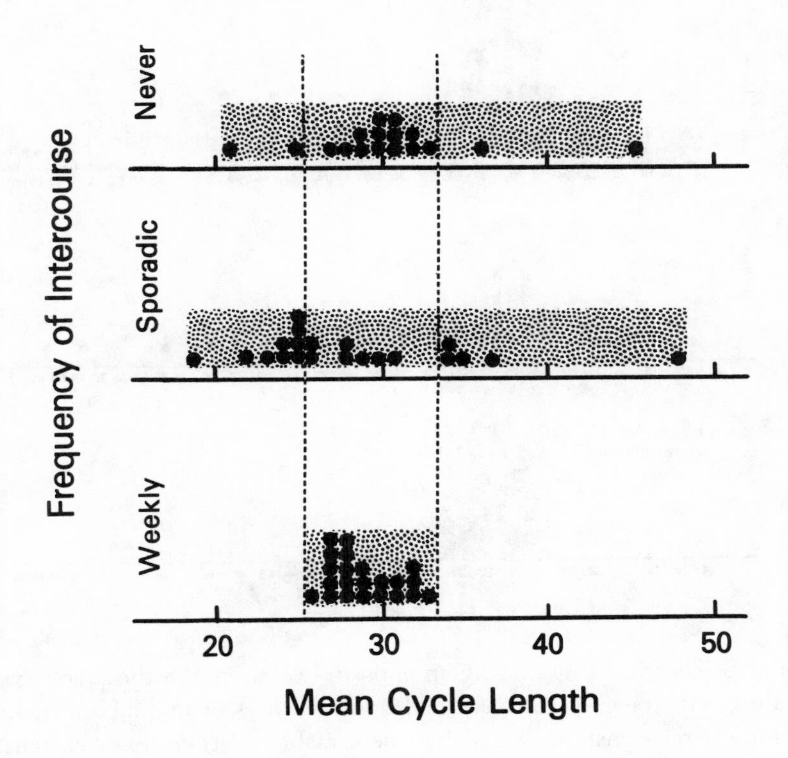

Each dot represents one woman's average cycle length. Look first at the twenty dots in the *weekly* category, the bottom grouping. You will see that every one of these women had an average cycle length that was in the fertile range (the vertical lines bordering twenty-six and thirty-three days). If you look above this at the *sporadic* group, you will see that more than half of the twenty sporadic women had cycle lengths that were outside this range, either shorter or longer. Then, looking above the *sporadics* to the *nevers*, you will find that fewer aberrantly short and aberrantly long cycles occurred in this group.

If once a week was good, would twice a week be better? It seems so, as you can see from Figure 1-2. None of the women who had sex at least twice a week had long cycles. Even though my intuition seemed right, more data would need to be gathered to test the question statistically.

We see that sporadic behavior is associated with certain detrimental effects that influence health and fertility. But what if regular sex was not available? I wasn't studying rats or monkeys, whose social interaction

FIGURE 1-2

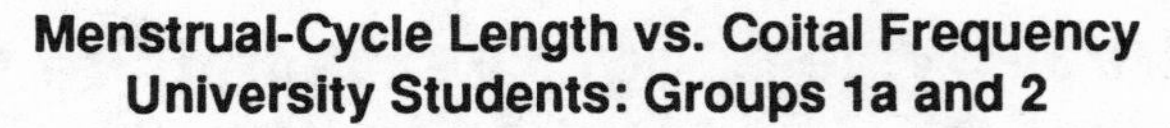

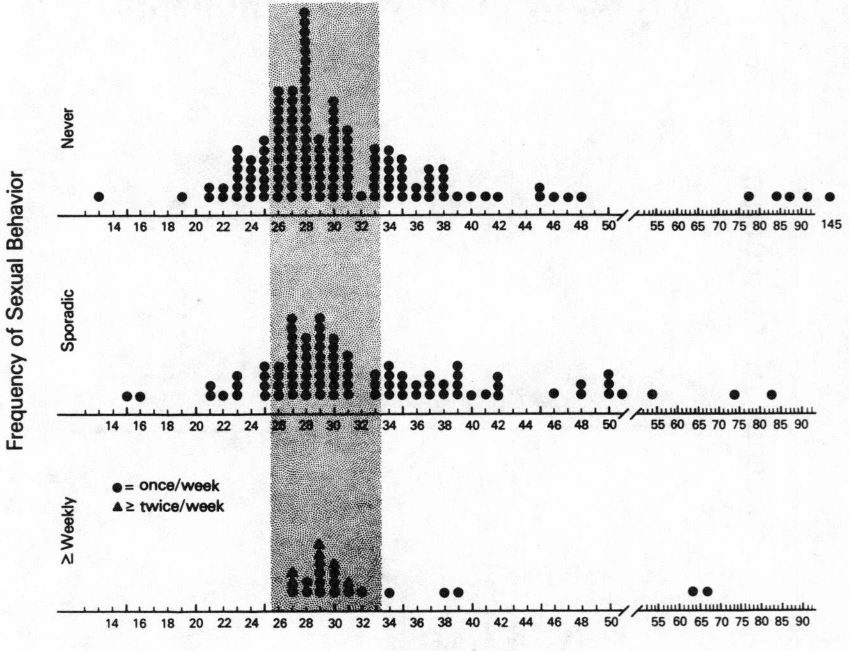

could be controlled by placing an opposite-sex animal in the cage. I was dealing with people, engaged in the complexities of finding and maintaining a relationship. If a stable, dependable relationship is currently unavailable, my studies suggest that celibacy is a wise choice.

THE POSITIVE SIDE OF CELIBACY, THE NEGATIVE SIDE OF PROMISCUITY

In addition to the potential for disease that accompanies sexual promiscuity, positive benefits emerge from abstinence. A truly intimate relationship requires knowing and being known. It requires ease and dependability as well as regular access to your partner. If regular sex is unavailable, my data suggest that it is better to be celibate. Celibacy seems to reduce the incidence of irregular menstrual cycles. For postadolescent women irregular menstrual cycles often indicate deficiencies in hormones that normally promote mental and physical well-being.

The sporadic pattern—concentrated bursts of sexual activity (sexual "feasts") interspersed with periods of abstinence (sexual "famine")—was highly correlated with subnormal estrogen levels and subfertile cycle patterns.

Sporadic sexual behavior in women in their midtwenties is associated with blood levels of estrogen as low as those that normally appear about a year before menopause, approximately age forty-nine. Not only do such low estrogen levels reduce fertility, they can also have serious health consequences. Diminished levels of estrogen can lead to loss of bone, gradually progressing to osteoporosis, and can lead to elevated cholesterol levels with increased risk of cardiovascular disease. Female athletes often stop menstruating normally. As their menstrual cycles become abnormal, these women lose bone mass—even in their early twenties. As with the female athletes, so it appears to be with sexually sporadic women.

JUST WHAT IS SPORADIC SEX?

Sporadic sex is *defined* in my scientific studies in a very specific way, and the definition underlies the principles that guide "a time to embrace." In the fourteen weeks of each study, *sporadic sex* meant having sexual intercourse in a pattern that failed to meet the "weekly" criterion. If a woman's records did not show sex *at least once* in each nonmenstruating week (i.e., once in each seven-day span) she failed to meet the weekly criterion. Regular sex seems to work in the body the way getting a weekly paycheck works in the household. Most people need their paycheck; they count on it. Skip one and they are in trouble. Bonuses are always welcome, but only on top of the dependability of the regular paycheck. A woman's body seems to need the regularity of the seven-day "receipt" of sexual contact in much the same way that the household needs to receive its weekly pay. Some women can handle sporadic activity. Some households can handle financial feast and famine.

HOW AND WHY RESTRAINT PAYS

Biology is teaching us an important lesson: Sporadic sexual activity is not good for a woman's endocrine system. Women who put sex before friendship, those who go to bed with a new man to see if something will come of it before they explore his character or his values, are more likely to have a sporadic pattern. Chances are such behavior will lead to lack of stability, frequent rejection, and the juggling of more than one man at a time in a doomed attempt to be certain of company. Like the problem of squeezing a handful of sand and losing most of it between the fingers, the behavior of "taking a chance" usually leads to islands of emptiness interspersed with brief periods of abundance. Most men don't want to form a stable bond with a woman involved with multiple partners. And vice versa.

Biology seems to favor the woman who is slow to consider sexual congress; the woman who waits until she really knows her partner. Consider the facts: A fertile-length menstrual cycle is most likely to arise when sexual encounters occur regularly. Regular sex occurs predominantly in relationships, not one-night stands. Multiple partners lead to fractionation of energy. Getting to know a man requires one to focus on him. The process allows a woman to assess his stability, dependability, and capacity for commitment. Once a woman is sure of the man and the friendship, then from the biological perspective it is appropriate to expand the relationship sexually. Until a woman meets such a man and gets to know him, celibacy preserves her hormonal system and her health. It is hard to be open to exploring a new man when one is intimately involved with another. I believe that dishonesty transmits a negative aura and that it drives away the really good candidate for monogamy. Restraint pays! What mother and grandmother preached about "waiting" has renewed validity—both in light of my studies and in light of the very real threat of sexually transmitted diseases.

Men and women who are currently celibate might think of a time without a partner as an intermission, a time of rest, renewal, and personal growth. Celibacy can offer a peaceful interlude in your life. And you will not be risking the potential harm that sporadic behavior promotes.

TIMING IS IMPORTANT FOR MEN TOO

Timing affects the male hormone testosterone. Excitement and desire raise the testosterone level. In preliminary experiments conducted by investigators at Stanford University, where men were put on deprivation schedules—in effect a "sex diet"—sexual potency increased at first. The volume of ejaculate increased as well. Past a certain critical period, however, delaying fulfillment is not helpful. If a man is kept waiting too long between encounters, potency declines.

Just what that critical period is has not been determined, but there does seem to be an optimal timing pattern that maximizes sexual health in men. Chapter 4 provides the details on the hormones and sexual physiology of men.

THE MORE SUBTLE IMPLICATIONS OF THE BENEFITS OF RESTRAINT

In my studies half the sporadic women showed fertile cycles, half did not. I wondered if the behavior differed between those sporadic women who were fertile and those who showed a subfertile cycle.[2] Did more sporadic sex make a woman more fertile? Or less fertile? Just how sporadic is sporadic when it comes to influencing a woman's cycle? Was there some magic number of monthly encounters?

What I discovered led me to conclude that if you're going to be sporadically active, it's better to have less sex, rather than more sex. Women who had a lot of sex in a sporadic pattern had a good chance of having shorter or longer (nonfertile) cycles. The "feast or famine" approach to sexual behavior seems to disrupt the body's endocrine balance. The sporadic women who feasted the most were the ones who got into hormonal hot water.

When I asked why these women had a feast-or-famine pattern, their responses tended to be "We just broke up," "There's no man around," "He travels a lot." These women weren't sporadic by choice. They simply did not have the kind of relationship that fosters weekly sex.

During the thirteen years in which I was presenting these discoveries to my scientific colleagues, I was frequently asked to speculate on whether hormones cause the pattern of sex behavior or whether the sex behavior was causing and promoting the hormonal environment of the woman. Merely demonstrating an association does not prove that one thing causes another. The question is, Does the estrogen promote the sex or does the sex promote the estrogen? Or is some third factor affecting both? Ideally in science one would want to tell people what to change in order to control their cycles. My experiments were not designed to manipulate behavior. After reading these studies, my scientific colleagues have related so many anecdotes indicating that at least some women and men are convinced they can manipulate their cycles by deliberately changing from celibate, to sporadic, to weekly and back again. Ideally studies should be done to follow up on these anecdotes.

As I considered the fact that sporadic women tended to say they wished there was more stability in their sex life, I came to understand that women were not deliberately choosing to be sporadic. I looked more closely at my data and discovered that the pattern of behavior would predict the state of the hormonal environment.

[2] I distinguish *subfertile,* meaning "*less* fertile" from in*fertile,* meaning "not fertile."

Weekly sex always produced a high likelihood of a fertile-type hormonal pattern. Likewise, sporadic sex always produced a high likelihood of a subfertile endocrine pattern. But the reverse was not true. That is, a fertile-type cycle length did not predict the sex behavior of the woman. Just looking at Figures 1-1 and 1-2 shows that. Take a look at the Sporadic group: Half of them are outside and half inside the fertile-type cycle range. Take a look at the Weekly group: Nobody is a short cycler; few are long cyclers. In other words, if you know that the behavior is weekly, you know what the cycle length will be. If you know that the behavior is sporadic, you cannot predict the cycle; it has a fifty-fifty chance of being subfertile in length.

Since the behavior predicts the physiology, but the physiology does not predict the behavior, I conclude that regular weekly sexual behavior promotes fertility.

Box 1-1: How to Measure Fertility by Degrees

Menstrual-cycle length is only one indicator of a fertile cycle. A more accurate method tracks the basal body temperature cycle *(BBT)*. Basal body temperature is the body's lowest temperature, and it can be measured, by mouth, immediately after waking up in the morning but before getting out of bed. It is a more refined measure, because it reveals what is going on inside the cycle itself. Your BBT typically varies throughout the menstrual cycle due to the shifting of the hormone levels.

From the onset of menstruation (Day 1 of the cycle) until ovulation (generally Day 14, 15, 16, or 17) the levels of one of the sex hormones, progesterone, are usually undetectable in most blood tests. The basal body temperature is also low—usually about 97 to 97.5 degrees. The fertile woman's basal temperature drops slightly just before ovulation and then rises steeply (up by a full degree or more) about the same time that she ovulates. Once up, the temperature tends to stay up until the cycle ends and the next menstruation begins. These elevations in temperature coincide with the presence of detectable progesterone levels in the blood. After ovulation, blood tests will continue to reveal progesterone levels until the next menstruation starts.

The *luteal phase* of the menstrual cycle is the term for this fourteen- to sixteen-day stage after ovulation and before menstruation, when temperature is elevated and progesterone levels are high. It is named after the *corpus luteum,* Latin for the "yellow body" that early anatomists saw when they studied the ovary during the two weeks after ovulation and before menstruation. The corpus luteum manufactures the progesterone as well as other sex hormones and is formed from the ovarian tissue that had surrounded the recently ovulated egg. After

(*continued*)

ovulation, although the egg has left the ovary, the remaining tissue re-forms into what is now transformed into a yellow body or corpus luteum. Yellow is the color of the cholesterol with which it is filled. This cholesterol is converted into our sex hormones. The luteal phase describes the time during each fertile cycle when the corpus luteum exists. In a fertile cycle one of the two ovaries will produce the corpus luteum, but only during the luteal phase, during the two weeks before menses. The corpus luteum appears after ovulation, grows to fill half the ovary, shrinks, and then disintegrates. As it grows, the blood levels of progesterone and estrogen rise. As it shrinks, the hormone levels fall.

Once there has been an *ovulation* (an egg moving out of the ovary and then into the fallopian tube) and a *fertilization* (sperm uniting with egg in the fallopian tube), the fertilized egg begins a three-day journey down the fallopian tube toward the womb, or uterus. In a normal fertile cycle that will lead to a pregnancy, as the luteal phase progresses, the *uterine lining,* or endometrium, grows thicker and thicker. Ovaries secrete the hormones responsible for this "thickening of the nest." As the nest thickens, the likelihood of the survival of the fertilized egg increases. If a fertilized egg enters the womb and can embed within a ready-and-waiting thick, plush lining, there is a very good chance that a pregnancy—and a baby—will result. If, on the other hand, the nest is thin, the egg will not embed, the pregnancy will self-destruct, and menstruation will ensue.

Proper thickening of the nest not only requires enough hormones but sufficient time for them to do their work. That is why the luteal-phase *length* is so important to fertility. It is probably the case that a woman could become pregnant on a ten-day luteal phase, but a twelve-day or longer luteal phase significantly increases the odds. The charting of the basal body temperature functions as a barometer. The chart reveals the number of days that the nest is being built. That number of days corresponds to the number of days that the BBT stays elevated, after ovulation and before menstruation.

WHY THE TIMING OF THE LUTEAL PHASE IS CRITICAL

Research has shown that anything shorter than a twelve-day luteal phase (the span from ovulation to the start of menstruation) tends to be inadequate for ensuring fertility. A woman may be ovulating monthly, may even be conceiving (sperm and egg uniting in the fallopian tube), yet still be infertile due to a timing problem—a "luteal-phase deficiency."

The luteal-phase length is just as important for the woman who does not want to get pregnant. When the luteal phase is too short, hormonal imbalances tend to follow. Fibrocystic breast disease as well as cancers of the reproductive system are much more likely to occur under these conditions.

HOW WEEKLY SEX AFFORDS PROTECTION

Once I had learned that regular weekly sex was associated with fertile-type cycle lengths while sporadic and celibate behavior often were not, it was time to look deeper. My mentor, Dr. Celso Ramon Garcia, was interested in the initial relationships I had discovered and agreed to guide my graduate studies. He suggested we look next at what the basal body temperature charts would reveal about sexual behavior and endocrine patterns. Dr. Garcia was constantly sought after by infertile women wanting help because of his eminence and great skills as an infertility surgeon and professor of gynecology at the University of Pennsylvania School of Medicine. Having been involved conducting the trials of the first oral contraceptives in Puerto Rico in the 1950s, he had an extraordinary fund of knowledge about hormones, fertility, and fertility control. As a scholar he encouraged his patients to keep careful BBT records and to record when they had coitus within the days of their menstrual cycles. He had a rare and accurate body of data that women with fertility problems had provided. He helped me to design a study to gather similar data of healthy women without fertility problems. Both sets of data turned out to tell the same story.

In our studies basal body temperature records of healthy twenty-three-year-old women were gathered. Dr. Garcia's records contained somewhat older women ranging from their early twenties to early forties. Both sets of data revealed that while most women did ovulate, many showed shortened luteal phases, time spans of elevated BBT that were too short for healthy hormone balance. In fact the most common problem in women who menstruated regularly was not a failure to ovulate but a failure to have an adequate-length luteal phase.

In the study comparing cycle lengths, sexual-behavior frequency, and basal temperatures, a pattern similar to my previous studies emerged.

Weekly sex was again associated with fertile-type cycle lengths (see Figure 3). True to form, women who were classified as sporadic or celibate had abnormally shorter and longer cycle lengths. Next we saw that weekly sex is also associated with fertile BBT patterns.

I was struck by the fact that there were very few deficiently short luteal phases in any of the women who had weekly sex. Almost 90 percent of women who had regular sex had fertile BBT rhythms. In contrast only half of the sporadic and the celibate women showed normal BBT patterns.

The shortened luteal phase, as measured by basal temperature graphs, was a common event in apparently healthy yet *sporadic* and *celibate* women of fertile age. In other words most celibate and sporadic women did

FIGURE 1-3

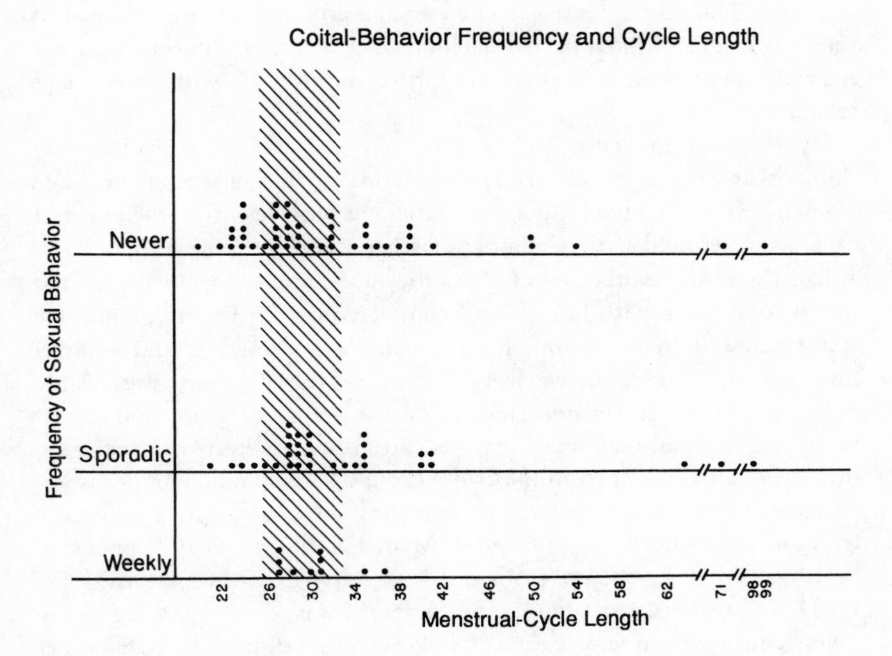

ovulate, but half were subfertile with a shortened luteal-phase length. Their hormonal timing was off. One other fact from the BBT graphs emerged: Short cycles were almost never normal; long cycles often (about half the time) were.

These data suggest a simple conclusion: Weekly sexual behavior promotes a fertile cycle, while less than weekly tends to disrupt it. Love cycles need to beat to a weekly rhythm.

THE ESTROGEN CONNECTION

Since basal body temperature reflects shifting hormone levels, it was interesting to examine women's actual hormone levels. In two of my studies, one at the University of Pennsylvania, the other at Stanford University, women gave blood along with their records. Levels of testosterone and progesterone did not differ among the groups, so the timing of intercourse did not seem to be associated with these hormones. (Women circulate about one thirtieth as much testosterone as men.) *The timing did relate to estrogen levels. The estrogen was nearly twice as high for women who had weekly sex than for women who did not.*

Women who have weekly sex apparently respond by secreting more estrogen. This fact emerged in the two age groups I studied, women in their twenties and those in their late forties. I went to California to gather the data about women in their late forties and to work with an eminent scholar.

By 1977 my graduate studies in biology and gynecology had already shown that weekly sex affected the endocrine rhythm of reproductive-age women. As I presented these results at national scientific meetings, I became acquainted with a number of other scientists. One of these, Dr. Julian Davidson, a professor of physiology at Stanford University, invited me to come work with him when I completed my Ph.D. in biology. He was interested in expanding his research focus from rats and squirrel monkeys and suggested we work together. His laboratory was highly respected for its neuroendocrine studies of sexual behavior and sexual hormones in small mammals. He was particularly interested in analyzing impotence in aging men in hopes of curing a very common and exasperating problem. I was interested in extending my knowledge. I wanted to learn more about hormones and I wanted to know what happens to healthy women as they get older. I was thirty-five when I earned my Ph.D. and realized I would soon be "older." I was ready—and eager—to investigate aging in female sexuality. Personally I didn't think that older women were going to be particularly interesting. I planned to study the end of reproductive life in order to expand my understanding of the years of reproduction. I had every intention of spending a little time and then returning to the interesting topic of young women. I turned out to be wrong. The mature woman had more fascination to offer a scientist than I ever expected to find.

Dr. Davidson and I agreed to work together to develop a research proposal and to seek funding from the National Institute of Aging. He would develop and write about the aging male: I would develop and write about the aging female in what was to become a long-term follow-up study of the sexual behavior and hormones as people age. We were awarded a five-year grant.

After setting up the Stanford Menopause Study, I became aware that estrogen levels normally decline as women approach menopause, but the link between sex and estrogen holds true in both age groups. Even as a woman approaches menopause, and her estrogen levels are declining, the decline is slower if she has weekly sex behavior. Those women who had weekly sex showed significantly higher levels of estrogen and fewer hot flashes than their less sexually active contemporaries.

Endocrinologists (hormone specialists) expect to see lower estrogen levels in older women. What was particularly startling was the discovery that half of sporadically active young women have estrogen levels as low as those of menopausal women. Although, in the young group, the average estrogen level in celibate women is the same as in sporadic women, there is an important hormonal difference between these two groups. The average level in sporadic women is composed of extremely high and extremely low estrogen values. The blood levels for celibate women were more uniform and exhibited no such extremes. In other words celibate women showed moderately low estrogen, but none of them were extremely low.

This is why sporadic patterns of sexual activity are so bad. This pattern has the power to disrupt a woman's endocrine functioning. Sporadic activity can cause estrogen levels to plummet. An apparently fertile woman in her prime reproductive years may be walking around with a menopausal level of estrogen. Although this group of women has not been assessed in scientific studies for menopausal symptoms, they may also be at risk of losing bone mass, developing heart disease, and being depressed. All three of these physical problems are associated with deficient levels of estrogen. To think that sporadic sexual behavior might have the power to promote osteoporosis, heart disease, and mental illness is astounding, but the data do suggest the possibility.[3] For an intimate relationship to be physically healthy, it must be stable and dependable. The science of intimacy shows it. A steady lover is good for your health. A sporadic, undependable one is really bad for you. But women probably know this anyway, just by the way they feel emotionally.

WHAT IS REQUIRED FOR GOOD, HEALTHY SEX?

Let us consider what kinds of sexual behavior had a positive impact on menstrual cycles. In the pilot study of sixty women intercourse was the only activity I had people record. My focus left out a range of sexual behaviors—masturbation, for example—and thereby excluded many women. In 1974, when I began these studies, many college women stopped short of intercourse as a way to retain their virginity yet still enjoy a sexual experience. In the pilot study those who engaged in extended foreplay without intercourse did not record that sexual practice. For the subsequent studies we gathered more detailed information, in-

[3] In most biological systems there are multiple pathways that lead to similar outcomes. Whether other conditions, such as depression, must accompany the sexual deprivation before endocrine health is compromised is not known.

cluding incidence of masturbation and foreplay. Foreplay was defined as "genital stimulation in the presence of a man," in an attempt to study whether it was the genital stimulation, the presence of a man, or intercourse that associates with the endocrine rhythms. I also considered whether orgasm is as necessary for a woman's fertility as it is for a man's. Since a man's biology requires orgasm for "firing" his sperm toward the egg and a woman's does not, my finding that orgasm was not a significant factor in producing a fertile cycle length is not surprising. That's not to say that orgasm is insignificant to a woman's sensual life, but simply that orgasm is not a prerequisite for controlling the length of a woman's reproductive cycle.

Trained as a biologist, I tend to see an underlying order, a kind of symmetry just waiting to be discovered. Particularly in my focus on human reproduction and sexuality, the design by which the continuation of our species works emerges as inherently elegant.

When I think about love cycles, I think about life stages. The early and mid twenties are biologically, and perhaps in terms of one's own subsequent career development, an ideal time for viable pregnancy. However, my research has shown that the continuation of fertile cycles by regular sexual behavior contributes to the biological well-being of the older woman as much as or more than the younger woman. While youthful lovers may have yet to develop their sensual skills so that they achieve the reflex of successful orgasm for the woman, the young woman's ovaries already contain her lifetime supply of eggs. Successful reproduction does not favor sexual prowess. In fact just the opposite may be true. People are supposed to be able to reproduce easily when they are young and innocent, but the late thirties and forties are the years when women often develop their richest sensuality. It is then that the capacity for sensual expertise most often expresses itself. By then men have become more patient, more capable of learning to control the timing of their own pleasure to enhance that of their partner. While a man needs orgasm for the couple's fertility, a woman apparently does not. Considering the tendency for raging sexual urges in young men, this gender difference in the requirements for fertility serves to promote the continuation of human life. As menopause approaches, regular sex helps to lessen the drastic age-related declines of estrogen. If so, a woman's sexual bloom occurs later, in her thirties and forties, this timing may work to her biological advantage. Sensual expression may serve to promote the continuation of the individual woman's fertile stage. Furthermore the blossoming of her own sensuality may serve to delay her aging by promoting estrogen production.

What of the woman without a partner? Does masturbation help her endocrine system? The research says no. Again, orgasm by itself provides no obvious benefit to the fertility of the female endocrine system. Masturbation did not have an effect on cycle length. What really seems to matter is the company she keeps.

IT TAKES TWO

When I evaluated the records that asked women about the frequency of their experience of genital stimulation in the presence of a man—not necessarily by a man—there was an effect on cycle length. Masturbation in its popular sense—an act performed alone—has no positive effect on cycle length. In fact it was no different than celibacy. Sexual stimulation by a man or with a man present does affect cycle length. A time to embrace requires a partner.[4] The science of intimacy is teaching us that an intimate partner provides some critical "essence" that a woman cannot provide for herself.

Consider the implications of this discovery. Throughout recent history, with the exception of the permissive seventies and eighties, women were counseled by their mothers and grandmothers to retain their virginity until they married. My results suggest that the old advice was probably good. Coupled to a romantic relationship, plenty of petting would promote the fertility cycles of a woman without jeopardizing her future fertility. The point is not whether data like these will compel women to go back to the standard of their grandmothers; the point is that those standards promoted, in many different ways, endocrine health and well-being while they preserved the viability of the next generation.

CONCLUSIONS

This chapter has covered a good deal of ground. It has introduced an important idea—that the poetry of intimacy has an underlying scientific

[4] Each study participant marked one of the four boxes each day. An *A* if there was no behavior, an *S* for genital stimulation, an *M* for masturbation, and *C* for coitus. Menstrual flow was also recorded. These calendars provided the raw data. In true "double-blind" fashion my research assistants could tabulate and analyze the data. The results allowed me to conclude that weekly sex in the presence of a man was closely associated with the fertile 29.5-day cycle. The presence of a man seemed to be just as good as intercourse in producing the fertile cycle length.

The term *double-blind* has a very specific meaning to scientific investigators. It means that neither the person who is evaluating the data nor the person who is the subject of the data knows what the hypothesis of the research study is. Double-blind studies help to inhibit the possibility that, unintentionally, the research result will "work out" in the direction expected. By having research assistants tabulate and analyze the data that subjects provide, a good scientific investigator protects herself from unwittingly misinterpreting the data she is recording. From the 1977 study onward all of my research studies involved this "arm's length" process because, by then I knew, in attempting to replicate studies, what I was looking for and I wanted to design pristine experimental methods.

basis. It has a cyclic harmony. And a "time to embrace" is built into the biology of the human species. To sum up what we now know:

- How often a woman engages in sexual behavior with a man strongly affects her endocrine system.
- Regular weekly sex is vital for maintaining a fertile hormonal environment in women who are of reproductive age. A fertile endocrine system appears to promote general health and well-being.
- Estrogen levels are higher in the luteal phase among women who have regular weekly sex than among women who are either sporadically active or celibate. Higher estrogen has been associated with better bones, better cardiovascular health, and a feeling of joy in life (what the French call joie de vivre).
- The luteal phase will be healthier—sufficiently long and hormone rich—in those women who are active weekly. Women who have a sporadic pattern may have an inadequate luteal-phase length, frequently rendering them subfertile. These women are likely to be the ones most vulnerable to fibrocystic breast disease, uterine cancer, and other maladies.
- As women in their forties begin to approach their menopause, regular weekly heterosexual relations can help preserve higher levels of estrogen and decrease the symptoms associated with the menopause, such as hot flashes and depression.

As my studies demonstrate, sexual behavior during a woman's reproductive years can have an enormous impact on her physical and hormonal health. There is a time to embrace (weekly) and a time to refrain from embracing (sporadically).

But sexual behavior is not limited to the childbearing years. A time to embrace includes the years before fertility has blossomed as well as those that follow them. Sex and the sensual arts play important roles in the biology of women and men. Chapter 2 discusses what is known about those relationships by focusing on the life cycle in the love cycle.

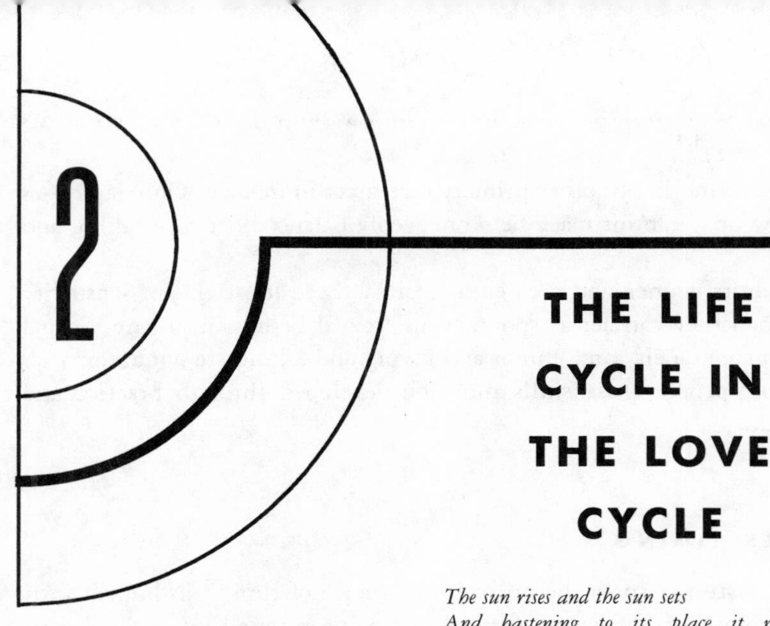

2 THE LIFE CYCLE IN THE LOVE CYCLE

The sun rises and the sun sets
And hastening to its place it rises there again . . .
That which has been done is that which will be done . . .
So there is nothing new under the sun

—ECCLESIASTES 1: 5, 9

The message of the last chapter was timing, specifically how the weekly timing of sexual intimacy enhances a reproductive-age woman's physical health, her childbearing capacity, and her emotional well-being. But weekly sex is just one aspect of timing, one type of intimacy, during one phase of life. Although the reproductive years are critical to the survival of the species, they cover only a portion of a woman's life span—twenty to twenty-five years of the average seventy she will live. The other forty-five years—those that surround the reproductive years—span fully two thirds of her life. When considering intimacy needs in men and women from birth to old age, the context necessarily changes, but the need remains.

THE EVOLUTION OF INTIMACY

When I consider the meaning of the word *intimacy,* I think of the dictionary definition in its reference to "an intimate act." And *intimate acts* pertain to "the fundamental, most private relationships we share with others." Our first intimate relationship begins in the womb. The second is with our mother or other primary care giver in infancy. Only later—as we grow up—can intimacy be more evenly balanced between taking and giving.

As adults women and men enjoy intimacy and its attendant sensuality best when they can act as spontaneous, sexual beings in private. Sexual skills of spontaneity and humor and giving and taking are not automatic. For most people these skills must be developed through practice and feedback.

IT TAKES TIMING

Scientific studies show the critical importance of timing in human intimacy. As infants, if we are fortunate, we are initiated into primal intimacy through out parents. The satisfactory baby-parent relationship is extremely sensual and is marked by intense physical closeness—breast-feeding, bathing, comforting, rocking to sleep, cuddling, kissing, snuggling, and playing. These sensual connections are vital for the overall development of the infant. Hold a normal child in your arms and the child will mold itself to your contours. Try the same thing with an abused or developmentally impaired child and you will get a different response. This baby will either hold itself rigid or will just flop. It won't mold itself to you. In order to learn to be physically intimate, a baby needs to get love and sensual warmth. Denied this necessary physical intimacy, a baby will probably develop a severe disturbance.

Intimacy needs change as we mature and our priorities expand. By the time we reach adulthood and the nurturing of children or career, distractions from intimacy are common. As child, career, family, and social demands all vie for attention, sex may temporarily take a backseat. Financing a loan, contemplating a career move, considering having another child—each can divert sexual energy, which can lead an individual or couple to trouble.

The awesome power of physical intimacy is a compelling force in human life. Nature's design is elegant: The human need and drive for physical intimacy promotes individual development, enhances physical functioning, perpetuates the species, and maintains quality of life. As a biologist, I was excited by the possibility of such a cohesive worldview. As a predoctoral student of experimental psychology and biology, I began to wonder about the science of intimacy. Was there a time element involved? Did certain sexual events have to happen in a certain order, within a certain time frame, to initiate and perpetuate the system? As it turned out, it appears that they do. The effects of intimacy are driven by timing principles. And science is teaching that we must either use it or lose it.

THE CRITICAL PERIOD—A WINDOW OF OPPORTUNITY

Biological research has thoroughly documented the importance of "critical periods" for the development of specific nervous system and hormonal pathways. Human intimacy is dependent on the normal development of both nerves and hormonal secretions. Nerve cells in the genitals and the brain, hormones in the gonads and pituitary gland—all intercommunicate in the service of intimate and erotic life.

One series of experiments that began it all was conducted by Drs. David Hubel and Torsten Wiesel in the 1960s. These scientists received the Nobel Prize in Physiology for their ground-breaking discovery of such a critical period. Working with newborn kittens, they figured out that the visual system of these little mammals had to be used in the proper way, at the proper time, for the electrical connections between eye and brain to develop fully.

Their experimental method was really very simple. Starting with Ping-Pong balls, which they sliced in half, they used sets of two halves to create "eyeglasses" for kittens. These opaque half-globe shades permitted some light to get through, yet blocked pattern recognition—the ability to discriminate vertical, horizontal, or angled lines. Using trial and error to determine the time when these abilities were being developed, they placed these "shades" on the kittens for two-week periods at varying times in kittenhood. Their discovery was relevant not only to kitten vision but to our broader understanding of biologically based critical periods.

The time points at which the "shades" had blocked pattern vision determined which ability would never develop in the adult. Drs. Hubel and Wiesel concluded that the visual capacities of a cat—the ability to see vertical lines, for example, a skill useful in detecting a prey moving across the landscape—developed in an orderly fashion at specific times in the early weeks of a kitten's life. Young kittens don't need to hunt, but they need to see patterns when they are young in order to be able to hunt successfully when they grow up. Their bodies require that they use their vision well before they actually need it.

SHEDDING LIGHT ON REPRODUCTION

The vision of kittens actually has a great deal to do with the science of intimacy. The proof that there is a critical period for the development of a particular part of the nervous system sets the tone for a rather universal principle in reproductive biology. There is a time for first use that *precedes* first need, and practice prepares the way. Nerve cells function something like telephone or electrical wires. They transmit electrical impulses. In the case of vision, photons of light (the energy source) enter the eye and hit the retina. This sets in motion the firing of an electrical impulse in a nerve cell, which connects to another cell, which connects to still another cell. The impulse gets relayed through these connecting "wires" to a specific location at the base of the brain. The anatomical pathway of these visual nerves goes from the source of light, to the retina, to the optic nerve. En route to the visual part of the brain, nerve cells send off peripheral signals to parts of the brain that, in some mammals, control reproductive development.

In a number of laboratory study animals, such as rats, the impulse of light and dark, the balance of day and night, seems to control the timing of the reproductive system. When the visual nerves are cut, these animals are blinded and their reproductive cycles cease. Without the "electric" signal of day and night passing into their optic nerves they become infertile.

Other manipulations of light and dark in the laboratory interfere with reproduction almost as much as severing visual-system nerves. If an animal is kept in a laboratory in constant light (or constant darkness), its reproductive cycling also stops. It becomes temporarily infertile. If the animal is later returned to a regular cycle of light and dark, its cycling resumes and fertility returns. Primates are different. Equivalent studies of different species of monkeys have shown that disruption of the visual pathways, through surgery or daylight manipulation, does not disrupt the reproductive cycle. The daily rhythm may change, but the reproductive aspects do not. Still, some influences of light on men's hormones have been demonstrated.

In men, in the months of September and October, the levels of the sex hormone testosterone rise in their bloodstream. The significance of this discovery has not yet been well developed, but it does intrigue me. Studies in both the Northern and the Southern Hemisphere have identified September and October as the annual peak elevation. It seems to suggest that even in men some aspect of reproductive-system hormones is influences by the movement of the earth in relation to the sun.

Comparative biology teaches that as you move up the phylogenetic tree, from mammals with simpler brain cortices to those with more gray matter, the control of the timing of the reproductive cycle moves from the visual part of the brain down into the gonads. In women the monthly hormonal cycle appears less influenced by light from the sun than by electromagnetic energy, which the moon reflects. Chapter 8 provides these details. Men seem to be more vulnerable to visual input than are women in their reproductive behavior. Chapter 4 explains the studies.

CONNECTING VISION TO SEX

In 1976, as I began my predoctoral studies of the reproductive system of women, I was absolutely fascinated by the research results about critical periods for the development of vision and the relationship of vision to reproductive cycles. I wondered whether there might be a critical period for the onset of sexuality in women, and if so, what role vision might play in the science of human intimacy. By 1990 I was able to answer those questions, but the answers provide just the barest hint of the power of vision on the inner workings of the intimate life.

At the University of Pennsylvania in 1975 a student in biology who was working toward her Ph.D. candidacy was required to do a pilot project, analyze the results, and propose an experiment either to expand upon this pilot project or to initiate something different. Figure 1-1 (page 9) showed my pilot data, that regular weekly sex was linked to 29.5-day menstrual cycles in reproductive-age women. I next proposed to expand on this finding with larger numbers of women and more rigorous methods of data collection. Thanks to the feedback and guidance of an extraordinary committee of University of Pennsylvania professors, the planned experiments expanded and led to the discoveries that were discussed in Chapter 1.

In the biology department of the University of Pennsylvania a Ph.D. is awarded only when the student is able to make a *contribution to knowledge.* If my project on weekly sex and fertile-type cycles could be replicated, and all other conditions were met, the committee told me that the summary of this contribution to knowledge would be sufficient to earn a doctoral degree, but they considered it unlikely that my results would replicate in the larger study we planned. They suggested that since my plan might not yield positive results, I should also do a back-up, fail-safe project that would contribute to knowledge regardless of the data outcome. My advisor and mentor, Dr. Celso R. Garcia, had an idea. As an infertility expert, he had amassed a large amount of data over the previous thirty years. He offered me access to these data in order to let me document the length of human gestation. In 1974 he believed that no published English-language study had actually calculated the exact length of human pregnancy. Dr. Garcia had the raw data that could be studied to provide some answers. These answers, regardless of outcome, would provide a contribution to knowledge.

I began poring over these records in order to plan a method for extracting the pregnancy data. In the records of each of the infertility patients were hundreds of bits of information, gleaned from questionnaires and interviews with Dr. Garcia. As I read the charts, I noticed something interesting. It seemed to me that these infertility patients reported later ages at their first sexual intercourse than what I perceived to be the experience of the typical young woman of the 1960s and and 1970s in Northeast America. I wasn't certain, but I had a hunch.

Having just been exposed to the notion of critical periods, it was natural for me to be thinking about them as I studied patient records. Could there be a critical period for starting sexual contact that, if missed, could impair subsequent reproductive-system function? I suggested this to Dr. Garcia, who liked the idea. He proposed I extract the information on age at first coitus (intercourse) as I was analyzing gestational length, to see whether my intuition was right.

SEXUAL INITIATION AND INFERTILITY

In order to extract the data, I needed help, and there was plenty at Penn. I formed a group of ten undergraduate students in biology who were interested in working on a research project with me. Next we made arrangements with the gynecology department. Once a week my ten students and I would troop to the hospital gynecology department and take over the conference room for a few hours. Because I knew my hypothesis and could inadvertently "taint" the data, my helpers divided massive quantities of patient records among themselves and each one plotted and recorded the data of a given batch of records. They did not know what they were looking for but were given the job of carefully transferring bits of information from record charts to the equivalent of ledger sheets for later analysis. After about eight weeks we had extracted and tabulated the data from the medical records of 792 patients. Once analyzed mathematically, the data produced evidence of a "critical period" for starting intercourse.

What emerged was a kind of stepwise progression: Groups with greater infertility problems showed later first-coital ages than groups without or with less serious infertility problems. In Figure 2-1 the results are organized for the type of gynecology patient and the average age of first coitus for that particular patient type. The routine-gynecological (RG) patient population comprised 92 women. This group showed the youngest average at first coitus—18.3 years. I went looking and learned, from other published studies, that in the demographic group represented here, 18.3 years at first sexual intercourse fit the national average. These 92 women had no fertility problems. The next step on the graph comprised women who were not infertile but who had some sort of gynecological pathology other than infertility. These 220 women also showed a relatively young age at first coitus—about 18.6 years. When I plotted the first-coital ages of women with infertility, the delayed first-coital ages became apparent.

Infertility can be classified as either primary or secondary, on the basis of severity—whether there has *never* been evidence of a pregnancy (primary infertility) or if there has *ever* been evidence of a pregnancy (secondary infertility). Among those diagnosed with secondary infertility, there are two subgroups:

- the women who have prior parity (*PP*) or prior births, but who now cannot successfully complete a pregnancy; and
- the women who can get pregnant but cannot bring the pregnancy to fruition (the "habitual aborter," or HA).

On a scale of relative fertility, the women with prior parity are better off than those who can get pregnant but who then suffer miscarriages. It means that their reproductive systems can nurture, maintain, and carry a pregnancy to term. Similarly these women with secondary infertility are all better off then those with primary infertility. Primary infertility is the most devastating type of fertility problem.

According to my data, the worse off a woman was in terms of her fertility, the later her initiation into intercourse was likely to have been. Put another way, the more fertile the group, the younger the first-coital age of the group.

As I thought about these data and prepared them for scientific presentation, I also reviewed the literature to see what other researchers had published. I learned that with the single exception of humans, reproductive behavior (coitus) begins at the same time the mammalian reproductive physiology matures. It is only in the human that coitus is conventionally delayed beyond the onset of the first ovulatory cycle or the first menstruation. This occurs for social, economic, and religious reasons rather than being due to the natural biology of behavior.

Clearly, delaying teenage sex serves important societal and economic functions. Yet I must state the lessons biology has to teach us. My results

FIGURE 2-1

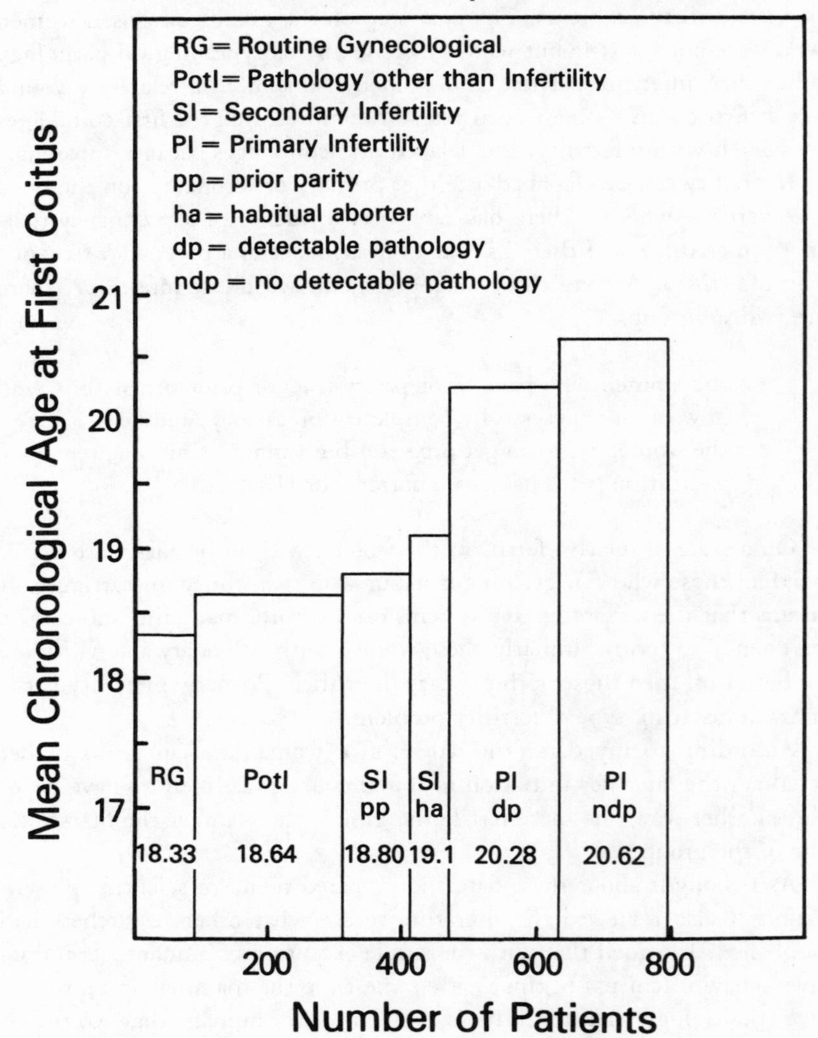

The 792 patients are grouped according to the six categories listed in the upper left of the graph. The routine gynecological patient is a perfectly healthy woman who is coming to this infertility service for routine health care because she happens to be in the hospital area. The width of the bar indicates how large the group is. There were 92 RG patients. The height of the bar indicates the average first-coital age of the entire group.

The order of the groups goes from the healthiest (reproductively) to the most infertile.

suggest a critical period for the onset of coital behavior in the promotion of fertility. Young women, I believe, need to be sexual in order to enhance their later fertility. How early in adolescence? To approach an answer to that question, the scientific discoveries of Alan Treloar provide some insight.

THE NATURAL HISTORY OF THE MENSTRUAL CYCLE

Dr. Alan E. Treloar captured the largest accurately recorded set of information on the normal variation in menstrual cycles ever reported. Published in 1967, his work helped to educate us about the biological meaning of menstrual-cycle length. Working on this project for more than thirty-five years, Dr. Treloar amassed the menstrual records of thousands of healthy women who were willing to keep a record of each time they menstruated.[1] Eventually daughters and even granddaughters contributed their records. When he finally put it all into order, he had tabulated twenty-five thousand years of accurate menstrual-cycle data.

To get an idea of how enormous a body of information twenty-five thousand years represents, consider that if one woman kept her records continuously for ten years, she would have provided ten years of data. If a hundred women recorded their data for ten years, that would supply one thousand years of data. In like manner a number of women at the different stages of life provided a total of twenty-five thousand years. Figure 2-2 shows what he found.

Dr. Treloar divided the stages of a woman's menstrual-cycle life into three large categories, because the normal variation from one woman to the next is different in these three life stages:

- The seven adolescent years
- The reproductive years (on this graph, ages twenty to forty)
- The seven pre- or perimenopausal years

Although he probably didn't realize it at the time, what distinguishes these three phases of menstrual life is how fertile the women are in each stage of reproductive life. In the middle phase, the reproductive years,

[1] This work was begun in 1934 at the University of Minnesota by Miss Esther Doer, a graduate student at the university. It was continued by Dr. Alan Treloar working with a great many other scholars. As of 1972 the first granddaughters were enrolled in the program. As of 1980 the data had been moved to the Center for the Advancement of Reproductive Health under the direction of Dr. Gary S. Berger (MRH Program—Menstruation and Reproduction). By 1980 computer systems had been developed to provide access to data that are now widely available to interested scholars.

FIGURE 2-2

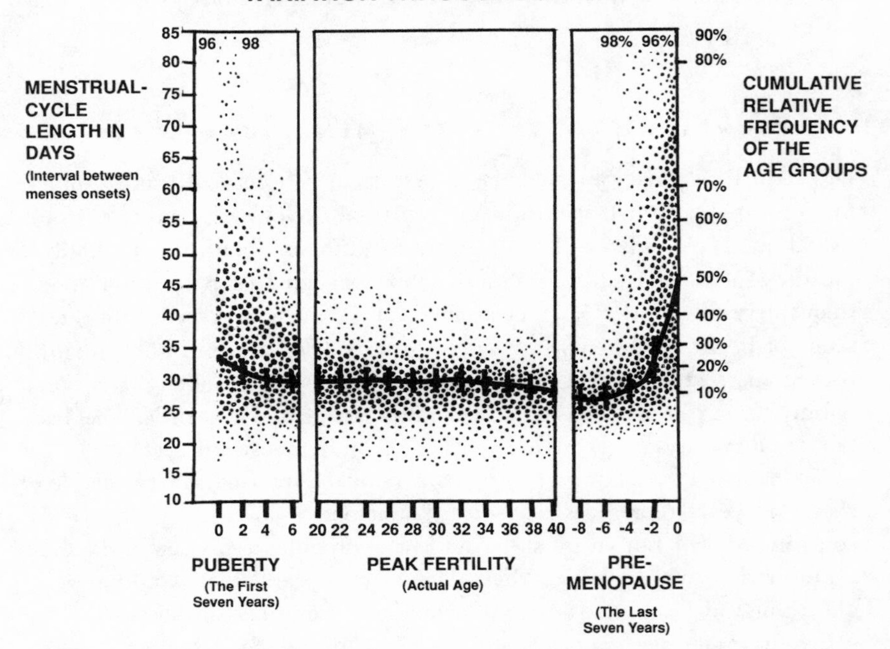

When you look at this graph, on patterns of timing in menstrual bleeding, you first need to know how it is labeled in order to make sense of it. *Look first* at the left side where the vertical axis shows a set of labeling numbers. The numbers on the far left, from 85 at the top to 10 at the bottom, and the labeling represent menstrual-cycle lengths and show the variation in cycle length from every ten days for the shortest to every 85 days for the longest cycles. *Look next* at the numbers at the far right, from 98 at the top to 1 at the bottom. The words on the far right of the graph ("Cumulative Frequency") refer to percentage of women. *Look finally* at the labels and numbers across the bottom of the graph. Three sets of numbers, each with its own label, show the three stages of menstrual life.

most women have cycle lengths that tend toward 29.5 days (plus or minus three days on either side of this). In the adolescent and premenopausal years a kind of "mirror image" emerges. In the first and the last seven years of a woman's menstrual life, many of the cycle lengths appear more and more erratic and extreme. Many women show short cycles; many others show long cycles. The farther in time from the reproductive years, the more aberrant the cycle length is likely to be.

From Dr. Treloar I learned that it is normal to see a wide variety of extremely long and short cycle lengths in the first and the last seven years of menstrual life. I put this fact together with Dr. Vollman's research (see page 5) on the reproductive meaning of these cycle lengths. Cycle lengths that are either short or long tend to be subfertile. This means that about half the women in their first and their last seven years of menstrual life are relatively infertile. And about 85 percent of women in the other years of menstrual life are highly fertile.

OUR BIOLOGICAL IMPERATIVE—SEVEN YEARS

The fact that Dr. Treloar had taught that it was seven years of aberrant cycles at adolescence, not six or eight, caused me to turn back to my data (shown in Figure 2-1) and focus on my "critical period" questions. First I replotted differently. Instead of chronological age, as shown in Figure 2-1, I plotted gynecological age. *Gynecological age* refers to the number of years a woman has been menstruating. (A gynecological age of six means she has been menstruating for six years.) The same pattern emerged. The later the gynecological age at first intercourse, the greater the infertility.

Next I asked whether there might be a critical seven-year period for the start of sexual life in promoting fertility. I also wondered if regular sexual behavior served as a cultivator—a cultivator that brought the cycles from these wide ranges of variation into the pattern characteristic of reproductive-age women. After all, my pilot study of sixty women seemed to suggest that regular weekly sexual behavior normalized menstrual-cycle lengths. I returned to the data of the 792 women (Figure 2-1) and tried to discern whether there was a critical period of seven years.

I asked two questions:

1. What percentage of women whose first sexual intercourse occurred more than seven years after menarche (first menstruation) were infertile?
2. How does this percentage compare with the percentage of women whose first coital experience occurred earlier (e.g., less than seven years after menarche)?

There was a big difference. In this predominantly infertile population a larger group—50 percent—of those who waited more than seven years past menarche to start sexual intercourse had primary infertility as compared with a smaller group—30 percent—who did not wait.

The results were highly significant, suggesting that these were not

FIGURE 2-3

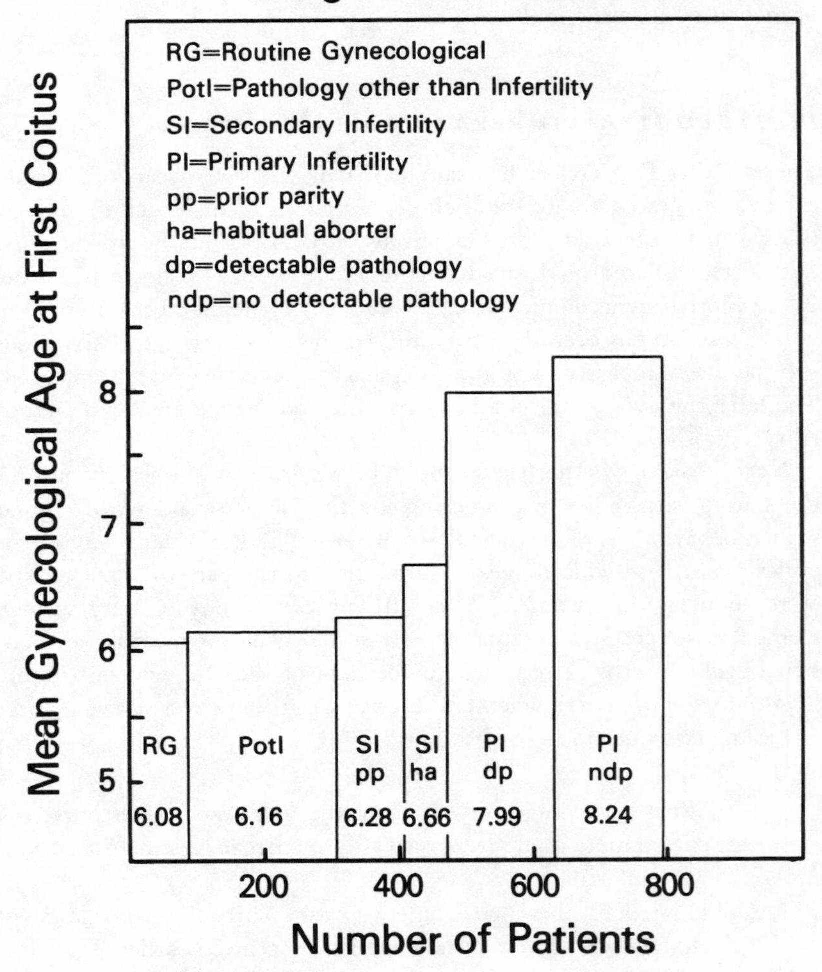

chance events, but that there was a real biological, statistically significant difference in the incidence of infertility between women who waited and women who did not.[2] What kind of results could we expect in a different

[2] "Statistical testing" is a field of scientific analysis that asks the question, Could results this extreme be likely to have occurred by chance more than, say, 1 percent of the time? The mathematical

group of women, say women not seeking help for infertility? Such data do not exist directly, but indirectly we can guess that among fertile women the age at first coitus would be younger, and less than seven years after the menarche. The reason is embedded in epidemiologic research. Demographic data for large groups of American women in the mid-1970s showed that first intercourse occurred on average at age 18 for white women and 17.6 for black women. The menarche age then was between 12.6 and 13. This means that first intercourse was widely occurring before the seven-year window (after menarche) had closed.

My discovery that delaying intercourse increased the incidence of subsequent infertility coincided with the data of another investigator, who showed similar trends. Dr. Joseph A. McFalls, a sociologist working in Philadelphia in the early 1970s, was investigating the impact of venereal disease on the fertility of the U.S. black population in the years 1800–1950. The day I presented my final thesis defense to my committee, one of my professors handed me Dr. McFalls's recently published paper. His investigation, while primarily focused on the correlation between infertility and venereal disease, also found a similar link between less sexual activity and decreased fertility. Those who weren't using it appeared to be losing it.

One other piece of information helps to clarify the association among sexual activity, age, and fertility. Once a young woman/girl starts having sex, she tends to continue.[3] Biology does not seem to condone promiscuity, however. As my studies suggest (see pages 8–10), the healthiest way to have sex is in the context of a committed, monogamous relationship. In fact the more sex partners one has, the higher the incidence of sexually transmitted diseases. And certain sexually transmitted diseases, if undetected, can permanently impair fertility.

So much for scientific analysis. The real-life meaning of these data needs to be addressed in order to use scientific discoveries to improve the quality of our lives in every life stage. I think that biology teaches important principles about adolescence.

....

methods for analyzing questions of this sort and providing logical answers have been well studied and are generally agreed upon by the scientific community. For those who would like a more thorough resource to understand these methods, *Basic Medical Statistics*, by Anita K. Bahn (New York: Grune & Stratton, 1972), is particularly clear for those unfamiliar with the discipline.

[3] According to one study by Shanna H. Swan, Ph.D., and Willard L. Brown, M.D., M.P.H., the statistical measure of correlation coefficient (R = .80) indicated that, among the young women who were followed, there was a very close relationship between starting sexual intercourse and continuing it.

ADOLESCENTS: ADULTS IN WAITING

Teenagers need to fall in love, to experience romance and sexual attraction, and, in the context of a stable responsible relationship, to begin the initiation into physical intimacy. My perspective as a biologist convinces me that the critical period offers a biological window of opportunity. And the window can close. I think that physical intimacy itself has vital lessons to teach the adolescent. I suspect that the intimate behaviors experienced by a teenage girl serve to recruit (or develop) the nervous system "wiring" and bring it toward fertility. Teenage boys, while not under the time pressure of the biological clock, also benefit from intimacy "practice." As they develop sensual skills, teens learn important life lessons as well: for example, concern for another's pleasure in addition to mere self-gratification.

Learning the lessons leads to the four intimacies that are each part of a complete relationship.

THE FOUR DIMENSIONS OF INTIMACY

The *physical, verbal, action/companionship,* and *problem-solving* aspects of intimacy together make up the intimate life. For the adult *physical intimacy* includes not just sexual intercourse but also the pleasure of touch, a caress, a hug. The *verbal intimacy* of a good conversation, a private joke, and shared humor enhances well-being. And the *action/companionship intimacy* of doing things together—taking a walk, rearranging the furniture, or seeing a movie—can be gratifying. And finally one can enjoy the *pragmatic intimacy* of solving problems together—of juggling schedules, planning and living within a budget, building equity.

Ideally men and women find an intimate partner with whom they can nurture all four intimacies. They are vital to promoting quality of life. Sexual intimacy—no matter how skilled and erotic the pairing—is empty intimacy without these others. Adulthood and reproductive maturity require us to experience all four intimacies for the well-being of the self—and for the well-being of the next generation.

If a teenage girl is denied intimate learning experiences during this biological window, the stage is set both for fertility problems later and for other developmental problems. We know of the critical importance of hugging and physical massage to the development of the baby and its maturation into a stable adolescent. I have learned to believe in the critical importance of sexual activity in the late adolescent—to promote his or her healthy development into a reproductively mature adult.

In a sense the initiation into physical intimacy gets the system started. Once begun, that system must be used—regularly. The need for regularity transcends our life stages.

Shifting from adolescence into adulthood, we approach the reproductive years of Treloar's graph (Figure 2-2)—the time span that begins when years of adolescent menstrual life have passed and continues until a woman is seven years away from her last menstrual period. Typically the reproductive years span from age twenty to about forty-three.

INTIMACY DISTRACTIONS: THE CRITICAL PROBLEM OF THE REPRODUCTIVE YEARS

Having learned to value intimacy may have led to marriage and family, but "having it all" is both the challenge and the curse of these years. Born of the women's liberation movement, the phrase *having it all* signifies women's aspirations to pursue and enjoy the same careers men do and still bear and raise children as nonemployed women do. In short, women want to do what men do vocationally *and* do what men can't do: have babies and nurture them. Whatever the source of the imperative—societal expectations, personal agendas, peer pressure—these years are characterized by maximum striving by women with these goals. For those who embrace this goal, the price is stress, a stress that threatens to break apart the fragile "house of cards" so precariously held together.

A less stressful set of goals focuses more broadly and acknowledges that we can have, across the span of our lives, many different kinds of experiences, but there are only so many hours of energy in a day. The ancient philosophies of the East teach the need for balance and awareness of one's place within one's universe. The Tao teaches the art of appreciating where you are, even as you are moving somewhere else. Addressing the role of a woman on her life time line, the Tao suggests that the appropriate allocation for mothering is wisely limited to twenty years. Twenty of her seventy is 28 percent of her life. It makes good sense to me. The heirs of the women's liberation movement are confronted with a very different problem than those of my generation. Today many young women believe they can choose and seek a life at least as freely as men apparently can. What a contrast to my experience, turning twenty-one in 1965. In my early twenties such an idea would never have been entertained. The culture taught—wherever we turned—that it was "unfeminine" to try to pursue the kinds of goals that striving men pursued. It was essential to marry as soon as possible. To be a woman and labeled unfeminine was among the greatest of slurs.

Fortunately for us all, the women's liberation movement clearly changed this attitude. Unfortunately what it has not and cannot solve is the more fundamental problem.

THE REAL FEMININE CHALLENGE

Only we—the female gender—can gestate and give birth to the future generation. And only we can nourish the infant at the breast. It is we, the women, who initiate the intimate bonds with the babies because it is we who carry them in our womb and give birth to them. The responsibility is awesome.

The innate drive for the reproduction of the species seems to be almost always stronger than the drive for creativity (or career) of the individual ego. That is not to say that the two drives do not clash. As the popular terms *baby hunger* and *biological clock* suggest, many of us find ourselves overtaken by our biology when it comes to gestation.

For women the needs of the body for fertility, for pregnancy, for parturition, and for caring for infants produce profound demands, demands that consume vastly more energy than anyone who has not spent it can realize. I have a new niece, by marriage, whose story really tells it well.

Stacey is thirty-six. She graduated cum laude from Princeton in 1975. She worked to develop a profession. And then at thirty-two and again at thirty-four she willingly became pregnant. We visited Stacey, her husband, and their darling babies and she told me something that she found extraordinary. She said that she had close friends who had mothered babies, taking time off from fledgling careers with a joyous anticipation. She had spent many hours with them and their babies in social visits. She said that until she was faced with the day-to-day experience herself, and in spite of her high perceptual abilities and closeness to these friends, she never perceived the physical and mental exhaustion that mothers of young babies endure, day after day. She was amazed. She had seen it when she visited with her "new mother" friends. She had heard the stories—of sleepless nights, ceaseless cries, and endless calls for care, attention, love. She now knew that she had missed the meaning: the profound impact it would have on both her physical and mental functioning.

And therein lies a critical problem of today's young women. They are faced with choices that are dreadfully difficult. As if that were not enough, there is a time element involved. If they develop a career in their twenties, then just as they are launching themselves, they find the conflict of family life hitting them in the gut. If they do conceive a baby and manage to go to work every day—rather than rest and nap and nourish themselves—they are likely to run up against the exhaustion that Stacey discovered. And once having given birth, many women find the options even more difficult. Does a mother put her baby into the hands of someone else to raise? Does she stay home and suffer a sense of isolation from the excitement of professional life? Does she try to have it all in the same five years and struggle to juggle more than she can hold? The choices belong to the woman because, regardless of the man's willingness to help, it is her body and her mind that bear the lion's share. It is women who pay the direct costs of reproduction.

I believe that "having it all" can be debilitating in the face of limited reserves of time, energy, and interest. What "all" means varies from individual to individual. For some women it is career, husband, children. For others it is career ambitions or expansion. For still others the highest priority is a meaningful relationship or the full recovery from the emotional upheaval after the loss of a spouse from divorce or death. Men grapple with similar priority pressures—ascending the corporate ladder, amassing prestige, creating financial security. In the push-pull of competing roles physical intimacy may get put on the back burner, waiting for the time when the couple or the individual will turn on the gas. And, as Chapter 9 shows, sexual infidelity is another form of intimacy distraction, one that punishes all three members of the triangle.

Intimacy seems to be dispensable, but as my studies show, it is not. We need intimacy—often and regularly. Regularity in intimate contact, in sex, must be maintained in order for benefits to continue and accrue. Although definitive research has yet to be done, I can't help connecting the astonishing rise in infertility stories with the social trend of delayed childbearing. For many professional couples their careers are their first "children." In their focus to establish a career, the pattern of intimacy that is so crucial for optimizing fertility may not become established until too late. And another window closes.

Even a child can become an obstacle to intimacy. "Inhibited sexual desire" among new mothers is emerging in the psychotherapeutic community. It became a significant focus of a recent International Academy of Sex Research meeting. New mothers may shun intimacy with their husbands because they are so intimate (breast-feeding, cuddling, etc.) with their newborns and feel both exhausted and satisfied. Intimacy overload is probably not the real problem in most cases. Rather the need for physical privacy can lead a woman to reject further physical intimacy with her partner.

As Lao-tzu said,

> Better stop short than fill to the brim.
> Oversharpen the blade and it will soon blunt.
> (Chapter 9, lines 1–2)

Translated for the nineties, I'd state it a little differently: Sometimes it may be necessary to shift the priorities; lighten the work load; achieve what appears to be less in order to have more; find more time for reflection and purposeful appreciation of the quality of our lives and our relationships; reevaluate the limited resources of money, time, material goods, and quiet spaces and what they mean to us. One must make these selections based on carefully negotiated considerations of the legitimate needs for intellectually satisfying work for both adult members of a family and consider how these balance against career demands and time requirements.

For women a different awareness of the life time line—with a well-orchestrated spacing of its intimate elements—can sometimes provide a better life-style. In other words, I suggest that women, especially, make plans based upon active choices. Effective birth control and limiting the size of the family are fundamental to this planning. Even men with positive intentions rarely carry even a third of the burden of child care, and this is important to recognize in the planning of family size and timing. Once there are children, the knowledge that the first few years of life move fast and that once past, the early-childhood years are gone forever—almost before we have the chance to adjust to them—can help modify the understandable sense of urgency for full-time work beyond the demands of childrearing.

I do not believe in the truth of the untested myth that "quality time" in our important relationships can substitute for quantity. That is like saying that quality time will substitute for the quantity of time we need to give to our other important life pursuits, such as career development and artistic growth. We need both. Quality is the nutrient but quantity is the water, both of which are required for the garden to grow. I believe adults should make active choices rather than fall ignorantly into the romantic notion that "we don't need to plan and everything will turn out all right."

For the sake of the young family at least one of the adult members of the family should be available in the home part-time. Day care, even when it is excellent at protecting babies from endless infections and immediate emotional stress, cannot provide two of the critical needs of the young: constancy and long periods of quiet. The constancy of one person and the stability this brings is replaced with an ever-changing array of "personnel" because day care workers have a very high turnover (i.e., burnout) rate. Each baby is receiving much higher rates of stimulation by being exposed to the noise of many babies and the detailed planning out of the day. Peace, quiet, and resolving boredom may turn out to be important developmental needs. The capacity to handle quiet time may be important to the development of a strong self. No studies have yet addressed these questions. My intuitions suggest that the choice of full-time day care is suboptimal for parents and for babies. These problems are tough; let us at least acknowledge them. I look to a time when a thirty-hour workweek will allow parents to be home with their babies for at least a few daylight hours a day. The years of our family's childhood are short, and a broad perspective may help to maximize them.

MATURITY: THE POTENTIAL FOR SEXUAL LIBERATION

Let's move forward and look at the years after childrearing and menstruation have ended. For the woman who limits her biological work of childbearing and child care to the twenty years recommended by the Oriental philosophies, her time line now offers new opportunities and challenges.

The reproductive years end with menopause—the time after the final menstruation. But before menopause, usually beginning at age forty-one or forty-two, a seven-year transition period occurs. The menstrual cycles become increasingly more variable, gradually becoming either longer or shorter in length. In the seven menopausal transition years it is not uncommon for a woman to menstruate every seventy or eighty days, or every twenty. And these changes in cycle timing signal the underlying changes in the hormone production of the ovaries. A reduction in fertility follows. During these last years of menstruation pregnancy is still possible, but as each year passes, the danger of pregnancy increases, both to the mother and to her baby. The incidence of congenital abnormalities rises drastically. Down's syndrome (a disease of the child) increases fourfold in the years from forty to fifty. Total chromosomal abnormalities increase threefold.

During this perimenopausal period the symptoms of the onset of menopause disrupt the well-being of 85 percent of all women. Hot flashes, hormonally driven mood swings, the aging of skin, and the drying of vaginal lubrication are all part of the normal experience of women who do not take hormone-replacement therapy after age forty.

Scientists do not know a great deal about the intimacy needs of the menopausal transition, but we know enough. In the Stanford Menopause Study, which investigated several hundred women during their menopausal transitions, I found clear evidence that weekly sex is good for premenopausal women. Women in their late forties who had regular weekly sex with a man showed almost twice as much estrogen circulating in their blood as those women who were either celibate or sporadically active. Women who had regular weekly sex showed other benefits as well. They reported much less symptom distress. Hot flashes were rare, and if they did occur, tended to be milder than in women who had sporadic exposure to sex life. I also found a vibrant sexuality in these women.

I closed my 1987 scientific paper on perimenopausal sexuality with the conclusion that women in their late forties are fully sexual, showing high levels of sexual desire, sexual response, and sexual satisfaction when they have intimate partners. These women who were using it were not losing it.

What most of the California women studied wanted was regular sex. And they were quite clear in revealing the reason why they were often experiencing only a less-satisfactory sporadic sex life. As one woman said, "No man around . . . how sad!"

INTIMACY OPPORTUNITIES: ONLY A FEW GOOD MEN

This woman summed up the experience of many of the unmarried women in this study. In their late forties California women were finding it difficult to locate a suitable, eligible man. That was in 1979 and 1980. By 1990, with the specter of AIDS and particularly its high frequency in California, we can expect that it is becoming even more difficult for these women. Perhaps because of the ripening awareness of the sexually transmitted disease epidemics, many people have retreated from what seemed to be sexual liberation and have become more careful in their selection of partners. As discussed earlier, when a woman is slow to expose herself to sex, the trend is that she is more likely to be either celibate or to engage at least weekly. Probably the friendship that flowers more slowly has a better prognosis for stability as well.

HORMONAL MIRROR IMAGES

During menopause, when menstruation ceases altogether, the hormonal environment changes significantly. Hormonally, transition years are somewhat like the adolescent years. The peaks and plummets of sex hormones—the tremendous changes from one day to the next—can upset the emotional life of both the adolescent and the premenopausal woman. Usually after age fifty the hormonal levels tend to even out. Hormonal secretions tend to stop their "mad" surges and purges, and things tend to settle down, although at lower levels of estrogen than were characteristic when younger.

Hormone-replacement therapy has come on the scene as a tremendous asset to many women in these years. I think appropriately dosed hormonal therapy provides a marvelous option for many women, but you must be informed about which hormones are safe and which ones are dangerous.[4] With hormone-replacement therapy the physical body and emotional life retain their youthfulness far longer. When the dosage is correct, the incidence of breast cancer is reduced to below that of women who take no hormonal therapy (see Box 3-2, page 80).

NEW SENSUAL RICHES

With or without hormonal replacement, as women and men age, some fortunate individuals enter their "lush years," the time of the full blooming of their sensuality. During these years sex as procreation is no longer on the biological agenda. Sensuality can come most fully alive. The transition to this life stage is complete. The intimacy distractions of the reproductive years have ended. The career, the baby, the child demands, the sometimes awesome financial responsibilities, and the tremendous work loads of the reproductive years have generally been resolved. Now there remains experience and knowledge, and maybe even wisdom. Even to those who gave it little thought before, the finite nature of the life span becomes obvious. It is also in these years that more people come to realize the importance of appreciation in promoting love and intimacy. Being able to appreciate what one has magnifies it. In younger years, in our naïveté, it is common to depreciate what we have. By the time we mature, most of us have learned better. We have discovered our bundle of opportunities and know that whether we enjoy or decry them, our lives slip by anyway. As we age, we tend to have less capacity, but with luck we will learn to appreciate what we have. Scientific research appears to support my intuitions.

[4] Hormone-replacement therapy is thoroughly discussed in my previous two books: *Hysterectomy Before and After* (New York: Harper & Row, 1990) and *Menopause: A Guide for Women and the Men Who Love Them* (New York: Norton, 1991).

INTIMACY AND FANTASY: IT'S THE THOUGHT THAT COUNTS

The research of Dr. Julian Davidson teaches that for men, starting in their thirties or forties there are declines in the experience of orgasm, in the frequency of morning erections and in the incidence of sexual thoughts and sexual urges. These physical changes may be linked to declines in the male sex hormone, testosterone. Chapter 4 provides the details.

What I find particularly interesting about these documented declines in reproductive physiology is that capacity—specifically, the capacity to appreciate sensuality—does not follow this precipitous decline. Men continue to enjoy sexual life as they age. In fact sexual enjoyment is not related to any hormonal variable yet investigated. Sexual destiny is not defined by any particular level of testosterone, whether high or low.

Equivalent kinds of data from the Stanford Menopause Study show aging patterns in women to be similar to those in men. Working with Dr. Norma McCoy, a psychologist at San Francisco State University, and Dr. Garcia in Philadelphia, I evaluated the answers to questionnaires about how often the women had sexual thoughts. Figure 2-4 arrays the data collected from perimenopausal women. Each dot shows the pattern of one of the women. This figure shows that many of the women experience sexual fantasy at least once a day.[5] The amount of menopausal distress they experience (which is known to reflect the levels of estrogen) appears to be independent of frequency of sexual fantasy. What do these data suggest to me? Some aspects of mental function have nothing to do with menopausal distress. Estrogen levels in a woman do not predict the amount of sexual fantasy in her life.

Despite inevitable physical aging, men and women can still act sexually and indeed continue to enhance their sexual intimacy if they continue to consider themselves sexual beings. Part of this ability is set by precedent, arising from the patterns we set in earlier life stages. Those who had a more fulfilling sex life during their childbearing years appear to be more sexually active as they age, according to several surveys. If the precedent was not set in earlier years, you are not, however, doomed. The function of our predominant sexual organ, the brain, remains vital as long as we can think clearly. And we always have the option of engaging it.

[5] The age of the women in the Stanford Menopause Study averaged forty-nine. Some women were much younger and others somewhat older, but all had in common that they were still menstruating and were now showing either irregular periods characteristic of Treloar's premenopausal years or the symptoms of approaching menopause, such as hot flashes.

FIGURE 2-4

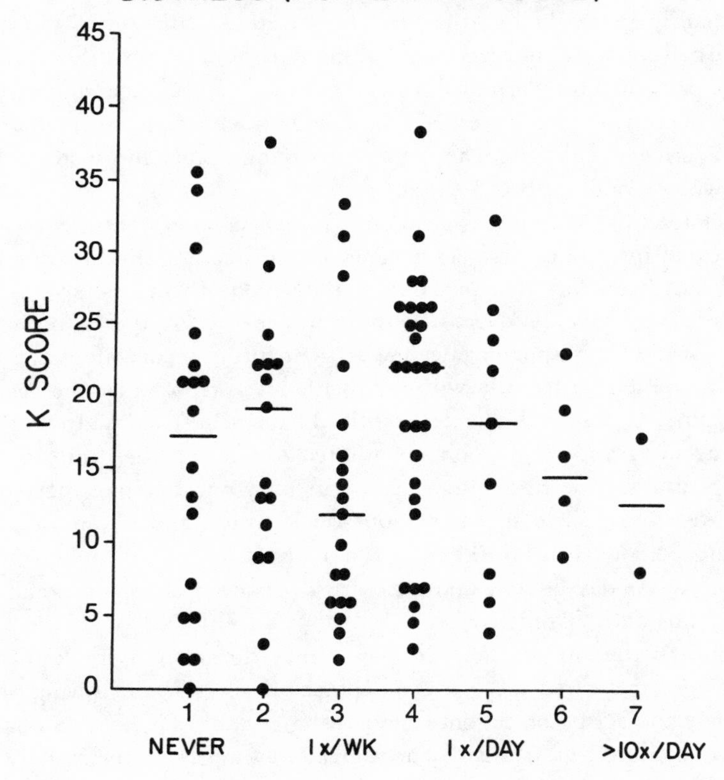

Each dot on this graph represents the answer that one woman gave to the question "Give a rough estimate of how many times you have sexual thoughts or fantasies during the last month." One hundred women answered this question. As the dots array on this graph, one can see that nineteen women said never, seventeen said something less than once a week, and the largest number (twenty-nine) of the women said something between once a day and once a week. The vertical axis, the K-score, is the Kupperman Score, which reflects how much menopausal distress (hot flashes, vaginal dryness, and other symptoms of menopause) that woman was experiencing. The higher the score, the more the distress. If you look at any one dot on the graph, you can tell what that woman's answer to the sexual-fantasy question was and what her K-score was calculated to be.

SEX GROWS UP

As people mature, their sexual behavior changes. Some men peak at twenty, others at thirty or forty, but almost all reach a point where they showed more sexual behavior in their previous years than in their later years. Women tend to peak later than men. For those who do bloom into a resonantly sexual being, the peak may come at thirty-five, forty, or fifty. I believe these differences in timing emerge from sex differences in the life pattern of the hormonal increases, peaks, and subsequent declines. I am also convinced that women who take hormone-replacement therapy will experience a different and more satisfying sexual life history than those who do not (see Box 3-2, page 80).

After reaching their peak, men and women, as they continue to age, have sexual intercourse less often. They experience orgasm during intercourse less. They find that sexual arousal overtakes them less powerfully. For those who have discovered the pleasures that an overpowering eroticism provides, this apparent decline in their future sexual life may seem undesirable. Those actually within this life stage report that the change is welcome. Consider the letter I received from a husband and wife who had read my *Hysterectomy* book and wanted to offer up their information to help others. The husband wrote, "I'm forty-one but like most men who love women, I've still got a 'one-track' mind, and like most of us I'm 'horny' more than I'd like to be (unsatisfied)."

Perhaps one day he will find a closer fit between his appetites and his satiety. The odds favor this.

Although the amount and the power may diminish, I think that as people age, they need more sensual contact. It is in the maturing years that body contact becomes, once again, as important as it is for an infant.

There is more time available and a greater awareness of the many ways to use "timing" to enhance sensual experiences. Details given in Chapter 7 describe the mechanics. Body contact, not necessarily genital contact, is what very old people say they want. Perhaps this is a good thing because the ratio of women to men continues to increase at each advancing age because women live longer; our endurance is greater. By the later years, age eighties and beyond, there are two women living for every man. In retirement centers the ratio is even steeper: six women to every man after the age of eighty.

We know something of the intimacy needs of people in this age bracket, thanks to Dr. Norma McCoy. Dr. McCoy was the first scientist to publish a study that actually asked old people, in a systematic way, what was going on in their inner, sensual lives. Her results are important because they teach us about the sensual spaces to which we are all headed.

In 1988, working with a graduate student, Judy G. Bretschneider, she published her study of the sexual interest and behavior of healthy 80- to 102-year-olds. These upper-middle-class men and women were living in retirement homes and agreed to be studied. There was a very high percentage who fantasized about being close, affectionate, and intimate with the opposite sex. Eighty-eight percent of the men and 71 percent of the women reported having these kinds of fantasies. Compared with fantasy, the frequency of sexual intercourse was much lower, but the married men and women were twice as likely to have intercourse as the group that was not married. Fourteen percent of the women and 29 percent of the men were married, but twice as many—28 percent of the women and 58 percent—of all the men reported having a sexual partner.

There were sex differences in the current sex activities, with the frequency of intercourse, masturbation, touching, and caressing differing in the men and the women. There were no sex differences in the reported past enjoyment of such sexual activity. However, the *present* enjoyment was much lower for women than for men. Why would old women enjoy sex less? We cannot be sure from the data presented. Maybe men are easier to please; or maybe men are less giving than women are. In either case Ms. Bretschneider and Dr. McCoy found that the importance of sex in the past was highly correlated with its present importance.

What can we take away from this study? If sex is important when she's younger, it will continue to be important to a woman when she's older—but it will be less available. For men, if sex was important when they were younger, it will also continue to be important when they're older, and it will probably be more available. Despite the seeming disparity between the sexes, remember that statistics don't tell the whole story here. Men may enjoy better odds of connecting with an eligible woman, but it is no fail-safe guarantee that they will. Similarly women may forge a meaningful relationship in spite of the odds against them.

THE POWER IN A CARESS

Older people need someone to hug, to put their arms around, and to be held in return. We all need to find a living creature we can touch. If we don't have one as a sex partner, there are alternative ways in which to fulfill these basic needs for physical intimacy. Perhaps the work of Dr. Erika Friedmann provides the best example.

Erika Friedmann was a graduate student in biology at the University of Pennsylvania at the same time I was. We became friends as well as colleagues. Every week, for about two years, she would drive from Philadelphia to Baltimore, Maryland, for her study at Johns Hopkins University Hospital. There she would go into the stroke-victim section to discover who had had a stroke since her last visit, to enroll them in her study, and to collect data. She wanted to know what would help a person survive a stroke. She followed each poststroke victim for twelve months. Among the ones who lived, she searched their records for a pattern of behavior they shared, something that was different from the pattern of those who had died; and her results made international news. She found that it did not matter whether the stroke survivor had a lover, a partner, or a family at home. What mattered was whether there was a pet at home. A pet to come home to helped provide a little insurance against another subsequent stroke in the ensuing twelve months.

Somehow a pet seemed to help "immunize" the person and increase his or her odds of survival. I have often asked myself why it was a pet, not a spouse or significant other, that immunized the stroke victim. I do not know for sure, but I have thought a lot about it. I can only conclude one thing: A pet represents what no other relationship represents. The pet brings unconditional love, uncritical hugging every time, something even the closest and most compatible of human partners cannot always do.

For both men and women, intimacy needs will probably increase as we grow old, and a good many of these intimacy needs are going to shift from sexual intercourse to touching, caressing, caring, and listening. For many elderly people there will be ways to meet their needs for intimacy if they remain flexible in their approach to a solution.

The need for physical intimacy continues, and these needs are real. The greater health and endurance of women produces more older women than men. The older the woman, the greater the disparity in the sex ratio. Are older women doomed to loveless, empty lives? I think not. When I study Dr. McCoy's results, I am struck by a notion that reassures me. The percentage of older men who want sexual intimacy with a woman is higher than that of older women, which means that most of the older men are sexually available. Most of the older women are no longer interested. Assuming compatibility, there is a man for every woman who wants one.

I don't see any data to prove that heterosexual intimacy is what older people require. I do see data (both from Dr. Friedmann's discovery and from numerous anecdotal comments) that physical intimacy with a living being—presumably in the form of hugs and nuzzling—is vital to well-being. Is it possible that aging women can find, in their loving connections to pets, grandchildren, and other people, the physical intimacy so needed by living beings? I think so. The inherent symmetry in nature suggests that what is generally available in the biological system is usually sufficient. And the idea fills me with hope.

CONCLUSIONS

Intimacy is a real, biological need, a driving force in human life. We need and benefit from appropriate exposure to physical intimacy from our earliest experiences in life to our very latest in old age.

Timing counts. Opportunities for the experience of intimacy occur throughout life. The intimate acts in which we engage affect both the development and the functioning of the sexual reproductive system. If the intimacy need is met in each life stage, the progression to the next life stage is enhanced.

The identifiable life stages and their intimacy requirements form a continuum as follows:

- *The first years.* Even before birth, intimacy occurs in the womb. A pregnant woman who wants the child and strives to make her pregnancy tranquil probably enhances her offspring.[6] In childhood, care givers foster physical, intellectual, social, and emotional maturity, laying the important groundwork for the following years.
- *The adolescent years.* Ideally within seven years of her first period, a young woman should experience her initiation into sexual intercourse. (This presupposes appropriate levels of emotional, physical, social, and intellectual maturity.)
- *The reproductive years.* Regular intimate contact helps optimize a woman's reproductive capacity and her general health. Finding ways to maintain this regular pattern despite the distractions of children, career, and evolving self-image help to ease a woman's transition through the years to come.
- *The perimenopausal years.* A habitual pattern of regular intimate contact helps a woman maintain more steady-state hormonal levels, so she is less likely to be affected by hormonal peaks and plummets. Staying sexually active helps keep intimacy a priority in the later years.

[6] Controversial data have begun to suggest this.

- *The later years.* Particularly in the face of the physical changes associated with aging, we can remain sexual beings. Once again, the precedent set in earlier life stages helps perpetuate vital intimacy. And physical intimacy in one form or another immunizes against loneliness and certain diseases.

The need for physical intimacy is very important throughout the life span. For the person who has missed out, who was unable successfully to conclude a developmental stage because of a stressed mother, a late first-coital experience, a sporadic sexual relationship, or the absence of a partner, the studies provide some reassurance. In my studies of infertility and age of first coitus it seems that half the women with delayed first coitus were fertile anyway. The same held true for the women with less-than-weekly sex; half showed fertile-type menstrual-cycle patterns. For the other half the future may be less bountiful.

Having traced the life cycle in the love cycle, we focus next on the biological context—the internal chemistry, the organs, and the corresponding sex hormones—that forms the basis of intimacy.

3 THE HORMONAL SYMPHONY OF WOMEN

We shall not cease from exploration
And the end of all our exploring
Will be to arrive where we started
And know the place for the first time

—T. S. ELIOT, "Little Gidding," *The Four Quartets*

Now the focus shifts inward, to the monthly synchrony and daily rhythms of a woman's hormones—the very stuff of sexual and sensual chemistry. As a scientist I've studied hormones for fifteen years. As a woman I live and experience their effects. As a biologist involved in sex research I perceive their magic. With a basic overview of the hormones and organs involved, you can begin to appreciate the biological music they create. Understand your hormones and you've taken a significant step toward better health, optimal fertility, and richer sensuality. Hormones are an integral part of each of these desired qualities.

THE NEW FACTS OF LIFE

Hormones are tiny substances that are produced in glands and released into the bloodstream. Derived from either cholesterol or protein in healthy women before menopause, they circulate through the body within the bloodstream in a continual process of supply meeting demand. Hormones act as chemical messengers, setting in motion many different biological responses.

Exactly how hormones work is not understood, but a great deal is known about them. Since the 1970s scientists have been able to measure their concentration in blood. The sex hormones include those manufactured in the gonads (ovaries and testicles) as well as others produced in the pituitary gland and brain. A competent laboratory can determine the circulating concentrations of the estrogen, progesterone, FSH, LH, beta-endorphin, and other hormones if a woman or man provides 20 milliliters of blood (perhaps a teaspoonful) drawn from her or his arm. In women these all relate to her intimate life.[1] We know that after hormones appear in the bloodstream, they travel about until they find cells with appropriate receptor molecules. Hormones attach somewhat like a key into a lock to their waiting receptor molecule "mates." Once attached, the hormone-receptor complex activates a chain reaction. The chain reaction produces effects that can be measured. From my own study of the research papers, I have pieced together data to form a mosaic. This "tapestry" of information reveals a series of relationships:

- When sex-hormone levels change, changes in physiology, spatial and motor skills, mood, and behavior follow.
- Changes in sex-hormone levels affect our capacity for sexual pleasure and our chances of fertility.
- Changes in sex-hormone levels predict or reduce the likelihood of certain diseases.
- Estrogen and testosterone, two of the sex hormones, affect skin —its beauty, tone, resiliency, and tendency to break out in pimples and blackheads.
- Estrogen and testosterone affect the growth pattern and the type of hair that covers the skin.
- Women who work have higher levels of androgens than women who are housewives; and women with higher-status occupations have higher levels than those with lower-status jobs.
- Even the prenatal hormones[2] affect the offspring's predisposition to engage in culturally defined male- and female-gender role behavior.

Although investigators do not know exactly how hormones work, they know enough to realize their profound importance in the everyday life of women and men.

[1] Throughout this book I use the word *estrogen* to denote the total estrogens produced by the body, most of which are contributed by the ovaries. Three principal estrogens have been identified that have slight variations in their chemical makeup. These are usually abbreviated as E_1, E_2, and E_3 and stand for estrone, estradiol, and estriol.

[2] For example, testosterone taken during pregnancy to prevent miscarriage.

THE DYNAMIC INTERACTION—AN OVERVIEW

In the body of the reproductive-age woman there is a dynamic interaction between the hormonal secretions of the uterus, the ovaries, the brain, and the pituitary gland. The ovaries produce the estrogens, progesterone, androstenedione, and testosterone. These sex hormones move into the bloodstream, circulate throughout the body, and cause the cells of the brain and the pituitary to respond with the secretion of their own hormones: the luteinizing hormone releasing hormone (LHRH), the luteinizing hormone (LH), the follicle-stimulating hormone (FSH) and the beta-endorphins. These in turn circulate in the bloodstream and stimulate the release of sex hormones from the ovary. The estrogen and progesterone released from the ovaries also stimulate changes in the hormonal secretions and fluid protection within the linings of the uterus and the vagina. The cervical glands of the uterus produce a hormone called prostaglandin, which although found elsewhere in the body, locally appears to affect the cyclic functioning within the ovaries. The changes in these ovarian hormones in turn affect the thyroid and parathyroid glands, changing the secretion of calcitonin and parathyroid hormones. These hormones control bone metabolism. Sex hormones from the ovaries appear to affect other glands throughout the body as well (parotid, pineal, apocrine, sebaceous). The scientific community continues to discover further interactions between secretions of one part, as investigators undertake further research of the body, and control of another part. There is a continual "cross-talk" going on among the various parts of the body, a cross-talk that creates the smoothly running machinery of human biology. The nature of this "cross-talk" appears to change as a girl moves from her childhood into her maturing years. Although the process is continuous, it is convenient to consider five major hormonal stages of female life. The reproductive anatomy and the specific hormones related to sexuality and intimacy dynamically change as women move from one stage to the next.

HORMONAL PASSAGES

The previous chapter highlighted three of the five stages of a woman's hormonal life: the twenty reproductive years, flanked by adolescence and the premenopausal transition. Figure 3-1 depicts the stages of the entire life span. On either side of the three stages of the menstrual years is a nonmenstruating life stage: childhood (before the menarche) and maturity (after menopause has been achieved). Prior to 1974 the age of first menstruation marking the passage from childhood to adolescence had been becoming progressively younger around the Western world. In 1944 it was 13.9 years. By 1974 it had stabilized at about 12.6 years of age. Although some girls do start menstruating a bit younger, the average has not moved lower in age. It appears that a girl's body requires at least twelve and a half years of growth and development before she is ready to enter her menstrual stage.

The timing of the end of the reproductive life remains unsettled. The average age at menopause (the last menstrual flow) had been hovering at about forty-nine or fifty throughout the seventies and the eighties. Recently, with the growing worldwide acceptance of intelligently dosed hormone-replacement therapy, menopausal ages seem to be shifting later. Hormonal therapies are delaying the aging of the women who take them.

FIGURE 3-1: The Five Life Stages of Woman

Stage	| Childhood	*|*	Adolescence |	Reproductive	| Transition	+|+	Postmenopausal	
Age	0	12/13	20		43	50	70	90+
Span	| 13 yrs.	|	7 yrs. |	20 yrs.	| 7 yrs.	|	20 to 45 yrs.	

| (Menarche—the first menstrual flow)
+|+ (Menoparch—the last menstrual flow)

SHOW THIS TO THE MAN IN YOUR LIFE

In order to understand the science of intimacy, an understanding of the female pelvic anatomy is crucial. This knowledge is vital for both women and men. A man cannot have intimate knowledge of a woman if he thinks of her sexual organs as a place "down there." Intimate knowledge is a fundamental basis of a rich sensual life. What the Bible refers to as "And he knew her" has a new resonance. Figure 3-2 shows the internal pelvic anatomy of a reproductive-age woman. The uterus, ovaries, fallopian tubes, and vagina are all drawn in relation to one another to show the complexity, size, and proportion of these vital feminine organs. Notice the interconnection of the parts. They are not separate entities but physically continuous, flowing from one structure into the next. They are also connected in three other ways: through a nerve network, through a localized blood-circulation system that shunts blood within the region, and through a complex hormonal interaction from one organ to the other, transmitted by the *local* blood supply. An ovulating egg journeys from the ovary into the fimbria of the fallopian tube, up the tube, down to its

FIGURE 3-2

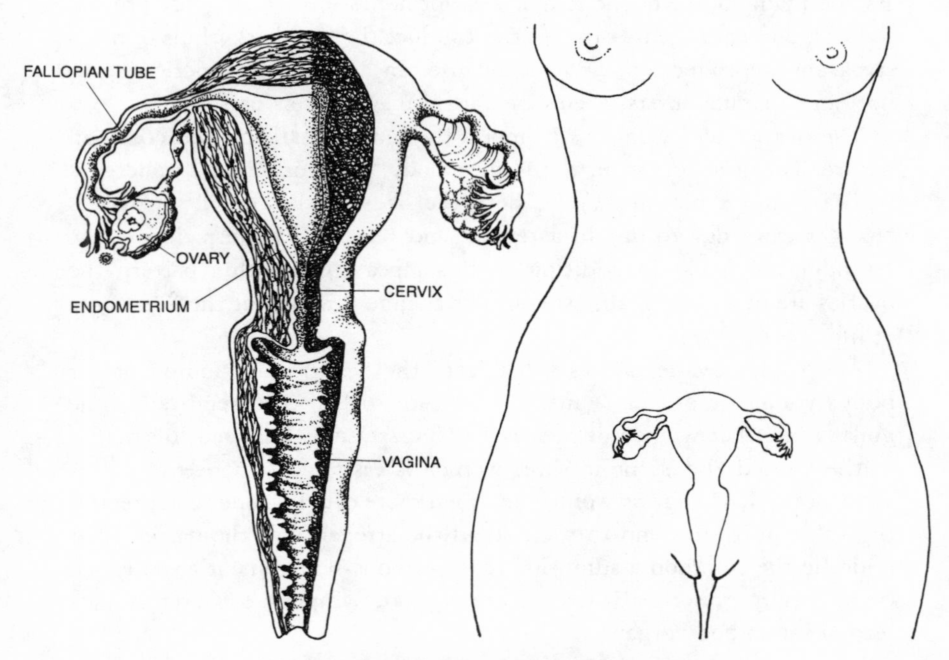

THE GENITAL REPRODUCTIVE ORGANS

reproductive destiny in the uterus, either to embed for the nine months of pregnancy or to be sloughed away in menstruation out into the vagina. What an elegant physical system! And think about the same structure at sexual congress, where a penis fits into that vagina, its coronal ridge tapping against and stimulating the vaginal walls and its tip tapping against the cervix. At ejaculation the penis directs sperm toward the cervical opening for its journey into the uterus, and a combination of sexually responsive uterine contractions and sperm tail flicks propel the sperm up into the fallopian tubes. Sperm meets egg in the fallopian tubes if fertilization is to take place. Chapter 7 details the exquisite design of the female anatomy and how it is shaped to fit the male genital anatomy.

THE DUAL-PURPOSE OVARIES

As the main source of the female sex hormones—the estrogens, progesterones, and androgens—the ovaries conduct the female rhythms of menstruation, reproduction, and menopause. In addition to functioning as hormone manufacturers, the ovaries are also egg warehouses.

The ovaries of a woman are among the most remarkable structures in nature. No other organ in the human body, male or female, undergoes such a complex monthly cycle. Both the size and the content of ovaries changes from day to day in a regular and repeating pattern during the reproductive years. In addition to this repeating monthly pattern the ovaries are in a state of almost constant change throughout the five stages of life.

A woman's ovaries are as valuable to the healthy functioning of her body, young or old, as a man's testes are to his. The removal of the gonads in either women or men is called castration. I am sad to say that in the United States one in three women is castrated. I hope to see this awful statistic change as women and men more clearly come to appreciate the value of ovaries and to understand the arrogance or the naïveté that underlie the common assumption that hormone-replacement therapy can be medically prescribed to equal the natural, symphonic, secretory pattern of intact body organs.

OVARIAN EVOLUTION

In an average thirty-year-old woman each of her two ovaries is about the size of a small egg, housing between 250,000 and 1 million eggs. Such little structures. So powerful their influence. Throughout the course of the menstruating years of her life, some of these eggs will, in regular cyclic fashion, be recruited to swell. One, the egg of the month, is destined to ovulate. Simultaneously a coterie of eight to ten more also swell, though not quite so fulsomely, and form a kind of "ladies in waiting" cohort. Each of the eggs is housed in a enveloping sheet of flat cells.

This whole structure—one egg covered with its sheet of cells—is called the follicle. This structure is the hormone machine. The follicles and their by-products manufacture most of the estrogen and progesterone that circulate in the blood. In order to understand how hormones work, it is useful to take a look at the ovaries and the follicles they contain at different stages of life, since the composition of ovarian follicles looks so different across the life span. Figure 3-3 shows a slice of the ovary of a baby girl magnified several hundred times.

When you look at this figure, you can see a number of oval and circular structures with the largest one toward the bottom near the center. Differ-

FIGURE 3-3: Ovarian Slice from a Baby Girl

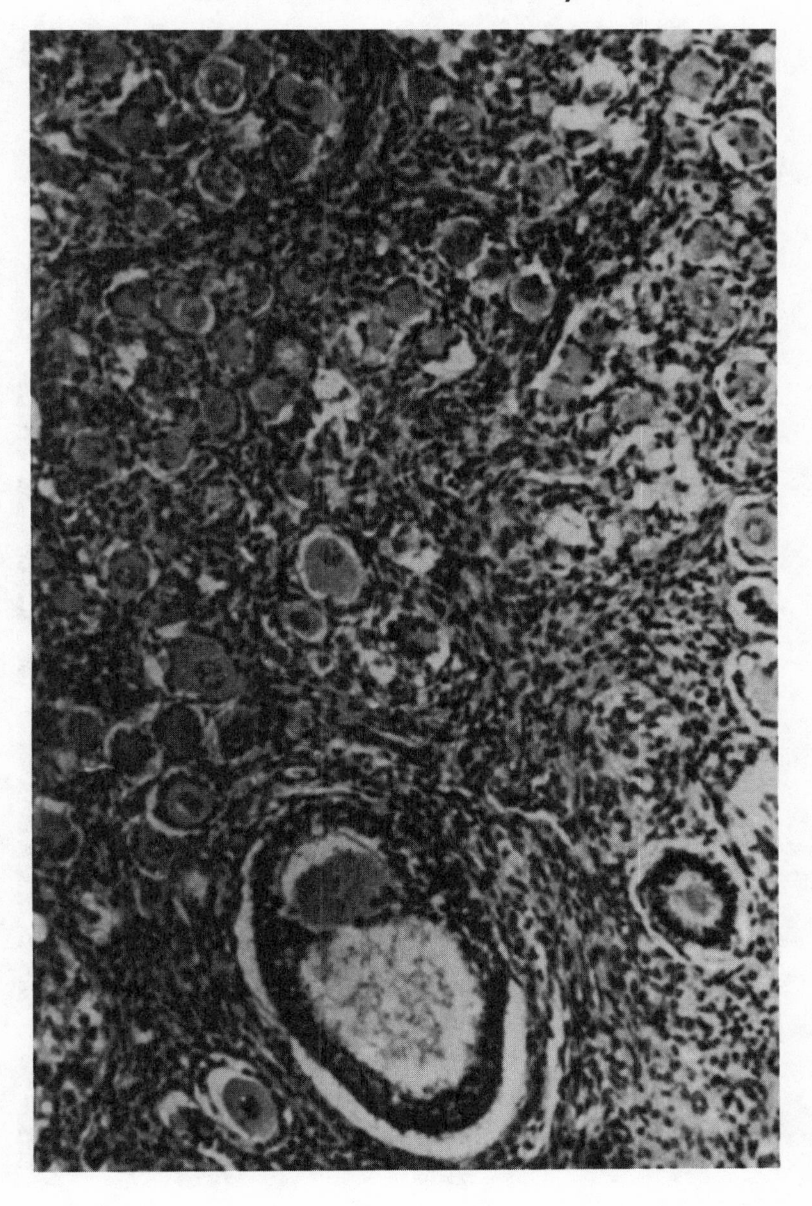

ent-sized follicles, in different stages of their growth cycle, are revealed. Some contain many more layers of cell structure than the original single layer that initially covered each egg. Most of the others remain in their primary state. In Figure 3-4 a comparable slice from an older ovary is shown. Here the woman who contributed this healthy ovary (at a hysterectomy with ovariectomy) was in her thirties.

FIGURE 3-4: Ovarian Slice from a Woman in Her Thirties

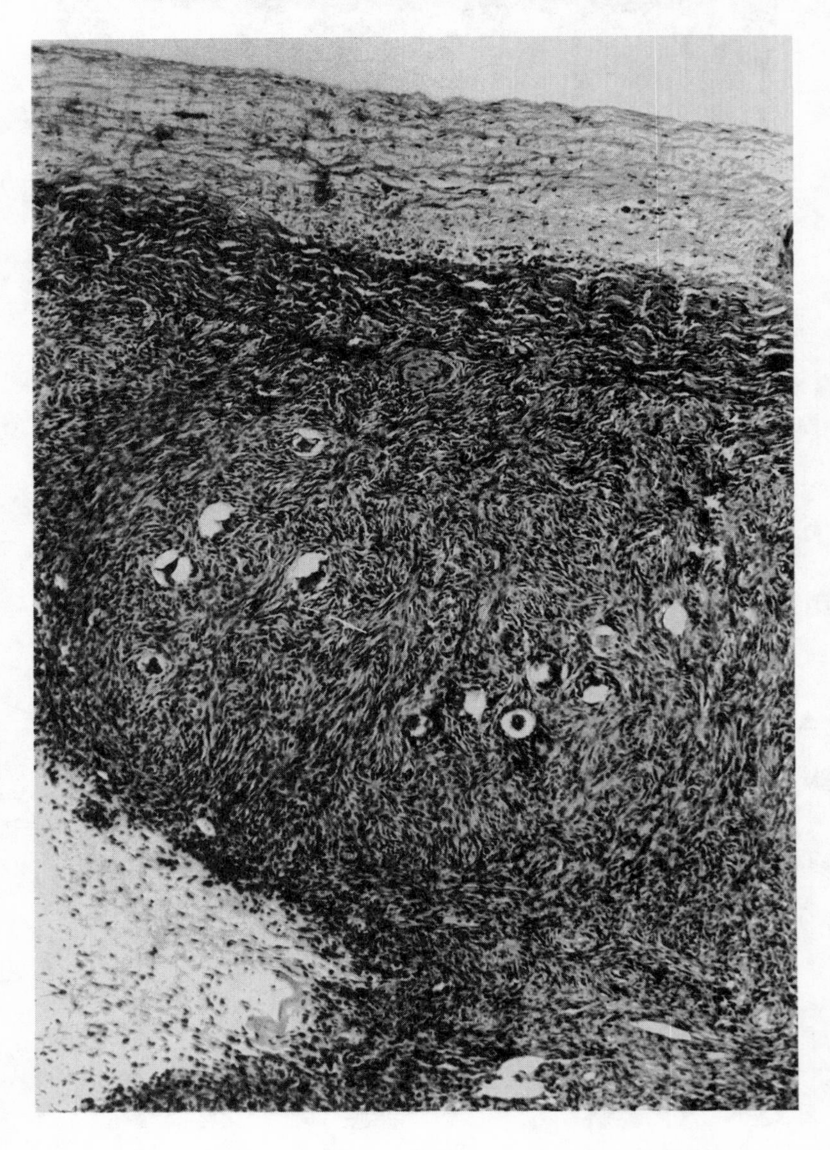

When you compare the ovary of a woman in her thirties with the ovary of a baby, the mature ovary has the less orderly structure. As women grow older, more of their follicles have been spent. Figure 3-5 shows an ovary of a woman in her menopause, past age fifty. The tissue within this ovary is even less orderly in its appearance.

The ovary changes throughout life, and as these three figures show, it

FIGURE 3-5: Postmenopausal Ovarian Slice

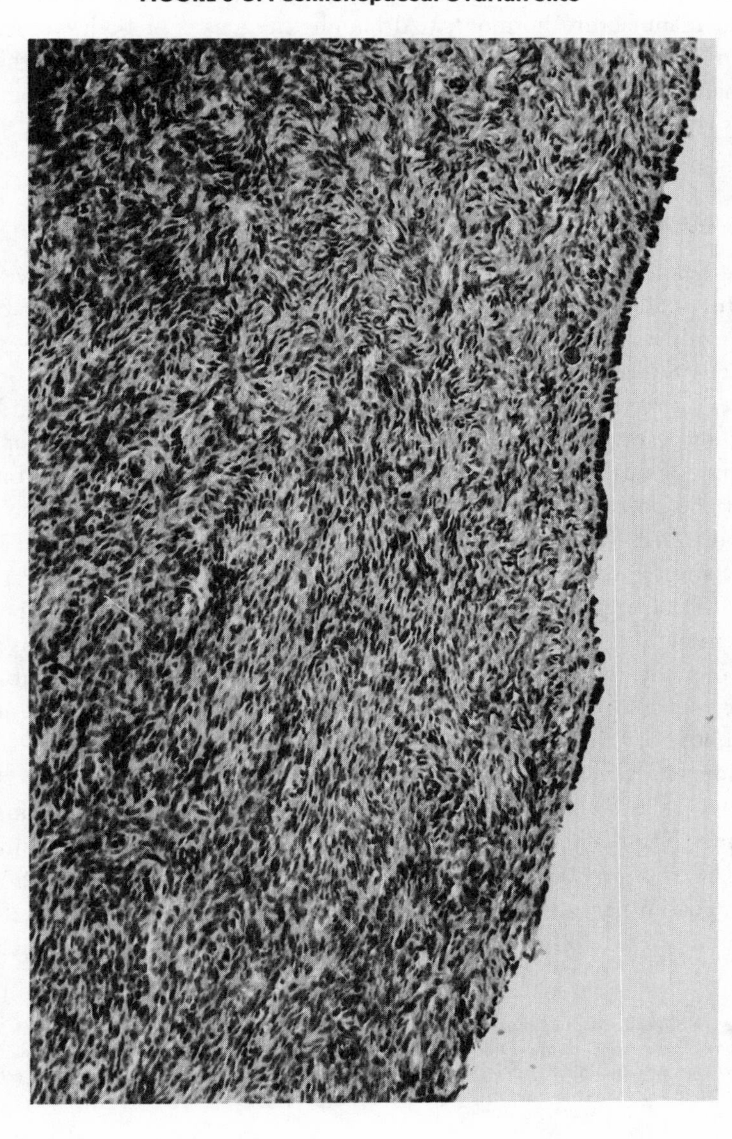

goes from a highly ordered state at birth to an ever less ordered state. It starts out small, grows larger throughout the reproductive years, and then, around age forty, starts to shrink. The observation of these changes in anatomy throughout life helped earlier scientists figure out what was going on in the ovary; and the most recent studies have finally proved what has always seemed obvious to me—that the ovaries are valuable at every stage of life.[3] An old woman's ovaries are as important to her well-being as an old man's testes are to his. The gonads of both women and men manufacture hormones. Although the gonads of both women and men also contain the genetic materials that will form the fetus of the next generation, there is a profound difference between men and women in this regard.

WHY MEN CAN LEAVE BUT WOMEN MUST STAY

The ovaries of a baby girl contain all the eggs she will ever have. The testes of a man do not. He will continue to produce freshly manufactured sperm just about as long as he lives. An old man can often supply sperm that is healthy. After her late forties a woman can almost never provide an egg that will be healthy. The Oriental twenty-year rule for childbearing and caring closely parallels the anatomical twenty-year rule for optimum egg health. Once again it seems to me that the principles of biology and the poetics of Eastern philosophy converge. And it seems to me a good thing. When a man inseminates with fresh sperm, he can set limits to his responsibility with that single coital action. He can disappear from the life he created and the one who houses it. When a woman permits insemination, she is biologically bound. For the next nine months and a subsequent twenty years chances are that she will be responsible for carrying out the biological mandate. The demands of pregnancy are so arduous that the body's increasingly greater inability to permit the condition tends to serve the woman's health. Life without eggs does continue. There's a great deal more to a woman's life than childbearing and caring. The twenty to forty postmenopausal years provide the forum.

The three ovarian slices show that, with increasing age, the ovary changes. What was in infancy a huge quantity of tightly ordered eggs housed within follicles eventually becomes the less-structured tissue of

[3] The scientifically inclined reader may wish to read my 1984 paper (coauthored with Dr. Celso R. Garcia) "Preservation of the Ovary: A Reevaluation and Reappraisal." The paper implored the gynecological community to change its then current practice of removing healthy ovaries at hysterectomy for women over age forty.

the older ovary. The unused eggs gradually disintegrate over the years; the follicular structure changes into an amorphous tissue, called *stroma,* that progressively takes up more and more space in the shrinking ovary.

THE HORMONES PRODUCED BY THE OVARIES

The eggs for the next generation are important to the species, but even more important to the woman and her intimate relations is the hormonal factory that her ovaries provide throughout the entire span of her life. As the structure of the ovary changes, the hormonal output changes.

Scientists have not yet identified the full range of ovarian hormones manufactured. New, previously unsuspected hormones of the ovaries continued to be discovered. Even so, what reproductive-system scholars already know describes a complex and beautifully rhythmic hormonal story. At different stages of life and, during the reproductive years, at different times of the monthly ovarian cycle, the ovaries manufacture and secrete into the bloodstream harmonically varying amounts of sex hormones. The cascading effects of this variation produce the hormonal symphony of woman. This variation in hormones *defines* what it means to be an adolescent girl, a reproductively mature woman, a perimenopausal woman in her transition years, and a maturing woman moving through her twenty to forty postmenopausal years. Each of these ages has its distinct hormonal tone, yielding predictable differences in appearance, strength, energy, emotion, and freedom from or propensity for disease. It is no accident that old women are the ones who suffer from cardiovascular disease, osteoporosis, wrinkling skin, and muscle wasting. These conditions emerge out of the shrinking ovary and its declining output of estrogen and progesterone.[4]

HORMONAL PATTERNS DURING DIFFERENT LIFE STAGES

Think of the ovaries as miniature warehouse-factories, housing eggs and manufacturing hormones that enter the bloodstream and circulate throughout the body. As Treloar's Figure 2-2 (page 32) showed, after a girl begins menstruation, about seven years of erratic menstrual patterns are likely to follow.

....

[4] For details of the aging ovary and its hormonal mandates, *Menopause: A Guide for Women and the Men Who Love Them,* 2nd ed. (New York: Norton, 1991), which I wrote with Dr. C. R. Garcia, provides a comprehensive resource.

Until age eighteen or nineteen the menstrual cycles tend to be irregular and unpredictable. This adolescent phase is associated with surges and purges of hormone secretions. Since the levels of sex hormones can surge high and plummet low from one day to the next, they produce the undertone for the adolescent extremes of emotionality. I think of this as a kind of biological programming for separating their lives from their parents'. As teenagers head toward their independent years, their hormonal surges and purges and consequent disruptions help the family to accept this inevitable parting and even feel relief when it arrives. By about age twenty (and coincidentally once the young adult is usually gone), hormonal output from the ovary has matured into a predictable ebb-and-flow pattern each month. Now the disruptive patterns change to harmonious flows—what can be a symphonic variation in tone.

During the reproductive years, estrogen levels rise and fall twice during each monthly cycle. Working together, the three principal estrogens—estradiol, estrone, and estriol—begin each menstrual cycle at relatively low levels and continue to rise until the time of ovulation approaches. Ovulation refers to the release of the egg from the ovary into the fallopian tube. Ovulation is schematically pictured in Figure 3-2. In that figure, on the left side, you can see the egg leaving through the surface of the right ovary (viewer's left) and about to be picked up by the fingerlike endings (fimbria) of the fallopian tubes for transport toward the uterus. The ovary itself, viewed from the surface, resembles the surface of the moon with craters where recent ovulations have occurred.

Estrogen levels rise continuously in the blood each month, from the onset of menses to the time of ovulation about twelve days later. After ovulation the estrogen levels fall slightly for a few days and then begin to rise again, reaching a second peak of the month about seven days after ovulation. From this second peak of the month estrogen levels again begin to fall, dropping day by day until the next menstrual period begins. Figure 3-6 shows this double-pronged estrogen cycle of the reproductive-age woman.

Many women, and the people who live with them, notice a monthly

FIGURE 3-6: The Monthly Estrogen Cycle in the Reproductive Years

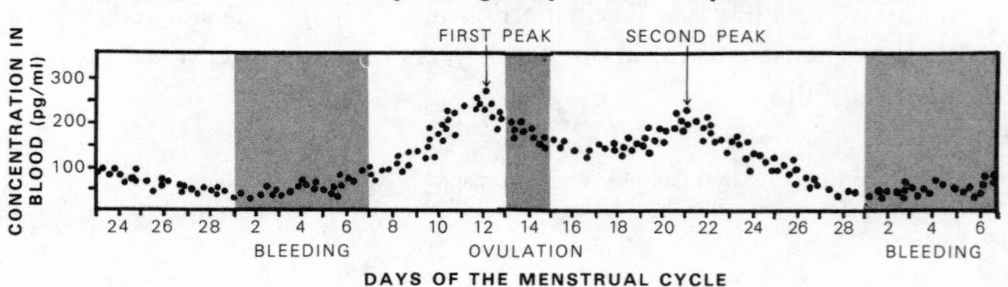

rise and fall in moods, a variation often timed to the rise and fall of estrogen. A lack of good cheer (when severe, known as premenstrual syndrome or PMS) during the premenstrual phase can be explained by the cascading effects of the estrogen cycle. Although almost all women experience and are are aware of this harmonic internal change, very few actually suffer such extreme surges and purges that their lives are disrupted by them.

It is reasonable to think of estrogen as the dominant female sex hormone. Estrogen is responsible for the feminine characteristics of skin, hair growth, breast development, curving hips and buttocks, and a certain capacity for receptivity that is labeled as feminine. The other female sex hormone, progesterone, also plays a critical role and seems to serve as a "temperance factor" modifying the effects of estrogen and protecting the woman against the female-system cancers and other biological excesses such as endometrial hypertrophy, excessive bleeding, and fibrocystic breast changes that can occur when estrogen levels are unopposed.

Progesterone follows a rhythm of its own somewhat different from that of estrogen. During the reproductive years progesterone has one peak each month—a peak that coincides with the second peak of the month of estrogen. Progesterone also plays a role in mood. For most women, as progesterone rises, it produces a narcotic effect, increasing the sense of pleasant sleepiness. When the progesterone begins its fall, simultaneous with the second estrogen fall of the month, a sense of mild letdown, a shifting from major to minor key, is common.

These two hormones, estrogen and progesterone, are manufactured by the ovaries and by the adrenal glands.[5] As women mature past their fifties, more and more of their circulating sex hormones are contributed by the adrenal gland as less are contributed by the shrinking ovarian tissue.[6]

Androgens, which include testosterone, dehydroepiandrosterone sulfate (DHEAS), and androstenedione, are also manufactured by the ovaries. Women and men both circulate estrogens and both circulate androgens. The critical difference is in the quantity. Women circulate between one twentieth and one fortieth as much testosterone as men do, but even this small amount has a powerful effect. Androgens play a major role in muscle strength, libido, and energy. For some women there is a monthly testosterone cycle, with peaks occurring simultaneously with the

....

[5] After menopause estrogens are also produced by the metabolic conversion of molecules of androgen into molecules of estrogen, a process that takes place in cells of fat. The fatter the older woman, the more estrogen she is likely to circulate in her bloodstream.

[6] The adrenal gland produces a more steady-state hormonal secretion but does undergo a single and permanent drop—around age fifty to fifty-five in both men and women. Called the adrenopause, it is usually noticeable only in castrated individuals.

estrogen peaks, but this testosterone rhythm has not been found for all women, who otherwise show the reproductive cycle of estrogens described in Figure 3-6.

WINDOW ON THE WOMB

The womb or uterus is a highly complex and remarkable physical structure—strong yet exquisitely sensitive. Mostly a mass of muscle, its inner cavity is usually flattened and holds about a teaspoon of liquid, yet it is expandable enough to house a growing fetus. More than just a baby carrier, the uterus contributes richly to a woman's physical and psychological health. It plays a role in sexual lubrication, hormone production, and orgasm. Its muscles contract at orgasm, producing intense, bodywide sensations. It has a muscle so powerful that it can expel the fetus at childbirth. Its glandular interlining, the endometrium, grows each month during the reproductive years into a nest and then, in a nonpregnant menstrual cycle, evacuates itself as waste. Once sloughed away, the endometrium re-forms into a new cycle of flat tissue in which another monthly-grown nest begins to form. When I look at these changes in the ovaries and the uterus, I see magic in the dance that they render.

THE MONTHLY HORMONAL SYMPHONY OF THE REPRODUCTIVE YEARS

In the Uterus

Figure 3-7 shows the orchestration of events that are occurring in the reproductive-age woman's endometrium, ovary, and bloodstream to show the hormonal symphony. If you look along the bottom line from left to right, the days of the woman's menstrual cycle are shown, beginning with Day 23 (six days before her menstrual flow will next begin). The first cyclically charted event at the bottom is the endometrial cycle, labeled "Endometrial Lining of the Uterus." On Day 23 of a fertile cycle the endometrium is composed of a thick nest. It contains blood vessels coiled within it and glands interwoven through it where hormones are produced and secreted. If you follow the uterus along the time line, you can see that as menses begins ("Day 1"), the thick endometrial lining has been sloughed off and a new flat uterine lining appears by the second day of blood flow. Beginning about Day 7 the nest once again begins to build and, by about Day 17, has reached its fully thickened size. Once again, at Day 28 or 29 of the cycle, the chart shows that the endometrial lining sloughs away as menstrual waste. Over and over again for twenty to thirty-five years this nest cycle will occur. Interrupted by pregnancy—when the nest serves its vital role in nurturing the growing fetus—the human monthly nest building is repeated elsewhere in nature on an annual cycle. Birds, rabbits, and men show an annual rhythm in their reproductive cycles. For women the dominant rhythm is monthly.

FIGURE 3-7: The Hormonal Symphony During the Reproductive Years

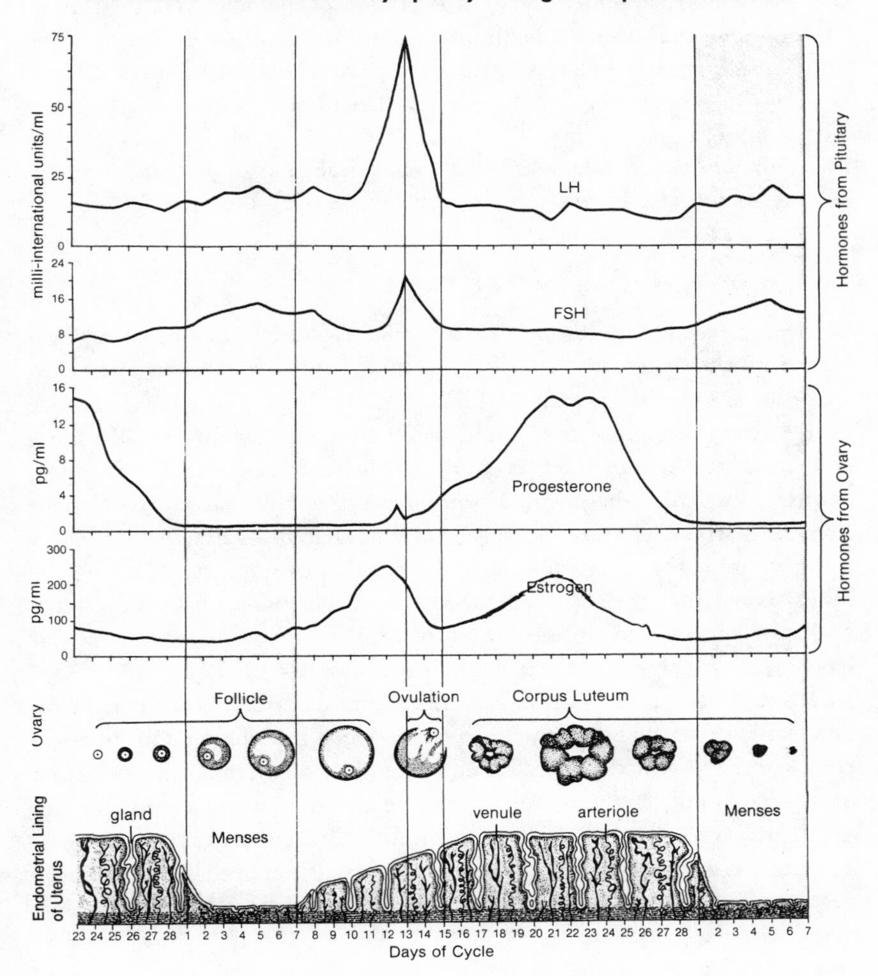

In the Ovary, Before Ovulation

Looking at the next level up on Figure 3-7, to see events within the ovary, the particular follicle destined to ovulate in that cycle is displayed. Although many other follicles are also swelling within the ovary, this figure shows only the one that will ovulate, or "the dominant follicle." Look at Day 23, where the follicle begins as a tiny speck with an egg in the center. By Day 27 that follicle has begun to enlarge as more and more layers of cells multiply themselves and build around this initial single-cell layer that surrounded the egg. By Day 2 of the new cycle the follicle has enlarged and is now beginning to form a hollow "antrum." The follicle continues to enlarge as the days pass. The antrum also enlarges. By Day 14 the egg has doubled or tripled in size, and the follicle has enlarged more than twentyfold. Then, on about Day 14, the follicle opens up at ovulation, to release its egg. This time line shows only the egg leaving its follicle. It does not show the egg leaving the ovary.

In the Ovary, After Ovulation

Within the ovary the emptied postovulatory follicle begins to re-form itself into a new structure. Called the corpus luteum, this structure now looks something like a crumpled doughnut with a hollow center. Notice on the ovary section of the graph that the corpus luteum initially starts out small (the size of the preceding dominant follicle without its egg), and then by Day 15 begins to grow, enlarging to reach peak size on about Day 22 of the cycle, around seven days after it formed at ovulation. It grows so large that it can fill half the ovary that contains it, crowding the other tissues into the edges. By about Day 23 of the menstrual cycle the changing corpus luteum has started to shrink. And by the next menses it has become much tinier. Eventually it fades away totally.

Women have two ovaries, and the events described are occurring in one or the other ovary at any given time. Much like the random toss of a coin that can produce five heads in a row and then two tails, the particular ovary that produces the dominant follicle and subsequent corpus luteum can repeat from the same ovary five times and then switch to the other. The process appears to be random. If one ovary is removed, the remaining ovary tends to take over and provide the full panoply of hormonal events described next.

These monthly events in the ovary (of follicle swelling, ovulation, corpus luteum formation, growth, and demise) are reflected in the changing levels and manufacture of estrogen and progesterone. These hormones journey out of the ovary to circulate within the bloodstream, chemically linking organs far removed from each other (ovary and brain, for example).

In the Bloodstream

Look at Figure 3-7 and the label on the right side titled "Hormones from the Ovary." Here you can follow the estrogen and progesterone cycle from Day 23 to a new cycle's Day 1, and on through one complete menstrual cycle to see the coordinated hormonal events and physical changes of the uterus and ovary each month. Here we see the estrogen cycle across the month, as was earlier described in Figure 3-6. Note also the progesterone cycle that was described but not pictured earlier. Looking at these two circulating hormones of ovarian origin, you can see why estrogen peaks twice each month. The follicles and the corpus luteum are the two "factory sites." As the factory site enlarges, its hormonal output increases. As it shrinks, the hormone levels decline. Estrogen has multiple factories; it is manufactured by the cohort of swollen follicles and by the corpus luteum. Progesterone has one principal factory site—the corpus luteum.

Two other sex hormones are manufactured in the pituitary gland, which is located behind the nose, hanging just below the base of the brain. These hormones are secreted into the circulatory system and are ferried around in the blood vessels. Follicle-stimulating hormone, or FSH, does just what the name describes. It stimulates the growth of the follicles in the ovary and was named for this role. Luteinizing hormone, or LH, graphed above it, was named for its role in stimulating the corpus luteum of the ovary. There is a dynamic cross-talk between the two sex hormones secreted from the pituitary and those from the ovary. They balance each other. When the level of one goes up, it stimulates a predictable change in the concentration of the other, turning secretion on and off. Somewhat like dancers who initiate and respond to each other's movements, in time to the rhythm of music, sexual glands initiate and respond to each other's secretions in response to the rhythm of cosmic music. While in this chapter our focus is on the hormonal dance itself, later, in Chapter 8, our outlook expands to reveal my perspective on this underlying cosmic pacesetter.

CONNECTING PELVIC ANATOMY TO THE OUTSIDE WORLD

The upper end of the uterus has two openings, the two openings that lead into the fallopian tubes. A number of external substances can move up the vagina into the cervix, up the uterus, and into the tubes. Sperm is the most obvious, but sexually transmitted microorganisms can also gain entry. For this reason women and men should protect the uterus, cherishing its value and treating it with great respect. How to protect it? With knowledge—of anatomy, physiology, and the spiritual principles of cause and effect. By the time you complete this book, you will be equipped to make knowledgeable judgments for yourself.

A NEW MEANING TO MENSTRUATION

The monthly event of menstruation is the culmination of this complex inner rhythm of hormones and organs. Menstruation also suggests some sensible principles of sexual behavior to the thoughtful. But first you need to know what menstruation is. Although most women and many men are acutely aware of the fact of the monthly blood flow, very few realize the true physiological meaning of this remarkable event. Knowledge of the facts can help the couple protect the woman's health. Here is what actually happens at menses.

HARMONY IN THE SYMPHONY: MENSTRUATION

On the last days of the twenty-nine-day cycle, the endometrial nest that has been built during the previous month loses its circulating blood flow as the declining levels of estrogen diminish their support to the local blood vessels. Anytime cells are deprived of the oxygen that is normally ferried in the blood, they die. The deprivation of the supply of oxygen to the endometrium, which results from diminished blood flow, leads to death of the nest tissue. As the tissue of this nest dies, it can no longer maintain its connection to the lining of the uterus. It breaks off. And small chunks of tissue become waste products to be removed from the body in the menstrual flow. As these pieces of endometrial tissue break, tiny blood vessels are torn. Much like any other cut that we sustain, menstruation is associated with blood flow from broken blood vessels. Within a day or two, in a healthy woman, the blood begins to clot, the broken tissue begins to heal, and bleeding slows down. Usually there are two heavy days of flow, and for most women these occur on the second and fourth days of the cycle. Something like the natural pruning of the limbs of trees, menstruation provides a monthly pruning of the endometrial nest. A D & C, dilation and curettage, is the surgical analogue, which includes the dilation of the cervix and "cutting" or surgical pruning of the endometrium.

DISCORD IN THE SYMPHONY: ENDOMETRIOSIS

When I think about the physiological events that occur during menstruation and the disease called endometriosis, it seems likely to me that there may be a relationship between the timing of sexual behavior and endometriosis. Endometriosis affects probably one in twelve women. For many the menstrual days are intensely painful as endometrial tissue that should have been removed from the body vaginally is bleeding into the peritoneal region (the internal spaces of the pelvis) in response to hormonal commands. I discussed these perceptions with Chung Wu, M.D., when we worked together in the late 1970s. By 1989 he reported that women who had sex during menses had a 70 percent higher rate of endometriosis than women who abstained.

Although many gynecologists are working to study it and to treat it, its origin is not well understood. There appear to be two components to the disease. First is the fact of misplaced endometrial tissue. Take a look at Figure 3-2 and imagine the endometrial nest tissue speckled throughout the surface of the uterus, the surface of the ovaries, adhering to the outside of the small bowel. How does the endometrial tissue get there? I suspect it gets there when the uterus contracts during menstruation and some of the material of the endometrial nest is forced up the fallopian tubes and squeezes into the internal pelvic space rather than down the vagina, where it belongs. If I am right, then logic would suggest that sexual intercourse and masturbation should be avoided during the first four days of menstruation. Until the menstrual tissue has passed out through the vagina, there is nothing to stop it from being propelled up the fallopian tubes if the uterus contracts during orgasm.

One has to consider why endometriosis is not epidemic. There are certainly women who probably do experience orgasmic contractions of the uterus during menstruation yet do not have the disease. I suspect that this has to do with the second component of endometriosis, individual variation in immune reactions. Autoimmune reactions that women show to the presence of endometrial tissue in their pelvic spaces occur in some women but not in others. The reasons for these differences are not yet understood. In addition, some women do not experience uterine contractions at or after sexual intercourse. Until we know more, it seems prudent to aim for prevention.

I suggest that women consider abstaining from orgasm during their first four days of menstruation to protect themselves. The idea of "a time to embrace and a time to refrain from embracing" has a biological as well as a spiritual basis.

THE REPRODUCTIVE GATEKEEPER—THE CERVIX

The cervix is the portal to the uterus. The uterus tapers off into the cervix, the inch-long doughnut-shaped neck of the uterus that dips into the vagina. Before the first pregnancy the cervix is about half as long as the uterus. After the first pregnancy it often shortens.

The cervical region has a unique structure. In contrast to the uterine body, which is composed of about 80 percent muscle mass (the other 20 percent is glands, blood vessels, and other tissue), the cervix has very little muscle. Mostly it is a region of glands, nerves, and blood vessels. The pattern of changes in the cervix is linked to a woman's estrogen levels. If a woman has high levels of estrogen traveling in her bloodstream, her cervix is richly supplied with nerve endings sensitive to pressure, such as a penis tapping. With women whose estrogen levels are low, such as a young girl, an older woman, or an athlete, these nerve endings tend to be absent. The underlying biology leads me to suggest that sexual intercourse can produce either no sensitivity, exquisite pleasure, or extreme sensitivity verging on pain. Likewise, surgical procedures such as endometrial biopsy that fail to provide anesthesia to the cervix can produce intense pain or no pain depending upon whether the nerves are there or not at that time. This anatomical information was well explained in the scholarly tome *The Biology of the Cervix* (University of Chicago Press, 1973). It is not generally taught in medical schools, and I find it is not yet generally known by many of the gynecological surgeons with whom I speak.

THE ACCOMMODATING VAGINA

The vagina, a muscular three- to five-inch tube, links the other reproductive organs to the outside world. It serves as reproductive exit (for menstrual blood, baby) and entrance (for sperm, penis). Amazingly, it is flexible enough to accommodate the varying widths of a tampon, a penis, a baby's head. The vagina not only changes throughout life, it changes monthly as well during the reproductive years. These changes are dependent upon the state of the hormone levels and the pattern of a women's current sexual activity. Childbirth through the vagina stretches and changes the shape of this barrel-shaped organ—for the better.

After vaginal childbirth the folds in the vaginal walls form a rich and undulating surface, somewhat like the interior of a many-petaled carnation. A vagina that has sufficient underlying muscle tone can provide an exquisite series of multiple sensations for the penis, much different and potentially richer than the smooth, albeit narrower, barrel of the woman who has not yet given birth. Chapter 7, "Sensuality Cycles," focuses on the relation of these vaginal components to the sensuality of the couple.

As women age and their estrogen levels decline, the vaginal lining tends to lose its thickness, its capacity for rapid lubrication, and its tone. These changes are almost inevitable with age, affecting up to 85 percent of women. Similar, reversible changes occur during periods of celibacy. Estrogen-replacement drugs, applied as a cream to the vagina or taken by mouth in pill form, reverse the problem within five or six days.

THE DISCORDANT NOTE OF SURGERY

With this anatomical overview completed, you can appreciate the intimate interconnection of the reproductive organs and their related hormones. Changes in one part of the system inevitably affect the other parts. Nowhere is that more evident than in medical interventions.

In the United States one in two women is told by her physician to have her uterus removed by the operation called hysterectomy (*hyster* for "uterus," *ectomy* for "removal"). In Western Europe about one fifth the number of women have this surgery. By the age of forty-four years, one in five American women no longer has a uterus; in Western Europe, one in twenty-five.

The hormonal symphony that is described throughout this chapter applies to the woman who has all of her parts in place and has not had her "tubes tied" (a tubal ligation).[7]

[7] For further details on the hormonal environment of a woman after hysterectomy, you may want to see my other book, *Hysterectomy: Before and After*. For the changes after tubal ligation, see Box 3-1.

Box 3-1: How Tubal Ligation Disrupts the Hormonal Symphony

About half a million women per year in the United States undergo tubal sterilization. In this process the fallopian tubes are blocked, by either cutting, clamping, or burning (coagulation) of the tissues. The surgery can be performed through an abdominal incision or through a periscope inserted through a tiny incision in the navel, a process called laparoscopy. Moreover in 1990, publishing in the *Journal of the American Medical Association,* Andy Stergachis et al. showed the results of thousands of post–tubal-ligation patients, comparing them to women married to men who had had vasectomies and to women who had had

(*continued*)

no surgery nor did their partners. The results were startling. For women who had undergone sterilization between the ages of twenty and twenty-nine, or whose husbands had, there was a 3.4-fold increased incidence of subsequent hysterectomy. The authors concluded that there was not any greater incidence of disease. Rather there was a positive attitude toward surgery that allowed women to accept surgery in situations where other women would have rejected such a prescription.

A series of research papers published throughout the 1970s and 1980s has provided the first intimation that the "tube tying" operation can affect the hormonal cycle of the ovaries. The kind of surgery, the skill of the surgeon, and other still-undefined factors affect whether or not a tubal ligation will affect the hormonal symphony of the woman.

In four different studies published in the early 1980s, investigators compared the different types of sterilization practices and their outcome on the hormonal cycling of the woman. Not surprisingly, certain procedures were associated with a high frequency of post–tubal-ligation complications. With one method, the unipolar high-frequency technique, 31 percent of women experienced changes in the menstrual cycle, and many of these had significantly lower levels of progesterone in their luteal phase. Severe menstrual pain was reported in 22 percent of the 1,700 women who were without pain preoperatively.

That is not to say that all tubal-ligation procedures are bad. When it goes well, 70 percent of the time it does resolve a contraceptive need, rendering a woman infertile yet maintaining normal hormonal cycles for her age. Unfortunately a number of negative postoperative effects are reported in about one third of the women of each study. The signs of these hormonal abnormalities are menstrual-cycle-length abnormalities (cycles become either shorter or longer than the normal twenty-six- to thirty-three-day length), pelvic pain and, in some women, menopausal-type complaints, such as hot flashes.

Some methods, such as the Hulka clip, produced results equivalent to women who had had no pelvic surgery (i.e., 85 percent with normal cycles and normal hormonal levels). Others, such as the high-frequency and endocoagulation methods, yielded lower likelihoods of hormonal normalcy. About 62 percent of the women of the group continued to have normal hormonal levels. Problems in hormonal deficiencies seem most accurately identified when twenty-four-hour urinary levels of total estrogens and pregnanediol are measured in the mid-luteal phase, about seven days before menstruation.

Women who have had a tubal ligation and later experience either pelvic pain or menstrual-cycle abnormalities should be alerted to see their physicians for either twenty-four-hour urinary collection tests (which are the most revealing) or a serial estrogen, progesterone, and gonadotropin test. This is a blood test that is repeated every two to three days throughout one's cycle. If the tests show that the young woman has hormonal levels equivalent to the perimenopausal transition, it is useful for her to know this and sensible for her to consider hormone-replacement therapy, a topic described later.

HOW THE SEX HORMONES CHANGE THROUGHOUT LIFE

Estrogen, the key feminine hormone of the ovary, has both a monthly cycle during the reproductive years and a life cycle throughout the lifetime of a woman. The estrogen levels in the blood reflect what is going on within the ovary because that is where most of them are manufactured, using droplets of cholesterol as the clay from which they are carved. For women in the reproductive years, Figures 3-6 and 3-7 show that the day on or before ovulation registers the highest estrogen level of the month. Notice that the middle of the luteal phase, about seven days before the next menstrual flow, contains the next peak estrogen level of the month. These dynamics change with the years.

As women mature through their twenties, thirties, and forties, the first (ovulatory) peak tends to get higher and the second (midluteal) peak tends to rise ever more shallowly. By the time women reach the premenopausal transition years, usually around the age of forty-three, they tend to circulate very high levels of estrogen as they approach the ovulatory time of the month (around Day 12 of a twenty-nine day cycle) and they tend to show ever-decreasing levels of estrogen in the luteal phase.

Meanwhile the FSH levels rise with the passage through the forties and beyond, a condition that stimulates an ever-larger cohort of follicles to join the "ladies-in-waiting court" of the dominant follicle. Each of these follicles manufactures estrogen, and the greater number of follicles leads to an ever-increasing blood concentration of estrogen in the follicular phase as women mature.

So, the highs get higher and the lows get lower each month as women move through their thirties and forties. With this change in the estrogen cycle women often report an increase in monthly premenstrual discomfort. This is in direct contrast to painful menstruation, the principal cyclic problem in young women, which tends to evaporate as the years pass. One of the most noticeable changes during the transition years, as revealed by the Stanford Menopause Study, was the frequent complaint that the premenopausal premenstrual days were very emotionally stressful. And most women commented that the distress was getting worse every month. My studies have led me to understand that these women experienced a reflection of the inevitable changes in ovarian hormone production. Without medical intervention the discomfort tends to increase each month—until the last cycle at the entry to menopause, when the follicles are, at last, exhausted.

The biological facts reveal why this is so. The stage is set by the combination of the increasing number of follicles that get recruited each month in the premenopausal transition years and the decreasing competence of a counterbalancing corpus luteum. The premenopausal equivalent of adolescent hormonal surges and purges leads most women (and their partners) to welcome the end of this monthly stress, an end that heralds the onset of menopause. For relief from this increasing premenstrual stress two approaches have been successful: Start taking hormone-replacement therapy during these transition years, and develop a physical-fitness regimen of aerobic exercises. For a research review of each, my book *Menopause: A Guide for Women and the Men Who Love Them* offers comprehensive facts.

ESTROGENS AND EMOTIONAL ENERGY: THE BETA-ENDORPHIN LINK

Rising estrogen levels promote a cascade of biological influences that increase a woman's sense of well-being. Newer research on a more recently discovered hormone family, the beta-endorphins, may expand to show that this hormone accounts for emotional well-being.

In the mid-1970s the first studies were published to show that hormones produced in certain nerve endings mimicked the effects of opium. The original investigators had gone searching for such an internal secretion because they reasoned that humans and other animals were responsive to opium. If the body had receptors for opium, logic suggested that the body also manufactured some substance that was opiumlike. As is often the case in biomedical science, once an idea is formed and an investigation undertaken, the search yields fruit. By 1990 hundreds of studies had been published to name, characterize, and analyze what came to be known as the endogenous (internally produced) opiates. The major class of such substances currently under study in humans is the beta-endorphins. Beta-endorphins are elevated in athletes after their athletic efforts. The more fit the individual, the higher the beta-endorphin response after a workout. Because of this, many runners and other athletes become "addicted" to their sports. They discover that exercise makes them feel good and that missing a day of training produces depression. They come to need their athletic behavior. I see this as a built-in mechanism that serves to keep us healthy.

In similar fashion women have been shown to have elevated levels of beta-endorphins when their estrogens are elevated and diminished levels of beta-endorphins when their estrogens are either in a state of decline or absent. Although other substances may turn out to be the final common path in the search for "the" internal chemistry in response to emotion, it is now clear that the beta-endorphins follow the rise and fall of estrogen levels. When estrogen levels are relatively high, the joie de vivre, of a woman tends to be relatively high. As her estrogen levels decline, her joy may also change into a more subtle refrain. Absolute levels of estrogen appear to be less important than the changing levels of estrogen.

The research community has shown that the three life events when a woman's estrogen levels drop acutely are associated with an extraordinary increase in the incidence of depressive events. The three stages occur immediately after childbirth; in the mid to late forties as the premenstrual levels of estrogen are declining; and after hysterectomy and ovariectomy, when the estrogen levels usually plummet. All three times are noted for their high incidence of depression. Close to half of all women are vulnerable to transitory emotional plummets at each of these hormonal shifts of gear. Medically prescribed hormonal replacement at each of these times usually helps to ameliorate this depression, unless its cause has a deeper component.

PROGESTERONE DECREASES WITH AGE

The other dominant ovarian hormone, progesterone, tends to decrease as women grow older. The corpus luteum, that structure in the ovary pictured in Figure 3-7, is the principal source of progesterone circulating in the bloodstream. It is also the main source of estrogen in the luteal phase, Days 15 to 29. The corpus luteum is the last part of the ovarian cycle to develop in adolescence and the first part of the ovarian cycle to show deficiencies as the premenopause transition years begin. Luteal-phase deficiencies, the inability of the ovary to form a corpus luteum or to maintain it with sufficient power to manufacture adequate hormones, begin to occur sporadically during the forties for almost all women. As you saw in Chapters 1 and 2, I believe that regular weekly sex helps to promote corpus luteum function and protect against this aging effect. Sexually active women in a stable relationship probably age more slowly. The higher estrogen levels that weekly sex promotes also serves the general sense of well-being.

SEX HORMONES AND SLEEPING

Women often report trouble sleeping as their menopause approaches, and this insomnia may have something to do with the changing levels of hormones, particularly the reductions in progesterone. Recent evidence by Drs. ElSayed Arafat, Joel Hargrove, Anne Wentz, and others reveals that when women take natural-progesterone therapy (a hormonal adjunct to estrogen that is just beginning to be prescribed in hormone-replacement therapy regimens), the progesterone therapy causes a pleasant sleepiness. If the dose is high enough, progesterone acts like a powerful sleeping pill. Synthetic progesterones, such as Provera, and other medroxyprogesterone acetates do not seem to promote this pleasant sleepiness. In fact the first three months of their ingestion are often associated with irritability. Natural progesterone therapy is not being routinely prescribed in the United States at this writing because the high doses necessary for adequate endometrial safety can have potentially adverse effects on the cardiovascular system. Research is under way to solve this problem, and once it is solved, natural progesterone may well become a better option for hormone-replacement regimens.

HORMONES AND THE DREAM DYNAMIC

Other relationships between progesterone levels and sleep were discovered earlier by psychologists. In 1974, when I was first studying menstrual cycles in graduate school, Dr. Jonathan Baron, an assistant professor of psychology at the University of Pennsylvania, suggested that I might want to read his wife Judith's Ph.D. thesis, which had not been published in a scientific journal. She had studied the changing pattern of dreams in reproductive-age women. She had not analyzed hormones but rather the effect of menstrual-cycle variation on dreaming. What she found in the early 1970s rather neatly parallels the 1990 work of Drs. Arafat, Hargrove and Wentz on progesterone therapy in menopausal women.

Dr. Baron showed that the pattern of dreams changes from the follicular to the luteal phases of the menstrual cycle. She found that in the luteal phase of the cycle (when progesterone is secreted) the pattern of dreams takes on a placid character. Women dreamers tend to be inside houses, inside closets, or inside other spaces (such as beds) peacefully overcoming chaos with orderliness. In contrast the follicular phase of the cycle, when progesterone is low, is associated with "outside dreams." In these dreams activity, action, leadership, and other events take the dreamer out into the larger world.

I have never since seen any citation of this work and wonder whatever happened to the follow-up of such studies. But it has lingered within me. And it blends with Ecclesiastes as it uses time to describe the nature of life. A time for inner reflection and planning is most effective if it precedes the action that follows.

Progesterone plays other roles in the internal machinery as well. When the progesterone levels are higher, a woman's metabolic rate increases by about 10 percent. During the luteal phase of a reproductive-age woman's life, she can consume about 10 percent more calories to maintain the same weight. Whether there is a relationship between a faster metabolic rate and sleeping better has never been systemically studied. Perhaps there is. Athletic activity increases the metabolic rate and seems to improve one's experience of restful sleep. Does female biology, particularly the luteal phase, serve such a pacesetter role, combining placid dreams with a faster metabolism and more energy to provide a time to plan and a time to act? I think it can if we allow it to; and if we do, it is a good thing.

MONTHLY SEX STEROIDS AND INTERNAL TIDES

The body's internal fluids rise and fall, shift and settle, like ocean tides during the course of the month. Progesterone is the prime mover. Since progesterone levels are rising from the onset of ovulation for about the next seven days, and then falling for the seven days before menstruation, one would expect other phased progesterone influences during the reproductive years as well. And there are. In the seven days before menstruation many women frequently report pelvic bloating. Studies show that the total weight does not change premenstrually. But the body fluids do shift some—from the general circulation into the pelvic circulation. The bloating is real. And it seems to prepare the body for its ensuing blood flow by distributing blood where it will be needed.

Other cyclic fluid changes are equally pronounced. Cervical mucus, a fluid that is secreted from the cervix and that lubricates the vagina, increases twenty- to sixtyfold around the time of the ovulatory peak of estrogen. Most women who keep a calendar can learn to discern the three days around the monthly midcycle time of ovulation by the copious vaginal fluids that result.

Other cyclic changes are more subtle. Irritability and depression are a common event premenstrually. Whether they should be attributed to the decline in estrogen, progesterone, beta-endorphins, or some combination remains unresolved.

MENOPAUSAL CHANGES IN SEX STEROIDS

By the time menopause has been reached, in the absence of any hormone-replacement therapy, both estrogen and progesterone levels will be much lower than they ever were during any of the days of the reproductive cycling years. Figure 3-8 shows this change in estrogens throughout the various pre- and postmenopausal years.

Note that two different estrogens are plotted, estradiol (E_2) and estrone (E_1). In the reproductive years when one of these is high, so is the other. As the figure shows, throughout the years of menopause they both decline, but then, after age eighty, the estrone levels may increase while the estradiol levels do not. Fat cells act as conversion factories, changing androgen into estrone. The old-age estrone increase may result from the weight gain so common in many old women.

HORMONAL REPLENISHMENT

Estrogen affects just about every body and mind system that has ever been systematically studied. When the sex-hormone levels begin to wane at the menopause, the bones, the cardiovascular health, the muscle tone, and the skin all show signs of aging, and this aging process increases with time. The vagina dries out, and the capacity for sexual lubrication diminishes. The longer a woman has deficient levels of sex hormones, the further deterioration in her bone mass and strength, muscle tone and strength, and cardiovascular-system health. Hormone-replacement therapy, when appropriately dosed, can retard these processes of aging.

There are those who suggest that hormone-replacement therapy should not be considered because it is not "natural." I disagree. I believe hormone-replacement therapy is equivalent to a number of other products of civilization that have allowed human beings to live far beyond their reproductive years. Heat and air-conditioning are unnatural, but they promote better quality of life. Packaged foods are unnatural, but they prevent the need for going out, hunting, and gathering for our nourishment, and in so doing increase the quality of life. Antibiotics are unnatural, but when an individual is in a state of serious illness, they may save her life. I see hormone-replacement therapy as the natural evolution of civilization. It serves to prolong and enhance quality of life. I am very much in favor of it for those who can safely take appropriate levels.[8]

[8] For a detailed discussion and tables of the various types of estrogens and progesterones, as well as discussion of the role of testosterone therapy for restoring absent libido, *Menopause: A Guide for Women and the Men Who Love Them*, 2nd ed. (New York: Norton, 1991), provides up-to-date information. For those who have had a hysterectomy, *Hysterectomy Before and After* (New York: Harper & Row, 1990) may better serve their needs.

FIGURE 3-8: The Estrogen Changes from Forty Years On

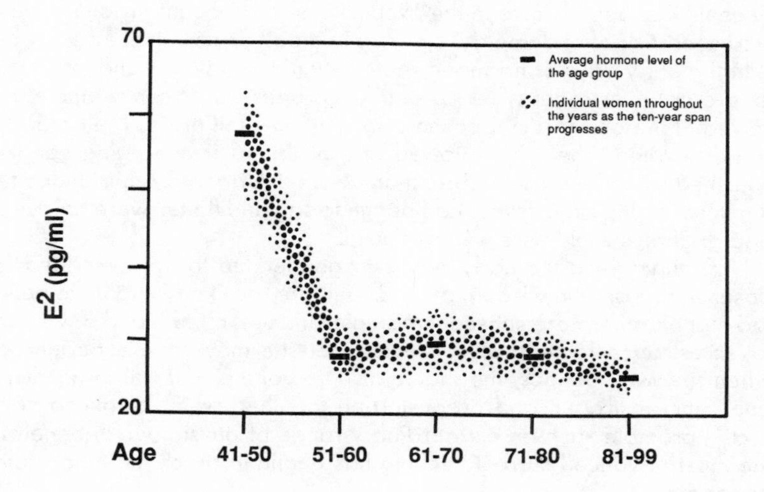

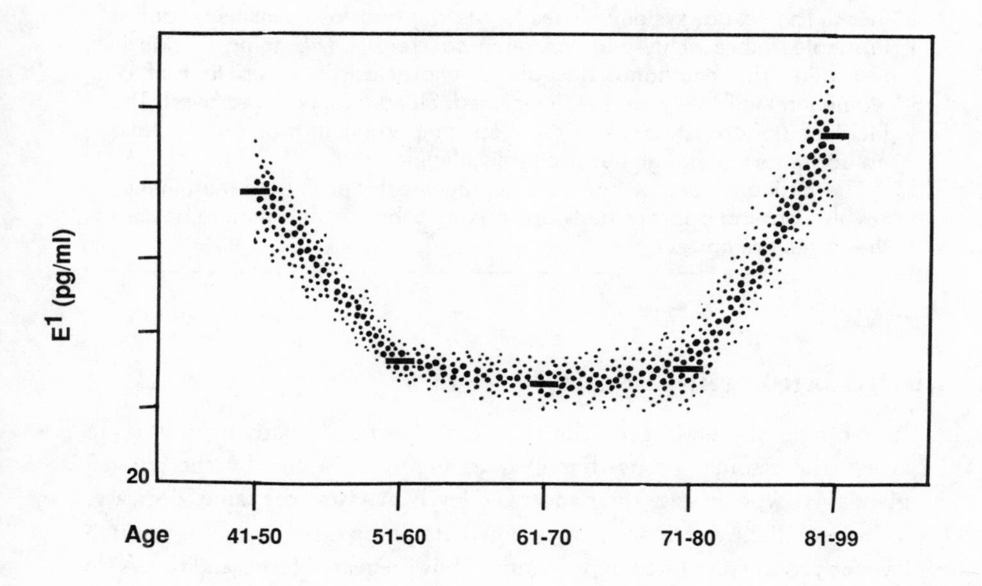

Box 3-2: Hormone-Replacement Therapy (HRT)

Hormone-replacement therapy in the 1990s has evolved enormously since its first appearance in the 1950s, when subsequent problems with increased risk of endometrial cancer put a pall on its value.

In the early days of hormone-replacement therapy the tendency was to prescribe estrogens unopposed by progestins. When women received high doses of unopposed estrogen, they felt great. Their moods were elevated. Their skin glowed and problems in the aging vagina vanished as an increased lubrication resulted. Urinary cystitis tended to diminish. Aging breasts that had begun to sag and flatten were full once again. Cholesterol levels were reduced.

Unfortunately in the early zeal estrogen tended to be prescribed in doses that, from the wisdom of hindsight, are now known to have been too high. Furthermore subsequent studies made it clear that if a woman takes estrogen to replace the levels that she may have experienced when she was younger, she should also be balancing it with progesterone. Since estrogen and progesterone together, when balanced correctly, promote such an extraordinary range of physiological benefits, the most advanced current thinking has begun to reach an initial state of consensus.

When estrogen is opposed by appropriately dosed progesterone, and when the estrogens prescribed are natural rather than synthetic, the overall health benefit to a maturing woman past the age of forty is extraordinary. Her cardiovascular-disease risk, the most common cause of death in women, is cut in half. The loss of bone mass, typically beginning in the early forties and continuing at a rate of 2 percent per year for the rest of her life, is halted. Bone may even gain a little bit of mass. The sexual system ceases to dry out and lose sensitivity, and it lubricates more richly and sensation can return. The aging of skin is retarded. The emotional despair so characteristic of up to half of women in their late forties is diminished. Sleep quality is improved. The incidence of breast cancer is reduced compared with women who take no hormones, although never entirely eliminated.

The critical issues of hormone-replacement therapy for the nineties revolve around appropriately dosing and subsequent testing to be sure the doses are correct.

AGING AND ANDROGENS

As we age, the androgens continue to be secreted, both by the aging ovary (the stromal tissue described in Figure 3-5) and by the adrenal glands. As women age, their androgen levels may stay the same, decrease, or increase. The difference may be a genetic group difference, but research has not yet answered why some women show increased levels and others do not. The effects of increasing levels of androgen can be seen in the facial hair that some women begin to grow, the deepening of their voice as they get older, and probably the better vigor that such women experience. Studies have also shown a direct but subtle relationship between the testosterone levels in women and their libido. More about this in Chapter 7.

Figure 3-9 shows the change in androgen levels with age, focusing on two of the androgens—testosterone and DHEAS (dehydroepiandrosterone sulfate). Both graphs begin with the premenopausal transition and show the variation from one woman to another.

Note that in this figure testosterone levels tend to show an initial postmenopausal drop and then begin increasing in the fifties after menopause begins. The other androgen, DHEAS, shows a steady decline throughout life. Science has not yet informed the biomedical community of the varied purposes and functions of DHEAS, but some understanding has begun. Chapter 6 reviews these theories.

PITUITARY SEX HORMONES SIGNAL THE MENOPAUSAL YEARS

Pituitary sex hormones (FSH and LH) are made of protein components rather than cholesterol components. These hormones are gigantic in size compared with sex steroids, and they work differently. They attach to the edges of cells and trigger chain reactions rather than entering them, as do the sex steroids. We know less about these than about the sex-steroid hormones, but what we do know is nonetheless quite interesting.

In contrast to the ovarian sex hormones, which tend to decline with the passage of our years, the pituitary sex hormones increase as we grow older. They increase even further after pelvic surgery, often shooting up to ten times the level that they were the day before surgery. Currently,their principal value to us is in their ability to inform us about where we are on the life cycle by a simple test that measures blood levels.

These pituitary hormones, FSH and LH, earlier shown in Figure 3-7 for the reproductive cycling years, are next arrayed for the life span.

As Figure 3-10 shows, LH levels tend to increase sharply about a year or two before menopause, to stay high throughout most of the menopausal years, and thereafter to drop in women who survive to their eighties. In contrast the other gonadotropin, FSH, shows a gradual increase beginning in the early forties, the time of the beginning of the perimenopausal symptoms, menstrual-cycle changes, and hot flashes.[9] For a woman who

[9] For details about hot flashes, the reader may want to look at *Menopause: A Guide for Women and the Men Who Love Them*, 2nd ed. (New York: Norton, 1991).

FIGURE 3-9: The Androgen Changes from Forty Years On

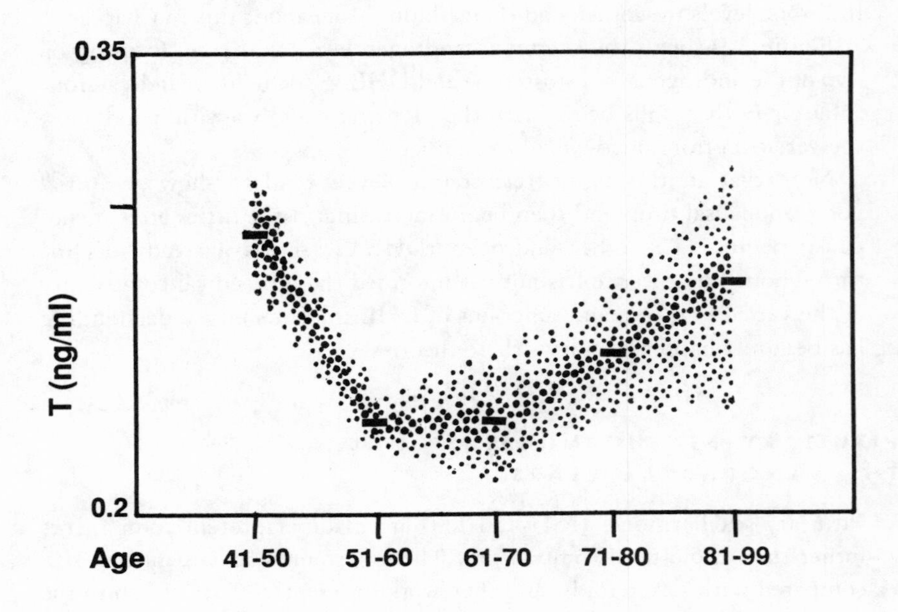

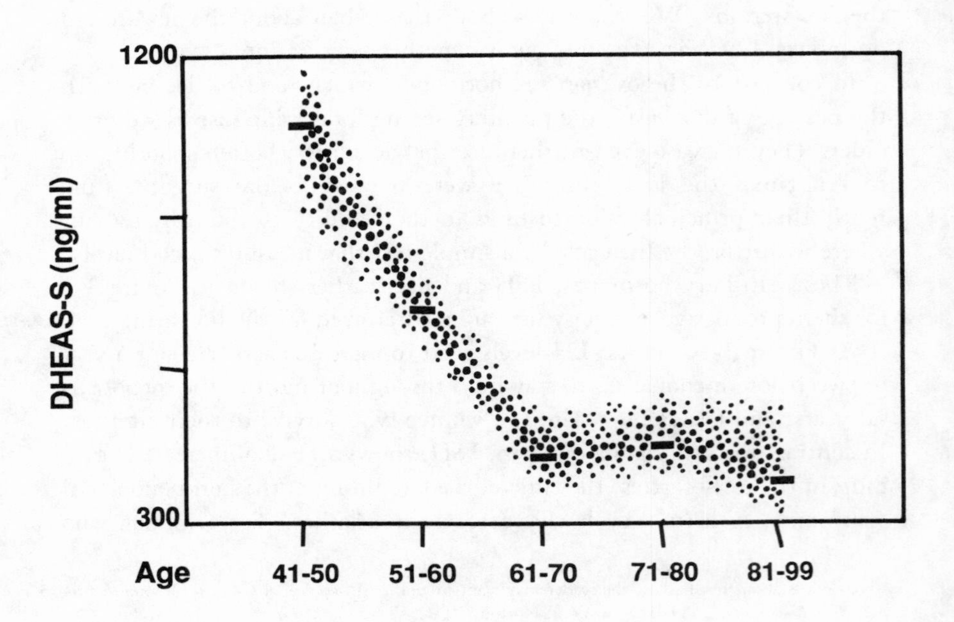

FIGURE 3-10: The Gonadotropin (FSH and LH) Changes Throughout Life

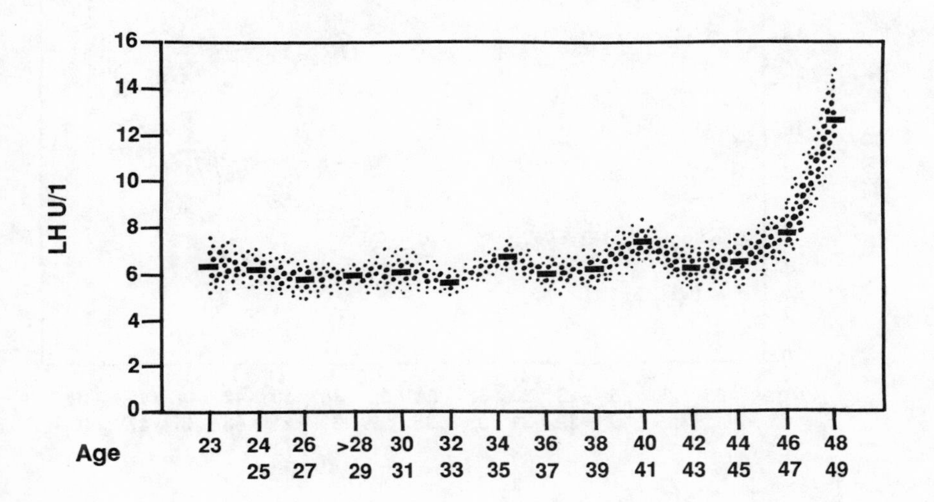

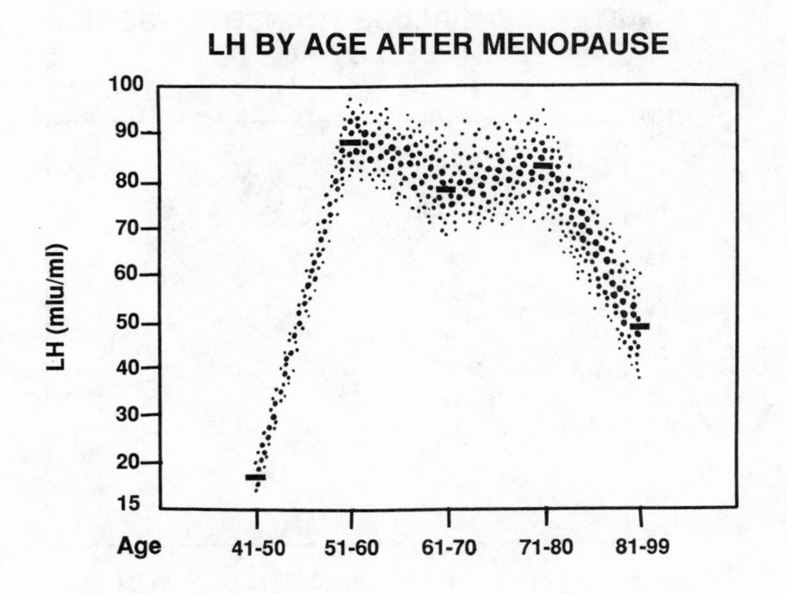

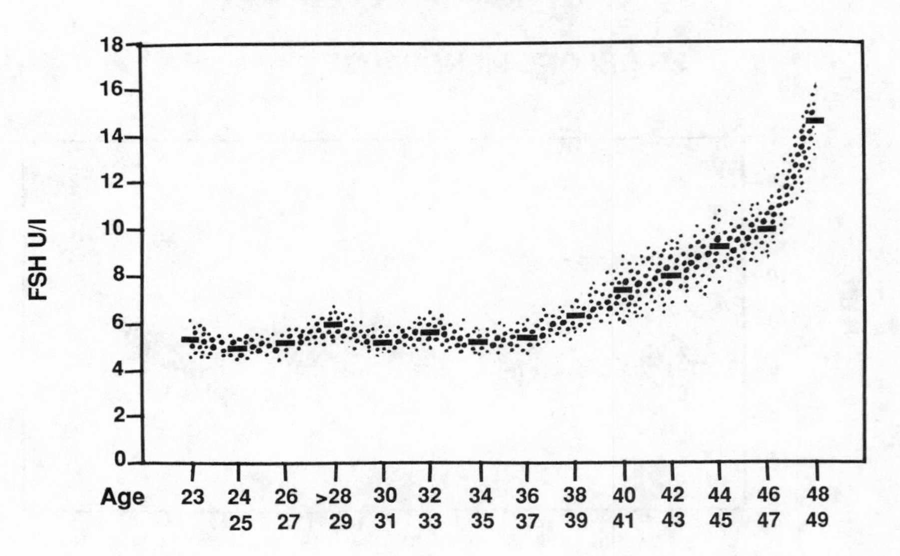
FSH BEFORE MENOPAUSE
FSH U/I
18
16
14
12
10
8
6
4
2
0
Age
23
24 25
26 27
>28 29
30 31
32 33
34 35
36 37
38 39
40 41
42 43
44 45
46 47
48 49

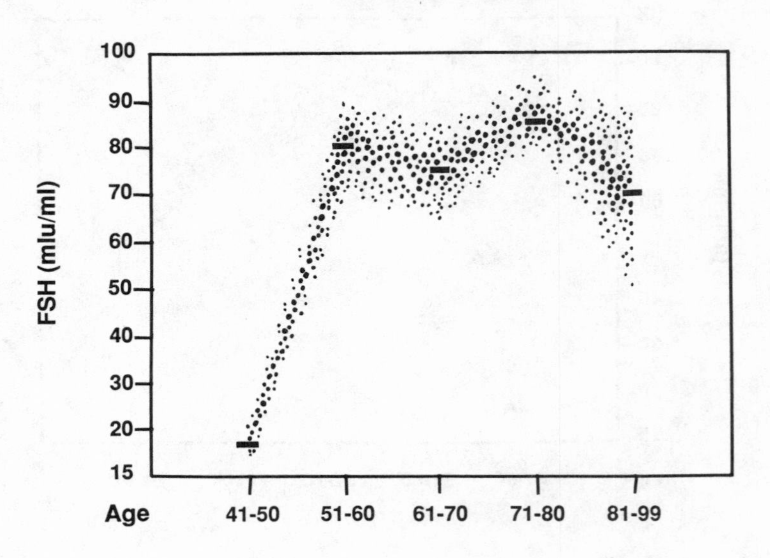
AFTER MENOPAUSE TRANSITION BEGINS
FSH BY AGE
FSH (mlu/ml)
100
90
80
70
60
50
40
30
20
15
Age
41-50
51-60
61-70
71-80
81-99

is experiencing irregular menstrual cycles and other premenopausal symptoms, a blood test for FSH levels can indicate if she is approaching her menopause. This is a particularly useful test when these menopausal symptoms begin in the thirties rather than in the forties. Perhaps 10 to 20 percent of women enter menopause seven or eight years earlier than the average age of fifty. If the last menses occurs at forty-three, then the transition years began in the midthirties.

I believe that the sum of the biomedical data from the international scientific-research literature strongly suggests that women will benefit enormously from taking hormone-replacement therapy during their menopausal transition years. Hormone-replacement therapy is of great value for those 10 to 20 percent of women whose menopause begins prematurely. For these women it would help to identify when these years have begun. The onset of the transition can begin as early as age thirty or as late as fifty-two. A blood test for FSH levels will help to pinpoint the transition. However, such blood tests can inaccurately suggest low levels (false negatives) and can lead to an incorrect diagnosis that the woman is not yet in her seven menopausal transition years. A woman who suspects that she is approaching menopause and has a test with negative results should know that three consecutive days of the same blood test may reveal what a single reading will not.

CONCLUSIONS

The hormonal symphony of women is an extraordinarily rich and complex subject. The chapter has covered a great deal of ground: naming the hormones, explaining how they change, explaining how the change in each derives from predictable changes in anatomy and physiology, and showing how hormonal changes in one region of the body inevitably affect hormonal changes throughout the entire body. For the purposes of *Love Cycles* I have focused attention on those specific changes and dynamics that I will be able to relate later to sexual response, sexual capacity, and sexual interest. In order to enjoy the intimate connections to which we are exposed, it helps to understand both ourselves and our underlying dynamics. Because of the complexity of a woman's body and the ever-changing harmony of its hormonal output, there exists an inevitable dynamic in emotion, capacity, energy, and quality of intimate experience. I believe this inherently feminine change produces a potential richness, which is what men, with their flatter hormonal patterns, are drawn to in women. We, the keepers of this symphony, do well to understand it. The hormonal symphony of women is magnificent, and we should relish the experience of being a female.

I have often heard the question, If women have a cycle, aren't they therefore less stable than men? Quite the opposite. Long before me, Heraclitus (500 B.C.) noted the law of stability and flux. He pointed out that the most stable structures are those that fluctuate. If flux leads to stability, then women are very stable, particularly when one considers the hormonal fluctuations inherent in menstruation, pregnancy, childbirth, and lactation, which women uniquely experience in addition to the events that men and women share in common, such as work and study, growth and aging. Although women are less strong physically by about 25 percent because of less muscle and bone mass, which are hormone dependent, their durability is nonetheless much greater. We have only to look at the longer female life span to see this.

There are those who suggest that because women have a hormonal symphony, they are therefore not fit for certain sensitive roles, such as in political and other human relations arenas. Some suggest that a woman should never be in the position of nuclear "button pusher." To these people I would respond, Yes, PMS (premenstrual syndrome) exists. It probably seriously affects about 5 percent of reproductive-age women. And it is true that those with severe PMS should probably not hold a politically sensitive job requiring emotional stability. The data show that the percentage of women with this debility is probably less than the percentage of men with similar inhibiting debilities of personality. For the broad majority of women the hormonal symphony they produce should enhance their perspective and increase their capacities to perform delicate and potentially explosive work. Because of a woman's experience of her own ebb and flow of energy and emotional volatility, there is the likelihood of having learned from the experience. When one triumphs over the highs and lows by developing a high tolerance for change, one has a greater depth of perception than one who has never had to develop such flexibility. In addition, a premenstrual decline in optimism may enhance judgment. In other words, to the degree that anatomy is destiny, the biological fact can be cast in a positive light.

A knowledge of the symphonic change and the ways in which hormones affect mood and energy should help us to know how to wait when the time of the cycle is not opportune for a particular dynamic. It should fill us with a positive outlook. When we know the inevitability of change, we can begin to plan and predict optimal times for action and goal achievement. The achievement of enhanced intimate relations should be a high priority. Let us turn now from the general relationship between sex hormones in women to the sex hormones and sexual capacities in men.

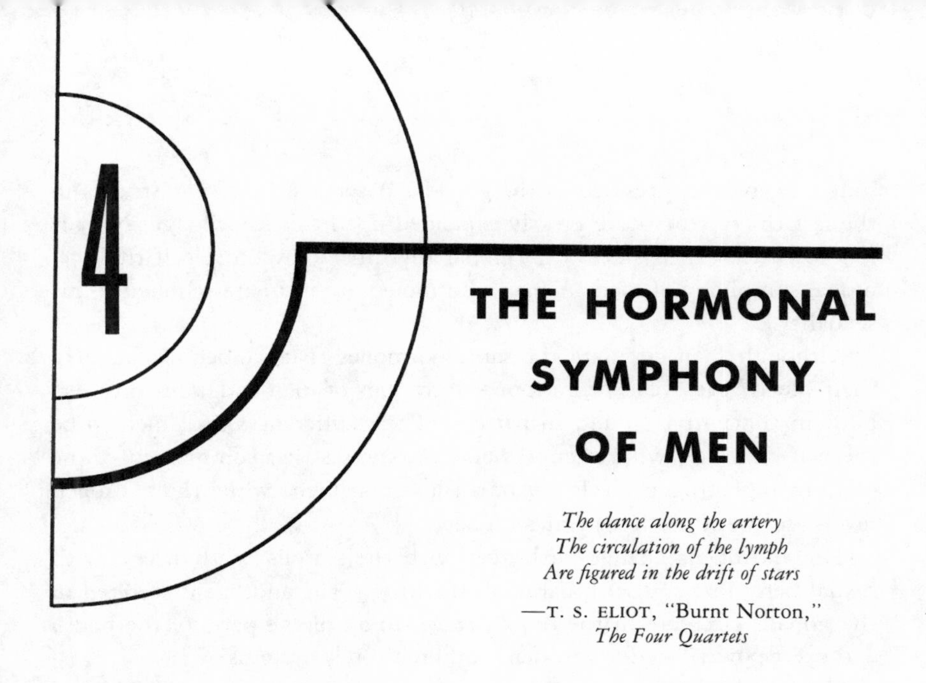

THE HORMONAL SYMPHONY OF MEN

The dance along the artery
The circulation of the lymph
Are figured in the drift of stars
—T. S. ELIOT, "Burnt Norton,"
The Four Quartets

THE MALE SEX HORMONES

Most educated people are aware that castration is devastating to the manly attributes of an adult male. Remove the testes and the man loses his deep voice, his need to shave his face, the power of his muscles, and his very virility. Precisely to deny virility to the men in their service, powerful kings in earlier times had certain of their male subjects castrated. Impotence was believed to render the eunuch safe to serve their women. Were the kings right? Was the eunuch rendered impotent? In some cases they were, and in other cases they probably were not.

The first "scientific" experiment designed to prove the value of the gonads was published in 1849. Physiologist Dr. A. A. Berthold experimented on cocks by castrating them. Inevitably and rapidly they lost their cockiness. Their male display structures disappeared. The combs and wattles, so pronounced in the male, lost their bold color and strength. They stopped strutting around the barnyard exerting dominance over the hens.

Drawing upon the knowledge about court eunuchs and adding the work of other researchers, reproductive biologists concluded that some source of maleness resided in the gonads. A series of scientific studies in the last thirty years have greatly expanded upon this knowledge. Now it is known that men show a hormonal rhythm, a rhythm I call the hormonal symphony of men. A man's hormones are intimately linked to his sexuality.

Although men circulate the same hormones that women circulate in their blood, the cycles of hormonal secretions of men and women differ, both in concentration and in pattern. These differences lead men to be the way they are, with strengths and weaknesses that complement those of women. Estrogen levels are much lower in men, while the androgen levels are twenty to forty times higher.

The androgens change with age, with the seasons, with stress, with sexual behavior, and with habits of smoking.[1] The androgens secreted in the gonads circulate in the blood, travel to all of the parts of the body, and exert powerful effects in many different body systems.

Meanwhile the two sex hormones of the pituitary gland, the LH and FSH, appear to be secreted in response to the androgens circulated by the blood.[2] The dynamic interaction between pituitary and gonads works in a very similar pattern to that of women. As men get older, their sex-steroid hormonal androgens tend to decrease and their pituitary sex hormones (FSH and LH) tend to increase.

Figure 4-1 shows individual testosterone levels in men as the levels vary with age. A total of 155 males provided their blood. The blood was tested for testosterone level, and the data were grouped by age. Each dot in this graph represents a hormone measure of one boy or man at one time of his life.

The figure also shows the average hormonal level at each of these ages, and the average is represented by a horizontal line at each age. Consider, for example, the "dots" of the men in their twenties and thirties, forties and fifties. The figure shows that their average testosterone level is 635 ng/100ml. But the figure also shows that this average level is only achieved by a few of the men in their fifties. Some men have levels as low as three hundred. Other men in their fifties have levels as high as one thousand. And that's a very important point of a graph like this. The average testosterone level declines from about age thirty onward in most studies, but at every age there are some men who are very high and others who are very low in their androgen levels.

---

[1] The term *androgen* is used to denote the general class of male sex hormones; most of the studies evaluate testosterone, which is the most powerful of the androgens.

[2] FSH, follicle-stimulating hormone, and LH, luteinizing hormone, exert actions at the gonads in ways that are similar to the actions they exert in women. LH stimulates steroid production and therefore testosterone synthesis. FSH is necessary to trigger the development of sperm.

FIGURE 4-1: Male Testosterone-Level—Changes with Age

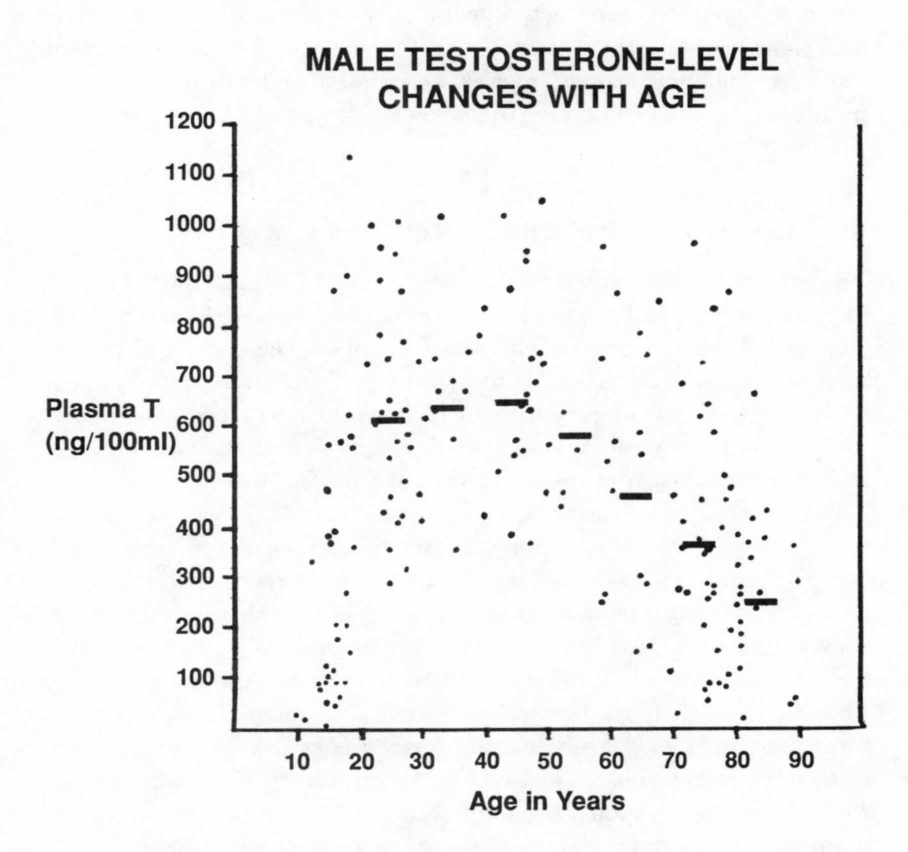

Is a decline in testosterone inevitable? Apparently yes. In all but one of the many studies (the Baltimore Longitudinal Study) this systematic decline with age has been reported. We have as yet no way of predicting how steeply a man's hormones will decline as he ages.

SEX HORMONES AND THE TIME OF DAY

If a young man has blood taken from his arm six times a day starting in the early morning and continuing every four hours thereafter, lab results will tend to show a rhythmic rise and fall of testosterone reflecting the time of the day. When he goes to sleep, his hormone levels will start rising hour by hour until, by the time he wakes, his testosterone levels will have reached the highest they will be for the nighttime sampling. By the early and late morning he is likely to show his peak levels. In the late afternoon his testosterone will usually fall.

The late-afternoon levels tend not to show significant differences between younger and older men, but the early-morning samples do. Older men show less testosterone than younger men when blood is tested in the morning. For these reasons, to evaluate testosterone levels in a man, one must consider the time of the day, when he slept, and his age.

SEX HORMONES AND THE TIME OF YEAR

The time of year is significant also. Studies conducted in North America in 1975, in Paris in 1983, and in Australia in 1976 have shown that men secrete their highest levels of the year in October. There was a 16 percent increase in testosterone levels from April to October and a 22 percent decline in their testosterone from October to the next April. Australia is in its springtime when Paris and North America are in their autumn, yet men in all three parts of the world showed a similar pattern.

Seasonal changes in reproductive rhythms have long been understood in a variety of mammals. Some mammals are known as "short day" breeders, others as "long day" breeders; this reflects the time of year when they are breeding their progeny. A great deal of scholarly research has shown that in such seasonal breeders the amount of light transmitted from the sun and the timing of it have a profound influence on these mammals and their sex-hormone levels. One hormone, known as melatonin, produced in the pineal gland inside the brain and elsewhere, appears to be responsible for influencing the seasonality of the sex-hormone rhythms in these seasonal-breeding animals. Human beings, as well as monkeys, appear to have escaped, for the most part, this light/dark-cycle control of their capacity to reproduce. Except for those humans who live near the North Pole—and who consequently experience extremely long nights (twenty-two hours) in winter and similarly long days in summer—human beings have the capacity to reproduce all year round, with only a minor cyclic fluctuation in the predisposition for one time of the year over the other.

This result leads me to suggest that the length of the day, reflected in the season of the year, is not what causes the annual testosterone rhythm, nor is ambient temperature. The position of the earth in its orbit around the sun seems to be a more likely cause. Chapter 8 expands on the subject of cosmic influence on the hormonal symphony of women and men.

For now let's look at a 1986 study, conducted in Germany, of thirty-three healthy young men, which showed higher testosterone levels than any other published study had reported. The authors suggested that the higher German levels reflected the time of the year the blood was tested, not the German nationality of their particular sample of young men. If the investigators had been less sophisticated, they might have concluded that German men show higher sex-hormone levels than the men of other nations. These problems are interesting to scientists and relevant to the public because they show how important it is to be alert to variables such as season or time of day when one is trying to elucidate principles of sex hormones and sex behavior.

SEX HORMONES CYCLE IN MEN INDIVIDUALLY

One rather detailed study in 1975 of twenty young men is worth reviewing. The investigators had these men provide blood every other day for two months. Twelve of the twenty showed a clearly discernible cycle of testosterone with regularly repeating rises and falls, but each man who did show a cycle had a cycle unique to himself and different from the others. In other words, men emerged as cyclic, but with unique individuality to their cycles. This pattern of individual cycles is very different from the pattern in women, where the 29.5-day cycle described in Chapter 3 is the universal optimum for fertility and is the average cycle of just about every large sample published. Thus males are hormonally individualistic creatures, whereas females are hormonally harmonic with one another.

A FUNDAMENTAL DIFFERENCE BETWEEN MALES AND FEMALES

Dr. Robert Sapolsky published a number of studies between 1982 and 1986 reviewing the findings of his field research in the wilds of Kenya. He would go annually to live among wild baboons. When he wanted to learn what their hormone levels were, he would use a dart gun to "shoot" one animal with a sleeping potion that lasted long enough to take some blood from the sleeping animal. The results of his studies appear in the next chapter within a discussion of the effects of stress and the patterns of dominance hierarchy on sex-hormone levels. One fundamental point is relevant here, though. Dr. Sapolsky was able to study the male levels, but he explained in a footnote that he was unable to evaluate female hormones. When he "shot" the male, no other animal came around to stop him from drawing blood from the sleeping male. When he tried "shooting" the female, he found his attempts futile. Immediately the downed female was surrounded by her "sisterhood," who would not let him get near her. This footnote intrigues me because I think it may reflect a fundamental biological difference between males and females. The physical demands of pregnancy and child care tend to require extended support, as well as economic help, to the mother or her substitute. Whether through the family or agencies, women tend to form the social communities that provide for the young, a process that requires cooperation. The biological underpinning—a harmony in the cycle of women in contrast to a cyclic asynchrony in the hormones of men—may be responsible. In any event the gender difference may be an important variation that serves to promote the viability of reproduction. The gender difference is similar to the biological difference in men's and women's hormone cycles. Fertile women cycle in a harmonic rhythm, in phase with the moon (see Chapter 8) and with one another. Fertile men apparently do not show this harmony with one another.

SEASONAL VARIATION IN THE REPRODUCTIVE CAPACITY OF MEN

The production of semen, an ejaculated fluid that contains sperm and a variety of other nutrient-rich substances, also shows a seasonal variation. Figure 4-2 depicts the percentage of the "adjusted total number" of births in three different parts of the United States. The birth rate is highest in the autumn and lowest in April in all three parts of the country, but in the hotter climates the lows are lower and the highs are higher.

Figure 4-2 was created from a sample of 1,200 births and reflects the discovery that conceptions (nine months earlier) are lowest in the summer and highest in the winter.

Investigators have also shown that the semen of men who work in the heat shows enormous declines in quality in the summer. The sperm concentration (total number of sperm per ejaculation) drops. The motility (how actively swimming the sperm are) also declines during the hot months among men who work in the heat. The motility is highest in spring, and a second peak of vigorous swimming shows up each November.

Other annual cycles in male sex hormones and sexual function have also been studied. One pituitary sex hormone, FSH, shows a peak in July, reflecting the relatively low levels of testosterone at the time of year. The other gonadotropin, LH, also shows an annual cycle. Each year the highest levels of LH in both elderly men and young men occur in the spring,

FIGURE 4-2: Annual Cycle of Births

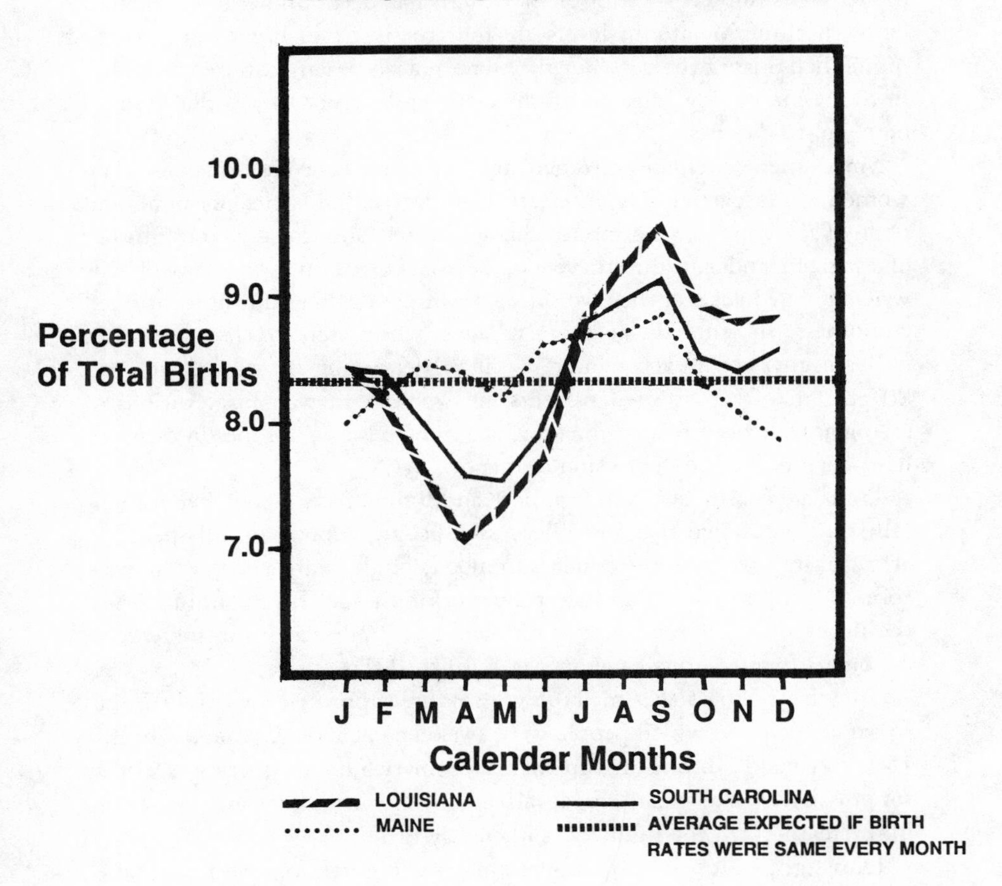

precisely when the men are showing their lowest levels of testosterone for the year.

SEX HORMONES AND SUNSHINE

Sunshine affects men, influencing their estrogen level. As the weather shifts from cloudy to sunny, estrogen levels rise in men. And when the sunny weather turns cloudy again, the estrogen levels fall. When men find that they feel depressed on cloudy days, this may help explain why.

No similar studies seem to have been done for melatonin, following day-to-day changes in weather and day-to-day changes in pineal hormones. The annual rhythm of melatonin has been studied and shows a daily rhythm. Melatonin levels are four to six times higher during the night than during the day, but they have not been demonstrated to relate, in the human being, directly to the estrogen levels or to weather changes on a day-to-day basis.

Since men circulate estrogen in much lower concentrations than women, it is particularly interesting to see that this sex hormone and none of the other sex hormones changes in response to sunshine. Studies of estrogen and sunshine have not been reported in women as of this writing, but I suspect they would show similar declines during continued cloudiness. In Finland's northern villages, where eight weeks of darkness (polar night) occur each winter, a winter depression is well known, and 80 percent of the affected patients are women. In women elevations of estrogen have been related to elevations in mood; and declines in estrogen have been related to depression of mood.

Light affects sex hormones in men; and light affects mood. Experimental data have shown that some depressed people respond to full-spectrum artificial light with improvements in mood. Night-shift workers can overcome deficits in energy and sleep by working under bright light during the night shifts. Is the effect of artificial light reflected by an increase in estrogen? Investigators have not yet published this answer.

They have shown that melatonin levels are suppressed by bright light. In experiments in which people were awakened at 2:00 A.M. and a bright 150-watt flood lamp shined in their face for two hours, there was a clear suppression of the melatonin levels practically instantaneously with the light treatment if the light was sufficiently bright.

Dim light bulbs did not suppress normal nighttime elevations of melatonin. But the issue is not so simple as it would seem. Simply suppressing melatonin, or making melatonin levels lower, provides no guarantee against depression. In fact studies in mental hospitals have shown that melatonin levels are *significantly lower* in men hospital patients who are suffering severe depression, according to a 1984 study by Dr. Johan Beck-Friis et al.

Therefore, although bright-light treatment does reduce melatonin levels and does help to reduce the problems of annual depressions (known as seasonal affective disorders), both of these effects transiently reduce the melatonin level. Light affects melatonin levels and light affects mood, but the complexity of the relationship probably involves other elements besides the ones so far researched.

THE EFFECT OF VASECTOMY ON SEASONAL HORMONE CHANGES

By 1988 investigators were showing that vasectomy abolished the annual cycle in male sex hormones. Within one to three years after vasectomy, men no longer show the normal annual cycle in hormonal production. We really don't know what this means for the health and wellness of men, but I am concerned. I perceive the cosmic control of the hormonal symphony of men and of women to be something akin to the music that sets the rhythm to which we dance. And it is good to be able to dance.

THE EFFECT OF SMOKING AND STRESSES ON MALE SEX HORMONES

Smoking increases the testosterone level by about 5 percent at every age. Surgical stress also affects the testosterone level. At the moment of incision testosterone level plummets and then quickly recovers. Other kinds of stress also affect testosterone. Parachute jumping resulted in a profound drop in testosterone level, but only after the first jump. When men returned and repeated the jump, they began to show elevated levels immediately after all the subsequent jumps. In other words, it appears that intense fear-producing stresses do affect testosterone level, but only the first time. Once men learn that they can (and will) conquer the obstacle, the effect reverses. Perhaps the surge in androgens after taking a risk is the hormonal "cause" of male daring. The results do suggest the possibility.

THE EFFECT OF PERSONALITY AND EATING HABITS ON MALE SEX HORMONES

Despite a systematic search for relationships between androgens and personality factors such as aggressiveness or the capacity to relax, no connections have been found. Neither has the usual pattern of a man's nutrition shown any relationship to his androgens. According to studies conducted in Belgium and Germany, men who lived on vegetarian diets, macrobiotic diets, and normal high-fat diets all showed similar age-related patterns in their testosterone levels (T-levels). Sudden changes in diet are different. When men go on a vegetarian diet for the first time, their T-level drops. These temporary changes do not continue over the long term. The relationship between hormones and nutrition in men is different from the pattern in women. Women who are vegetarians show lower levels of sex hormones than women who are meat eaters, and the effect remains over the long term.

MALE ATTRIBUTES AND MALE SEX HORMONES

Several other relationships between male sex hormones and aspects of a man's life have also been studied. The higher the level of testosterone, the more often a man needs to shave, the more muscle mass he can build, and the more "sexual" he is likely to be.

Box 4-1
Sex Hormones: Which Testosterone Fraction is Biologically Active?

In order to learn how sex hormones may influence sexual behavior and physiology, it is necessary to study the right hormones. A variety of "testosterone" components circulate in a man's blood, and only recently have the relevant ones begun to be identified.

Testosterone (T)is the most powerful of the male androgens. *Plasma* (or *total*) testosterone refers to all the T circulating in the blood. Plasma T is measured in units—nanograms (ng) per milliliter (ml) of blood drawn. But plasma T is an imprecise measure in that it fails to isolate the biologically active fraction that is traveling in the blood.

Biologically active means what it seems to mean—the part of the hormone that acts on the mind or on the body. The term *free testosterone* evaluates the hormone that is traveling in the blood free, not "bound" to a binding protein molecule. To get an idea of the intention underlying the use of this terminology, you might think of the difference between a man who is bound to a woman and a man who is unattached (free). The distinction between "free" and "bound" hormone was thought to differentiate what was biologically active from what was inactive—that is, held in reserve. Besides, bound versus free testosterone concentrations were relatively easy to measure by chemists who could separate the two fractions through a kind of strainer. The bound was bigger and could be strained out from the smaller (unattached) free hormone. Until about 1986 investigators exploring relationships between sex hormones and sex behavior tended to measure only the total plasma T and the free-T levels as they attempted to discern relationships between sex hormones and sex behavior.

(*continued*)

More recently biochemists have discovered that the biologically active component is not what they expected. It includes two different elements:

1. The testosterone that travels bound to one particular protein (albumin), but not the testosterone traveling bound to other proteins
2. The testosterone traveling "free" in the blood

Why is this important? Because the free T comprises less than 5 percent of the total T, but the total active portion—the albumin-bound and free T—together make up about 60 percent of the total. Because the measures of free T that were reported previously comprise so little of the total hormone that is biologically active, most of the studies published before 1989 could only crudely address the question, How does testosterone level in blood relate to a man's sexuality? As a scientist I provide these details because the reader needs to know that the available evidence uses data that are very limited.

The most recent research has shown that the crude measure of "free testosterone" is inadequate (see Box 4-1), but this was the common measure of analysis before 1989. Although some discoveries were made with these relatively crude "free T" measurements, it is likely that we will learn a great deal more in the next few years about how closely related the biologically active testosterone levels and the patterns of sexuality are in the men who are governed by their hormones. Even so, with only these crude hormone measurements a number of very significant relationships between a man's testosterone levels and his sexuality have already been demonstrated in healthy men as they age.

Figure 4-3 shows the age-related decline in orgasm, morning erections, and sexual thoughts (frequency of thinking about sex).

When you study the figure, you can see that men in their sixties show a much lower incidence of monthly orgasm than younger men, and that men in their seventies show an even more drastic reduction. Compare the decline in sexual behavior with the decline in the average testosterone level at these same ages in Figure 4-1. As you will see, although the mental part of sex, the thoughts and enjoyment, decline with age, cognitive aspects do not decline as steeply as physical capacity. This means that while a man's sexual thoughts can be active, his sexual organs do not have to be. Let us turn next to the subtleties of the hormonal symphony of male sexuality.

FIGURE 4-3: Sexual Expression of Men—Changes with Age

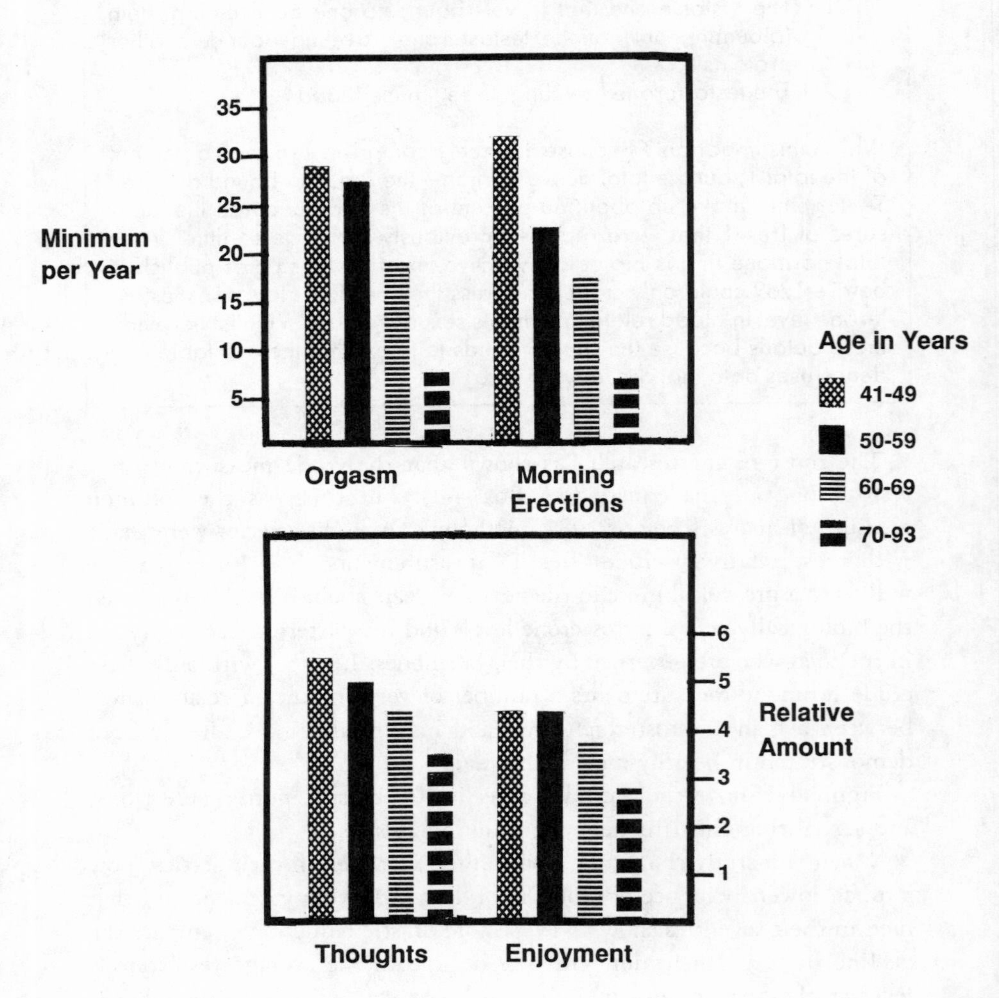

THE SEXUAL BEING—DO THE HORMONES CONTROL HIS BEHAVIOR OR DOES HIS BEHAVIOR MODIFY HIS HORMONES?

The science of intimacy for men leads us inevitably to the role of sex hormones and sexuality in men. Many different studies have shown that the relationship is powerful. But it is like the problem of the chicken and the egg—the question of which comes first, which causes what. The answer seems to depend in part on the stage of life.

THE ADOLESCENT BOY

To understand the science of male intimacy, and to consider the hormonal symphony of men, it makes sense to begin with the hormonal changes in adolescence. As Figure 4-1 showed, a startling surge of testosterone occurs sometime between the ages of ten and twenty in healthy young men. In 1985 the first scientifically sound study on adolescent male hormones and sexuality was published; it found a powerful relationship between the T-levels of boys and their sexuality. Drs. Richard Udry and Naomi Morris, working with colleagues, investigated eighth-, ninth-, and tenth-grade boys in two different school systems. What they found has an important message for us all.

They took blood from thirty-three eighth-grade boys and sixty-nine ninth- and tenth-grade boys. Through questionnaire and interview they asked a great many questions about the sexual behavior of each boy. They assayed the blood to measure the testosterone concentration and compared the testosterone levels to the age at which boys began their active sexual life, as well as the amount of energy spent thinking sexual thoughts; the ease with which they reported being "turned on" by a girl; the amount of kissing, hugging, and fondling they experienced; and their physical development. Their findings were clear. A strong relationship exists between the T-level and each of the sexual measures. The higher the testosterone, the more frequently the boy thought about sex, the more easily he was turned on, the more frequently he held hands, hugged with girls, and had intercourse, and the more sexually mature was his physical development. Even among the boys who were virgins (sixty-six of the group) the same relationship held. The T-levels showed a strong and close relationship to sexual experience such as masturbation, ease of being turned on, closeness and physical intimacy with a girl.

The effects of the androgens—in this case the strongest androgen, which is testosterone—on both sexual motivation and sexual behavior are very strong. As testosterone level in the blood increases, sexual thoughts increase proportionately, and the behavior follows.

THE RELATIONSHIP OF COURTSHIP BEHAVIOR AND TESTOSTERONE

Six years earlier, in 1979, Dr. Harvey Fader published his study of the testosterone level of courting ring doves. His results lend flavor to the human research. Ring doves coo when they're courting. The male performs a courtly bow in front of the female of interest. The couple preen each other, removing parasites from each other's necks with their bills, and in that way provide service to each other that they cannot provide for themselves. Male and female ring doves share parental responsibility too. They take turns sitting on the nest as they incubate their eggs. Dr. Fader showed that the T-level served as a kind of sexual barometer in the male. Before pairing, T was at the lowest level of the study. Four hours after the pairing the T-level had doubled. Three days later it had doubled again. Clearly T-levels are related to courtship and pairing in ring doves. Courtship behavior activates the secretion of sex hormones in ring doves. Something similar was shown in Atlanta, Georgia, in another species—rhesus monkeys. It seems that humans have certain similarities as well.

SEXUAL BEHAVIOR IN MEN IS MORE COMPLEX

In the study of sexual behavior in young, middle-aged, and old men, knowledge has now expanded to include at least fifty published scientific studies. The results reveal a tremendous concordance with the underlying principles. Still, the study of sexual behavior, sexual desire, and sexual response in men is complex. One leading researcher in this field, Dr. John Bancroft, a psychiatrist in Scotland, believes that three questions must be considered in order to have a reasonable understanding of male sexuality:

- Cognitive—discovered by asking the man "How do you think?"
- Affective—discovered by asking the man "How do you feel?"
- Physiological—discovered by asking, through actual scientific measurement, "How is the physiological response happening?" (e.g., the change in circumference of the penis during erection)

Scientific investigators, and most everyone else, have long understood that mood and sexuality are closely related. Men who are depressed tend not to be sexual during their depressed state. Depression tends to impair or abolish even the automatic penile erections that occur during sleep. In women similar reactions are said to exist. Both thinking and feeling alter sexuality.

Through the experimental method scientists ask and attempt to answer the question, Where in the body and/or brain is sexual function controlled? For example, scientists can separate the components of male sexual response to show that erection and ejaculation are two physically and mentally separate functions. Ejaculation without an erection can occur. Erection without ejaculation can as well. In the lower animals the control of sexual function appears to be located in specific regions of the brain; specific areas can be stimulated electrically to yield erection, while other regions can be stimulated to produce ejaculation. Recall, from Chapter 2, that by cutting the optic nerve of female rats, experimenters abolished fertility. Certain nerves that course through the brain are necessary for fertility.

But men are more complex, as the experiments in the mid-1950s of a neurosurgeon, Dr. Calder Penfield, showed. During surgery on the brain he discovered that he could stimulate different areas of the brain to elicit memory and smell, as well as entire sequences of filmlike arrays of visual memory, but he could not elicit an erection or even a mental image of sexual response in humans. What these experiments seem to suggest is that the neuroendocrinology of sexual response in men does not occur in the nerves of the brain, the way it does in the lower animals. As seems to be the case for women, where control of the timing of the hormonal symphony occurs in the gonads, so it may be for men.

Although brain stimulation does not increase human erectile sensitivity, hormones can. In fact the gonadal hormones directly influence the penis. Especially here, the hormonal symphony exerts its influence.

PENILE SENSITIVITY DECLINES WITH AGE

Three extraordinary studies of sensual response have led to the clear conclusion that the sensitivity of the penis declines as men age. And the decline is severe. In 1970 Dr. Herbert Newman published the first study. He evaluated one hundred men, all of whom agreed to have an electrode (a tiny piece of electrically sensitive metal) placed at the tip of the penis. Vibration was applied through the electrode, and the man would announce the moment he could first feel the energy. The experimenter continued to increase the amount of vibration voltage until the man reported he could feel the stimulation. The results help to explain the declining sexual function that is a characteristic of aging.

In young men the ability to feel vibration occurred at very low levels of stimulation. In other words the "threshold for sensitivity" was low. The older the man, the more stimulation was required before he felt anything. The threshold for sensitivity increased. Further discoveries followed. In each of the age groups the men who had had no sexual intercourse in the past year had *five times higher* thresholds for sensitivity than the men who were sexually active. In other words their penises were one fifth as sensitive to touch as those of their contemporaries who were sexually active.

Men who lack sensitivity at the penis tend to be men who are not sexually active. Another study a few years later reported the same finding. Every day, men ranging in age from nineteen to fifty-eight years old logged in a calendar whether or not they had had intercourse. Here again sensitivity at the penis tended to decrease with age, but this study added information. There was a close relationship between the amount of intercourse and the amount of sensitivity. The more sensitive the penis was to vibration, the more often the man tended to have intercourse. The author of this study, Dr. Allen Edwards, suggested that these results might account for several well-known phenomena in men:

- Unwanted spontaneous erections tend to be an adolescent problem. Dr. Edwards suggested that for the highly sensitive penis of a young man the inevitable friction of his clothing is sufficient to produce erections.
- Premature ejaculation tends to be a young man's problem as well. Dr. Edwards suggested that the hypersensitivity of the young man's penis provides so much stimulation that uncontrolled ejaculation is more likely to occur in youth.
- As the sensitivity decreases, the capacity for self-control seems to increase. As men mature, they are better able to delay their ejaculation in accommodation to their partner's pleasure.
- The age-related insensitivity may contribute to the impotence that becomes more common as men pass into their fifties. It may account for the sexual lure of irritants (such as flagellation, pornography, and others) for more stimulation to achieve the same sensual effect.

In 1989 a third study of sensory thresholds, this time comparing the penis and finger, was published, and again the same results were found. As the age of men increases, the sensitivity of the penis decreases. Younger men who have diseases such as diabetes show patterns characteristic of old men: The sensitivity is low. One further discovery was reported. Among the younger men there was a close relationship between how sensitive the penis was and how much sexual intercourse the man recorded. In fact the relationship was so close that it appears that just as a man was beginning to lose sensitivity (usually before age thirty-eight), there was a sharp decline in his frequency of sexual intercourse.

As men age, there is a decline in their penile sensory perception, and with this decline in sensation a decline in sexual intercourse follows. This might help to account for the midlife sex crisis so common around this age.[3]

The research results seem to suggest that either behavior or hormone change can trigger the other. The relationship is complex. For example, according to studies conducted in the later 1970s by Dr. Harold Persky and colleagues, how often a husband initiates sex with his wife was significantly correlated with the wife's responsiveness. In other words, if she receives him with a positive reaction, he tends to initiate again. If he is rebuffed, he tends to stop initiating or at least slow it down. Chapter 7 reviews some of the facts men need to know in order to reduce rebuffs.

Other hormone-physiology studies show further connections between testosterone and erection. In young men the higher testosterone levels in the blood are associated with both a faster erection and a slower loss of erection (detumescence) after ejaculation. Other relationships between the physiological response at the penis and the levels of testosterone in the blood have also been shown. But erection also occurs independent of testosterone control.

ERECTION—WHEN DOES IT ORIGINATE IN THE MIND AND WHEN DOES IT ORIGINATE AS A PHYSICAL REFLEX IN RESPONSE TO PHYSICAL STIMULATION?

There appear to be two types of erection response in men. One type is unrelated to androgens. This means that the erection can occur even in a man who has been castrated, has low levels of testosterone, or is having

....

[3] A similar decline in skin sensitivity has been reported in women where the ability to discern that two separate pencil points are on the skin (versus perception as one) has been shown to be directly related to sex-hormone levels (estrogen). Sensitivity improves when these women take estrogen treatments.

other androgen-deficiency problems. This androgen-independent erection has been shown to occur when there is sufficient visual stimulation, that is, an erotic film or other visual "turn-on." Even the eunuchs described earlier on would probably have produced erections if they saw enough stimulating displays of behavior.

The other kind of stimulation, that requiring the use of fantasy, or of emotional involvement, appears to require androgens. A man who lacks adequate testosterone levels appears to be unable to generate an erection through the stimulation of his cognitive awareness.

Recall that in Chapter 3 I spoke of the studies on vision and reproduction in rats. Cut the visual pathway and the rat became reproductively neutered. Put the rat in constant light or in constant darkness and in that way block the normal visual stimulation of the cycle of day versus night (or light versus dark), and this was temporarily equivalent to actually cutting the nerves.

Now we find that men who have insufficient testosterone circulating in their blood can use vision—but not the more complex mental function of fantasy—to turn on the erection reflex. The mind of a man is very complex compared with that of a rat. That which passes through his visual system appears to be operating at a primitive level, comparable to what we saw in the rats. That which passes through his cognitive function, the way he thinks or the way he feels, is different. Sexual creativity requires the presence of his sex hormones to get and keep it going. Pornographic magazines quite clearly capitalize on the uncreative, primitive responses.

For these reasons, the study of male sexuality has a level of complexity beyond that in all the other male animals that have been studied. I think we are dealing with a combined process. Men can be primitive or men can be elegantly complex, depending on a man's nature and character and on the quality of the relationship. At the crudest level his sexual interest is triggered by visual stimulation—watching women. And at this level his arousal is expressed in catcalls or other forms of behavior, where he intrudes his desire for connection without regard to his object's willingness or arousal. As the artistry of the man increases, he learns how to engender arousal and willingness through his skills at seduction or through his pursuit of courtship.

ASSESSING MALE SEXUAL BEHAVIOR: OR, IS WHAT THEY THINK THEY DO WHAT THEY REALLY DO?

Woody Allen appears to have been right. In his movie *Annie Hall* I was struck by a scene, a split screen in which a man and woman were each talking to their own psychotherapists about the problems of their relationship. They were seeking help. We, the moviegoers, saw and heard each of the therapists ask, "How is your sex life? How often do you have sexual intercourse?" Their response was classic: He whined, "We only have sex three times a week." She complained how relentless he was—"three times a week." He felt deprived, and she felt exhausted by his need. This couple had unmatched libidos and an unsuccessful negotiating style. Their problem is common. It fills the waiting rooms of psychotherapists and sexual counselors. Research has begun to explore how this universal problem can be successfully addressed.

In 1983 Dr. Anthony Reading published a study of the sexual-behavior patterns of men. He enrolled close to fifty men in their thirties and divided them into two groups. One group kept a daily diary card and were also interviewed once a month. The other group only had monthly interviews. First he discovered that the process of keeping a daily record produced impotence in about 10 percent of the men and caused them to drop out of the study. Somehow keeping a daily record interfered with the sex life of some men. Monthly records were less debilitating. None of the men who came in for a monthly interview dropped out or reported this impotence problem. Second, the written daily record showed a higher level of intercourse, morning erections, and sexual thoughts than the memory (of the same man) at the interview at the end of each month. In general, men were about 40 percent more sexual than they believed themselves to be. In other words their perception was of 30 percent less sex than they were actually having. For example, at the end of the first month the estimate of how many morning erections had occurred was ten; the actual number was fourteen. Likewise, at the end of the first month the estimate of how many times per day the man had sexual thoughts was four and a half times a day; the actual number they recorded was a little over six. Men believe they are having less activity than is actually happening. The same thing has been shown for dieters. People tend to think they eat less than their records reveal. Does this reflect that they want more? Probably yes. Whether we consider the male character in *Annie Hall,* or the men in Dr. Reading's study, the dissatisfaction represents a common male complaint.

THE TIMING OF THE CYCLIC VARIATION IN MALE SEXUAL BEHAVIOR

Each day the most likely time for intercourse is about 11:00 P.M., and the next most likely time is early in the morning. There are annual rhythms of sexual behavior also. The highest rate of activity, according to studies that were conducted among medical students, were in October; the lowest time of year was in February. The four-month span from January to April was the lowest; the three-month span from September to November showed the highest activity. What interests me here is the close relationship between these annual rhythms of sexual behavior and the annual rhythms of testosterone levels. These studies of behavior and sex hormones were conducted by different investigators in different places in the world, but the pattern is the same. When testosterone is highest, sexual frequency is highest.

As men age, they slow down. They have sexual intercourse less often, masturbate less, and have fewer morning erections and sexual thoughts. Different investigators, publishing in the late 1970s and early 1980s, reported the same thing. There is individual variation, and some older men are more active than some younger men. One study, by Dr. Clyde Martin published in 1977, analyzed the patterns of men who had been followed in a continuing study to see what happened throughout their lives. Dr. Martin discovered several principles:

- Men who have been highly active tend to remain relatively active throughout their lives.
- Men who have had their first coital experience at an early age tend to have more sexual intercourse per year for the next ten or fifteen years than "late bloomers."
- Masturbation is a significant sexual outlet in men until age thirty; after age thirty masturbation tends to continue but at a much lower level compared with the early years—probably because a partner is now available.
- Delaying the first coital experience doesn't cause a lasting effect; by age forty the initial effect of the late start has evaporated.
- There is an overall decline in the total amount of sex behavior as men age. By sixty-five less than 10 percent of them report as much behavior as was common in men in their forties.

Other studies confirmed these findings, both in women and in men as they age.

BUT LESS CAN WELL MEAN MORE TO HIM

The declining frequency is definitely not associated with a declining quality of sexual life. The universal conclusion has been that although frequency of behavior tends to decline, sexual activity continues to play an important role in the lives of aging men. Previous behavior helps to determine present behavior. If sex was important to a young man, it is going to be important to him when he is old. The same is true of women. In addition, the data have consistently suggested that healthy men with higher levels of testosterone tend to display more sexual behavior than healthy men with lower levels of testosterone at each age. Consider this example of the thirty-three young German men, discussed earlier. Each Monday, Wednesday, and Friday for two weeks the thirty-three men in their twenties came into the laboratory, gave some blood, and turned in their records about sexual behavior in the previous few days. The results were rather clear-cut:

- When sex did occur, it was usually not more than once in a day. Two of these young men each had an Olympian day in the two weeks. One had five orgasms in one day, not to be repeated; and the other reported four in one day, also never to be repeated.
- When the hormone levels were measured, the men with the highest levels of testosterone showed more orgasms than the men with the lower levels of testosterone.
- Likewise the men with the higher levels of testosterone showed more sexual activity (intercourse and masturbation) than the men with the lower levels of testosterone.
- When the scientists compared the behavior forty-eight hours before and then forty-eight hours after a testosterone reading, they found that behavior "before" is a better predictor of the subsequent testosterone level than testosterone level is of the behavior following. But both times were related to the increased testosterone. In other words the behavior might be stimulating the testosterone; but the testosterone also might be stimulating more behavior. What's the chicken and what's the egg?

SEXUAL BEHAVIOR AND SEX HORMONES—PUTTING IT ALL TOGETHER

There seems to be an elliptical process between male sex hormones and male sex behavior.[4] The androgens do stimulate sexual behavior, but

[4] If it were circular, there would be an evenness in the speed of relationship between hormones and behavior. Rather, it is elliptical, because sometimes the reaction can move faster than at other times.

sexual behavior in turn seems to stimulate the production of sex hormones. The testosterone levels before, during, and after sexual intercourse are usually higher than they are during times of celibacy. Masturbation does not alter sex-hormone levels in men, nor in women.

Other relationships between hormones and sexual behavior confirm these general trends. Similar findings are occurring in the Baltimore Longitudinal Study (referred to earlier) of exceptionally healthy men as they age, exercise, maintain their fitness levels, provide blood for hormone testing, and recall every two years their general pattern of sexual behavior. In this study of 183 men the least sexually active group contains the most individuals with very low testosterone levels. Likewise the group of men with the highest level of sexual activity shows the highest testosterone levels of the total study. However, these are only trends—averages. Even among some men in the "low activity" group, there are some individuals with very high T; and likewise even among the most active group of men, a few men show very low T. In other words, to reiterate, level of hormones and frequency of sexual behavior do correspond, but men are more complex than a simple "hormone equation" can describe.

The Baltimore Longitudinal Study has been showing a decline in sexual behavior at the same time it has been reporting that testosterone levels in aging men are not declining. Since this was the only study to conclude that testosterone levels do not necessarily decline with age, I was particularly interested to discover what may account for the difference between these results and the other six studies that found testosterone consistently to decline with age. The Baltimore Longitudinal Study took blood in the afternoon. The decline in sex hormones that have been demonstrated have been in morning samples. It may well be that if the Baltimore Longitudinal Study investigators started taking the blood in the morning, they would find the same decline as have others.

All of the studies tell the same story. Testosterone does not serve as an on/off switch. And sexual behavior, likewise, does not trigger a sudden change in testosterone. Instead there are general trends of testosterone secretion that respond to and predict general patterns of sexual behavior. The individual studies provide the subtleties.

In men, as in women, stable patterns of behavior seem to be reflected in the pattern of sex-hormone secretions. Dr. Harold Persky and colleagues studied testosterone in men involved in monogamous sexual relationships. They found that the testosterone level in men each month tended to reach its maximum simultaneously with their wife's ovulation. They found this to be true in twenty-five of the thirty-two cycles that they studied for eleven young couples. In 1976, when I first met with

Dr. Persky, he was immersed in the detailed analysis of his remarkable scientific study. Although his study was never repeated, the elegance of his research design coupled with the known integrity of the scientific team convinced me of the accuracy of the work. The cost—in money, scientific effort, and volunteer participation—was enormous. The results—that the hormones of monogamous men dance to the rhythm of their partner's hormones—is an important discovery. Indeed it makes a strong case for the value of male monogamy in leading to endocrine harmony.

THE EFFECT OF ALCOHOL ON MEN'S SEXUAL BEHAVIOR

Alcohol can inhibit a male's sex life. Men who drink more than four ounces of alcohol per day show an increased likelihood of being in the low-sexual-activity group as opposed to men whose drinking is more moderate. No difference emerges in the testosterone levels of the drinkers who drink more than four ounces as opposed to those who drink less. Yet a clear difference in sexual behavior does. Dr. Millicent Zacher, a reproductive surgeon who works with male infertility, suggests that the alcohol may affect the liver and its enzymes in a way that alters the binding proteins and that this change in blood chemistry may affect the biologically active testosterone that is circulating. Perhaps she is right. Since, as Box 4-1 showed, the biologically active testosterone comprises 60 percent of the circulating testosterone, I suspect the failure to identify a T-and-alcohol relationship may be due to the failure to analyze the correct fraction.

Alternatively alcohol dulls the coherence of a man's conversation, and it can produce an unpleasant physical aura. Gracelessness and bad odors emanate from a man who drinks immoderately. Excess alcohol consumption is usually a turn-off to a man's sexual partner because alcohol dulls his sensitivities. While alcohol may relax the uptight, inhibited fellow, it may also disable him if he drinks too much.

WHEN SEX BECOMES A PROBLEM

Perhaps more common than generally discussed, two classes of sexual problems are showing up, both in the scientific literature and in the doctor's office: (a) difficulties with forming an adequate erection, or impotence; and (b) hypoactive sexual desire, or low libido.

Eight studies published between 1979 and 1990 have provided some useful data that explore relationships between hormones and sexual dysfunction. In each of these reports the population studied is composed of men who actively seek help for what they consider to be sexual dysfunctions. Investigators have studied erectile dysfunctions using laboratory techniques. A strain gauge placed on the penis measures the size and quantifies the strength of an erection. Other investigations have examined problems of inhibited sexual desire in men with and without erectile dysfunctions.

In addition to men who actively seek professional help, a more pervasive problem also exists: Men who do not consider their diminished sexual desire a dysfunction may refuse to seek help. Others who discover a diminished erectile capacity may respond by lowering their expectations. In order to avoid sexual failure, they may behave as if they do not have desire, even when they secretly do. Fear of failure can disrupt relationships. If the fear is hidden, access to help is denied.

Figure 4-4 shows the age distribution of men who do seek help for erectile dysfunctions. Although this figure cannot show how often the problem occurs in the general population, it does show that it is in the late forties, fifties, and sixties that most of the erectile dysfunctions appear to be occurring.

Erectile Dysfunction

The first scientific study of the relationship of erectile dysfunction to testosterone therapy was published by Dr. Julian Davidson working with colleagues at Stanford University. In 1979 they reported an improvement in total erections for hypogonadal (testosterone-deficient) men who received injections of 200 or 400 milligrams of testosterone. Dr. Davidson reported that the higher doses of testosterone produced better results than the lower doses. He also noted that the dose wore off within three weeks and suggested that effective testosterone-injection treatment for erectile dysfunction required treatment every three weeks.

Another study produced in Dr. Davidson's laboratory, four years later, expanded on the subject. Dr. Marie Kwan and her colleagues reported that the number of nighttime erections (the scientific term is *nocturnal penile tumescence)* was lower in men who had very low plasma levels of testosterone. She reported that these nighttime erectile responses were significantly improved in response to testosterone-therapy treatment, but only in three of the six patients. The investigators were not able to explain why some men improved and others did not. All the men got the same treatment. Since the nature of the sexual relationships

FIGURE 4-4: Age Distribution/Impotence

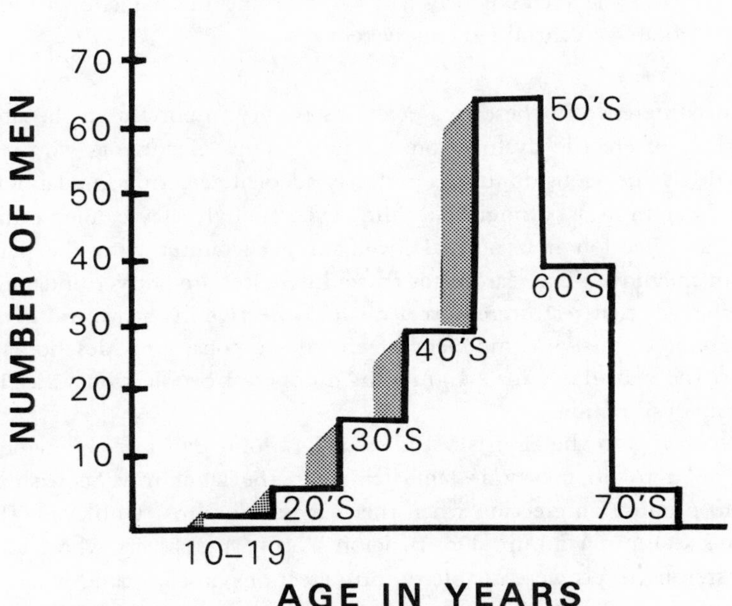

in their lives was not evaluated, I am left to wonder whether the quality of their interpersonal relationships helps account for the difference in success or failure. Unfortunately both hormone and sex research are expensive and time-consuming. Each study provides only a partial answer to the complex questions about male sexuality and its dysfunctions.

Some studies have confirmed in dysfunctional men what was described earlier for healthy, functioning men. However, there are limitations. A good scientific observation with a caliper around the subject's penis while watching pornographic film provides objective, if limited data. One of the critical problems in sexual-dysfunction research, which has not been addressed in the scientific literature, is how to factor into the evaluation either the quality of the relationship of the "dysfunctional" man or the general sexual attractiveness of his partner. No doubt both of these variables have a profound effect on the sexual function of a dysfunctional man. Even in men with erectile dysfunction the capacity to form an erection comes from either of two sources:

- *A reflex,* which is defined as an automatic response to a specific physical stimulus. Penis reflexes do exist. And most men learn to suppress them.
- A *central cognitive process cerebrally induced,* which refers to the fact that the erection originates from some mechanism within the brain or central nervous system.

The difference in these two erections is very important in the understanding of erectile dysfunction. Certain groups of patients who report erectile dysfunctions do form a perfectly adequate erection in a laboratory while watching a pornographic film, even though they cannot form an adequate erection in a normal lovemaking encounter with the partner whom they love. The reasons for these differences are not yet understood, but the fact that testosterone treatment is effective in the central nervous system or cognitive component of erectile response provides hope that help is now available for a significant number of people who suffer from erectile dysfunctions.

According to the Scottish studies of Dr. John Bancroft, hypogonadal men tend to do poorly at fantasizing. In the laboratory Scottish men would produce an erection when they saw erotic films. Unlike Dr. Davidson's California group, Dr. Bancroft's Scottish subjects who had low testosterone levels were unable to form erections on command when they fantasized about sex. When these hypogonadal men were given testosterone therapy, they experienced increases in the numbers of sexual thoughts, in the numbers of sexual acts, and in the intensity of sexual excitement.

In other words testosterone therapy is repeatedly shown to improve the sex life of hypogonadal men by improving the perception of pleasure. Hormone therapy for men seems to work in the creative elements of the mind when given to men with deficient natural levels of testosterone. Unfortunately, although T-therapy increased the interest in sex, it did not correct the erectile dysfunctions in Dr. Bancroft's study.

Hypoactive Sexual Desire

Perhaps the most instructive study in the series of research papers on sexual dysfunctions and hormone levels was published in 1988 by psychotherapy investigator Dr. Raoul Schiavi working with Dr. Patricia Schreiner-Engel at Mt. Sinai Hospital in New York. Their studies of men with hypoactive sexual desire were broken down into those who also had an erectile problem and those who didn't. They found that plasma testosterone was significantly lower in men with hypoactive sexual desire (whether or not there was an erectile dysfunction) than in age-matched men who did not have this problem. They also showed that both the dysfunctional group and the healthier group shared a similar rise in testosterone during their sleep, from hour to hour. As the night wore on, the total plasma-testosterone levels rose. The other sex hormones were not different between the groups, but each man's individual testosterone average for the night did reflect how severe his own dysfunction was. The men who engaged in the least amount of intercourse attempts or masturbation showed the lowest levels of testosterone; the men with more sexual behavior showed higher levels of testosterone. Some men with inhibited sexual desire did show "total plasma" testosterone abnormalities, and this was particularly noteworthy, since the more refined T-level (see Box 4-1) could not be tested with the assays Dr. Schiavi had available at the time.

The Mt. Sinai investigators also found major differences in the nighttime erections in the impotent versus the potent men with the low levels of desire. Men who had hypoactive sexual desire and also reported being impotent during the day did not show a normal quantity or duration of erections at night. Erections at night during sleep are a normal part of the male psychophysiology. Typically a man will have three to six erections during the night, and these will occur while he is dreaming, during the stage of sleep known as REM (rapid eye movement). Impotent men with hypoactive sexual desire did not show the normal incidence of nocturnal erection with their dreams. And subsequent work by these scholar/therapists revealed that men with hypoactive sexual desire tended to have a lifetime history of depression or other psychopathology.

Men with hypogonadal (reduced) circulating levels of testosterone produce normal erectile responses to erotic films. As in men with normal T-levels, input to the visual system can evoke an erection, and this appears to bypass the cognitive functions. Libido, the awareness of pleasurable reaction during sex, does seem to be related to the testosterone circulating in the blood and probably in the brain as well. T-therapy will probably increase the motivation.

WILL TESTOSTERONE IMPROVE A HEALTHY MAN'S SEXUAL PERFORMANCE?

Research has made it clear that if a man has normal testosterone levels, testosterone therapy is not likely to produce any change in his sexual capacity or his response to erotic stimulation.

Unfortunately we lack sufficient research to evaluate what the side effects of testosterone therapy are in men who already have adequate circulating levels. Studies of general endocrinology and cardiovascular health show a strong relationship between male sex hormones and cardiovascular-disease risk indicators. There is the potential for T-therapy to increase the (bad) cholesterol levels, to decrease the (good) HDL-cholesterol levels, and to increase the incidence of cardiovascular disease. For these reasons one cannot simply use testosterone therapy for dysfunction without considering its potential side effects and health risks.

NORMAL VARIATIONS IN MOOD NOT PREDICTIVE OF SEXUAL BEHAVIOR OR HORMONES

Although it is well known that when a man is depressed, his interest in sex will wane, studies that evaluated the mood of men showed that mood itself did not predict sexual behavior or reflect what the testosterone levels were. So apparently the normal fluctuation in mood is different from the experience of a serious depression. Does testosterone plummet during a depression? The research has not yet been published to tell us, but I suspect that it probably does.

THE EFFECT OF DRUGS ON MALE HORMONES AND SEXUAL BEHAVIOR

Many drugs and hormones used to overcome disease and dysfunction have been linked to changes in a man's sexuality. Let us look at the principal pharmaceutical agents.

Sex-Hormone Treatments to Control Deviant Sexual Behavior

In 1974 Dr. Bancroft published his study with imprisoned sex offenders designed to discover whether hormone therapies would reduce their deviant behavior. These twelve imprisoned offenders had committed atrocious sexual crimes.

The results were clear. An estrogen (ethynil estradiol) and an antiandrogen (cyproterone acetate [CPA]) were equally effective in reducing the frequency of sexual thoughts and sexual activity. Only the antiandrogen, CPA, also showed some effects in reducing erectile responses to erotic stimuli. Again we learn that androgens promote sexual function, this time by revealing how antiandrogens inhibit it.

Sex-Hormone Therapy Used as a Contraceptive

In 1982 the World Health Organization Task Force on Contraception evaluated the possibility of a male contraceptive. First they studied men in six regions of the world to learn whether men would be willing to take oral contraceptives. The results were clearly affirmative. Preliminary studies in Fiji, India, Iran, Mexico, and Korea showed that 41 to 74 percent of men, both rural and urban, would use the pill as an oral contraceptive if it were available. Armed with this information, a prospective study was undertaken evaluating a variety of hormone therapies to test the effectiveness of the contraceptive agents and how they affected the sexuality of the pill users. The hormones given were effective. Various combinations of antiandrogens or androgens opposed with progestins did inhibit sperm production, sperm motility, and the fertility of men. Unfortunately sexual interest, arousal, morning erections, intensity of desire, and self-reported quality of sexual activity were consistently and adversely affected by the antiandrogen. Again we learn that sex hormones and male sexuality have intimate connections with each other. Inhibit the androgens and you inhibit both the fertility of the man and his sexual functioning.

Despite problems, men hoped to see the pill developed. They said they would be willing to take the hormone in order to prevent unwanted pregnancies.

Prenatal Hormones

Clinical experience in which pregnant women were given hormones in order to help preserve pregnancies that were at risk have taught that these hormones can subsequently affect the personality of a young man as he grows. One synthetic estrogen was widely prescribed in the 1950s—DES (diethylstilbestrol). It worked. It stopped miscarriages. Meanwhile divergent reports from two different laboratories have evaluated "gender-specific" behavioral patterns. Dr. June Reinisch and her colleagues at the Kinsey Institute for Sex and Gender reported that the sixteen-year-old boys who had received these female hormones (DES and progestin) while developing in utero showed less aggression, less assertion, less athletic coordination, and less heterosexual experience than their age-matched peers who had not been exposed to these hormones during gestation. She also found more group dependence rather than independence in the teenage boys who had been exposed to female sex hormones. Another competent investigator, Dr. Heino Meyer-Bahlburg, working with colleagues, studied the offspring who had been exposed to lower doses of these estrogens. They did not find these effects in the boys they studied.

Hormone treatments do save pregnancies. We now know that some of them adversely affect the reproductive systems as well as certain personality traits of the "exposed" recipients.

Estrogen Treatment Used for Men with Cancer or Sex Change

Estrogens are sometimes prescribed to men as treatment for prostatic cancer or to transsexuals who are changing sex from male to female. If the dose is high enough, the men start to show breast development, a softening of the skin, a reduction in facial hair, a redistribution of their body fat, and increases in weight, blood pressure, and death from cardiovascular disease. Other hormonal changes also occur. When men take estrogens as adults, the hormones reduce their libido and reduce their capacity for an erection. If you turn off the male hormones (with the antiandrogens or the female sex hormones), you turn off some of the normal male patterns of sexuality.

Certain Widely Prescribed Drugs

Prescription drugs that show a negative effect on sexual life include, in addition to the antiandrogen sex hormones, certain classes of commonly prescribed cardiovascular and psychiatric treatments. A long list of drugs has been associated with unfortunate sexual-function changes. The list is long and is growing longer.

The most common prescription drugs that are reported with sexual dysfunction as a side effect are the following:

- Antihypertensive drugs (to fight high blood pressure)
  - Centrally acting antiadrenergic drugs
  - Alpha-receptor blockers
  - Beta-receptor blockers
- Diuretics (to release urine)
- Histamine-2 receptor antagonists
- Antidepressant drugs
  - Heterocyclic antidepressants
  - Monoamine oxidase inhibitors
- Antipsychotic drugs
- Antiepileptic drugs
- Opioid analgesics (painkillers)
- Sex steroids (female type, such as DES described above)

Adverse sexual reactions to these drugs are not universal. Between 10 and 30 percent of men who take them report difficulties of sexual function. The individual reaction to these drugs is quite variable. Perhaps the dosage of a useful drug could be modified so as not to upset sexual function. If you are concerned about such a problem, you should know that by switching to another drug of its class (i.e., a related compound) the sexual dysfunction may evaporate. Consult your physician.

CONCLUSIONS

The same general trend continues to emerge as I look at the studies of healthy, young, middle-age, and old men and note how these compare with men with sexual dysfunctions. There are relationships between the amount of testosterone circulating in a man's blood and the amount of sexuality reflected in his behavior. But one cannot simply score a man on his testosterone level and be able to tell what to expect about his sexuality. Still, the male sex hormones and male sexuality are intimately linked. As studies continue to refine the assay methods for the biologically active component of testosterone, results are likely to become linked with even greater precision. The hormonal symphony of men is different from that of women. As we understand the differences, we begin to learn how elegant is the dance that together they comprise. The general principles in men are these:

- Aging produces inevitable declines in the androgens and concommitant increases in the gonadotropins.
- Sex-steroid hormones show a seasonal variation, with peaks in September–October and troughs in July–August, regardless of one's location on Earth.
- Vasectomy abolishes the seasonal variation in hormones.
- Testosterone shows a twenty-four-hour rhythm—with highest levels in the late morning and lowest levels in the late afternoon.
- Drugs that inhibit the circulating levels of testosterone have a likelihood of inhibiting certain elements of male sexual response.
- Alcohol can inhibit male sexuality.
- Men with higher levels of testosterone tend to show more erections, firmer erections, and more frequent sexual behavior than men with lower levels of testosterone. However, exceptions are common.

- The capacity for self-control over erectile responses increases as men mature and as their penile sensitivity begins its age-related decline.
- The sex hormones cycle in men individually, as compared to the cycle in women, which tends to be more harmonic with the group.
- Sunshine increases male estrogen concentration, and cloudiness depresses it.
- With age there is a clear decline in the amount of sexual behavior in which men engage, but a much milder decline of their perception of pleasure in reaction to sexual life.
- Hormone-replacement therapy will not improve a healthy man's sexual performance. Many drugs inhibit a man's sexual performance, but substituting the drug with another of its class often corrects the problem.

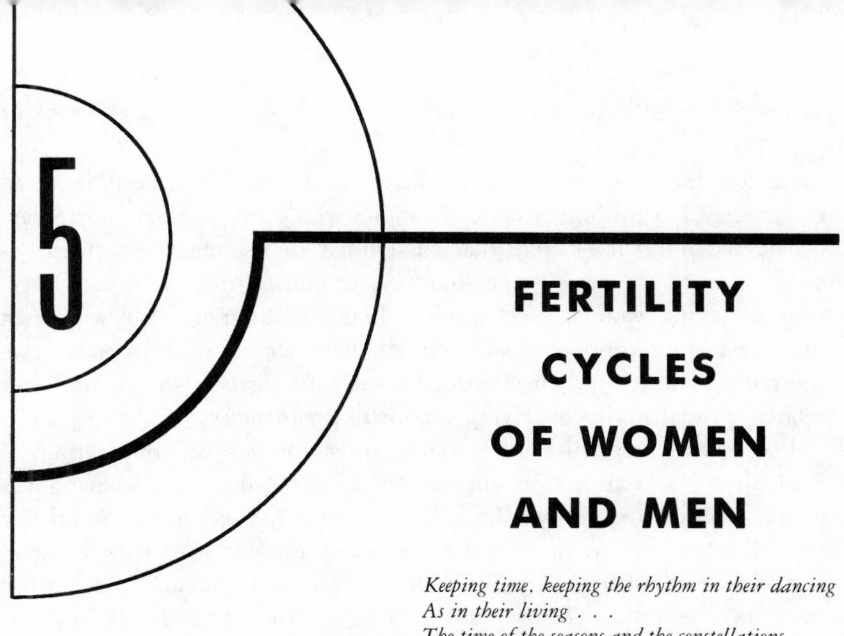

5 FERTILITY CYCLES OF WOMEN AND MEN

Keeping time, keeping the rhythm in their dancing
As in their living . . .
The time of the seasons and the constellations . . .
The time of the coupling of man and woman

—T. S. ELIOT
"East Coker," *The Four Quartets*

The individual hormonal rhythms of woman and man meet in this chapter as they often do in real life—within the context of a marriage or a long-term relationship. The unique hormonal, psychological, and physical cycles of the female and male partner detailed in the two previous chapters continue of course, but now each affects and is affected by the cycle of the other. A new rhythm, a new cycle is born—the one shared by the committed man and woman.

At first glance the most visible sign of cycle harmony would seem to be pregnancy, but fertility regulation is actually what is significant. What matters is not merely producing a pregnancy but having some measure of control over pregnancy. The science of intimacy offers a relevant perspective. New knowledge of sexual timing can enhance the couple's chances of creating a pregnancy. The same understanding can increase the couple's chances of avoiding a pregnancy.

THE INFERTILITY EPIDEMIC

Most couples assume that when the time is right, they'll have no problem getting pregnant. Unfortunately infertility is a fact of life for more and more couples today. At a recent congressional hearing, experts reported that nearly 2.5 million American couples who want to have children are considered infertile—a rate that has tripled among married women between ages twenty and twenty-four. According to other surveys 28 percent of young women of this age acknowledged more than a year of unsuccessful attempts at conception. At the other end of the reproductive spectrum, women in their late thirties and early forties also may find age-related difficulties in conceiving successful pregnancies.

People with infertility suffer. They suffer from stress, from feelings of inadequacy, and from the unpleasantness that goes with wanting yet being unable to conceive. They suffer when they go to the infertility specialist because the process of a medical evaluation is expensive, invasive, uncomfortable, and painful. A new and unfortunate situation has developed because of the availability of expensive and marginally successful reproductive technologies. Technology has dramatically increased the money a couple can spend without achieving the desired outcome. It is not unusual for a couple to spend $10,000 only to conclude that there is no point in continuing to try other available technologies that have low monthly success rates with fewer than 20 percent of the users conceiving. With monthly "medical assistance" often exceeding $3,000, the bills can mount rapidly.

Take a look at the growth in assisted reproductive technology as charted by the United States In-Vitro Fertilization Registry and redrawn from *Fertility and Sterility,* published in January 1991.

For each year one can see the number of menstrual cycles that were stimulated with hormones to produce ovulations, the somewhat lower number of oocyte retrievals (surgical entry and egg removal to a test tube), and the very low proportion of successful babies delivered as a result of these procedures. As the graph shows, the process is becoming slightly more successful as the technology is expanding the number of procedures conducted. However, the success rate remains very low.

Not all reproductive physicians focus on the critical role of sexual timing before initiating the couple into the medical merry-go-round of tests, shots, pills, possible surgery, and sex geared to the clock. Fortunately the weekly pattern of sexual timing allows a cost-free process all couples who want to conceive should try. Even if they are already working with medical professionals, this method may save them money and precious time.

I presented some of my scientific discoveries at the American Fertility Society Meeting in San Francisco in 1988, expanding on my previously published studies of how weekly sexual timing enhances the fertility cycle

FIGURE 5-1: The Growth of Assisted Reproductive Technology

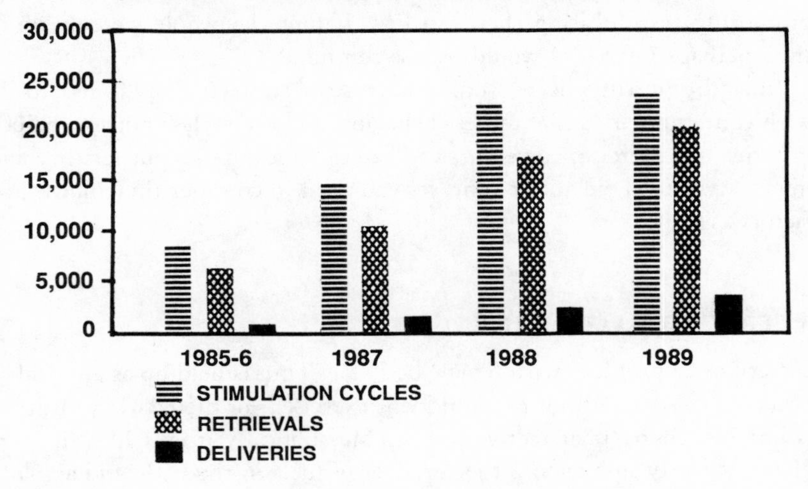

of women. I was surprised at what I learned. Several hundred fertility experts, mostly physicians and surgeons, approached me to learn of the newest scientific findings (discussed in Chapter 6). I kept hearing the same thing from the physicians I talked to: *They rarely discussed the intimate sexual life of the patient when she came for help with her infertility.* I was astonished. How could it be that questions of infertility did not automatically lead to an evaluation of the sexual life of the couple? The doctors had their reasons, ranging from personal discomfort, the potential for being misunderstood, and concerns about appearing prurient, to fear of malpractice assaults. It seemed clear—these patients need to know about weekly sexual timing in order to use their own behavioral practices to maximize their fertility.

SEX MAKES THE CYCLES

My own research (first discussed in Chapter 1) consistently revealed how the pattern of sexual activity, whether sporadic or regular, profoundly affects fertility through hormone levels, menstrual-cycle lengths, and the length of the luteal phase (crucial for implantation of the fertilized egg). Psychologists and biologists knew my studies, but the published results were not generally known by the general medical community—the doctors infertile women went to see. Many physicians today are exquisitely competent at using the extremely costly state-of-the-art technologies. What is generally not discussed with the patient, or even considered, is the critical role that the couple's pattern of sexual behavior plays in promoting or inhibiting their fertility. If more laypeople were aware of this, perhaps infertility would be less common.

Infertility/fertility is not only about sexual intercourse. It has to do with contemporary social mores, behavior, and life-styles choices, including diet and exercise, traveling, and career building. To put fertility and infertility into their fullest context, you need to consider these nonsexual factors as well.

THE FAT-FERTILITY LINK

Society has a problem with female body fat. Thin is held up as an ideal of beauty; fat is something to be hidden, exercised, or dieted away. But fat is not just inert, unattractive lumps. Metabolically it acts like an extra gland, actively influencing the turnover of fuels in the body and actually producing and storing estrogen. During the reproductive years fat plays a necessary role in allowing the estrogen supply to be adequate. After menopause a substantial percentage of a woman's estrogen comes from body fat. When a woman exercises heavily or greatly reduces body fat, as she might to conform to contemporary standards of thinness, fertility may be adversely affected.

It's not hard to understand why a woman who has gotten so thin that she has erratic periods or no period will not become pregnant. A more subtle problem is the woman who continues to menstruate but does not ovulate, or the one who ovulates but does not have a sufficiently long luteal phase. Below a certain "set point" of body fatness and estrogen level, the entire hormonal cycle is compromised. Surprisingly a shutdown in fertility can occur at a weight that's as little as 10 to 15 percent below ideal weight. Let's look more closely at diet and exercise—the two methods a woman can use to manipulate her weight.

FOODS FOR FERTILITY?

Women who diet sometimes become infertile. Several published studies in the late 1980s showed that women who lost as little as two pounds a week, on a diet of about 800 calories a day, had a disruption of their fertility cycles if their diets did not contain meat. These vegetarian dieters produced about half as much progesterone as the meat-eating dieters. Each dieter restricted caloric intake to lose two pounds a week, but only the vegetarians showed reductions in their fertility. Of the thirteen vegetarian dieters in this study, seven never developed a dominant follicle—that is, one that could ovulate. Four of the thirteen women did ovulate, but after the ovulation their luteal phase was inadequate to support a pregnancy, both in its length and in its hormone production.

Studies of less severe diets (1,000 calories a day) also showed that a vegetarian diet could disrupt fertility; and it was most disruptive in the youngest women. Women younger than twenty-five years of age were much more vulnerable to the disruption of their luteal-phase length and hormonal secretions than older women.

If you are a young woman who is trying to become pregnant, these findings suggest that this is *not* a time to be a vegetarian. Research has consistently shown that vegetarian women (even those who aren't losing weight) circulate lower levels of sex hormones than meat-eating women.

EXERCISING YOUR FERTILITY

Athletic men seem to stay fertile. For women the news is not so good. Endurance-trained female athletes tend to have reduced fertility. For example, one study published in 1988 compared endurance-trained athletes (who had been in training for one to two years) to age-matched, nonathletic women. The results were shockingly clear. Luteal-phase abnormalities were common in the athletes. These women were not amenorrheic—they were still menstruating and ovulating. In the luteal phase the athletes showed abnormally low levels of both estrogen and progesterone. Their luteal phase was shorter than normal as well. These athletes may have been in shape for physical activities, but not for pregnancy. Not every athletic woman is affected. Those who want maximal athletic conditioning without compromising their endocrine system need to know the underlying principles in order to tell when they have crossed the limit of their individual exercise tolerance.

Athletic conditioning reduces the fat and increases the muscle on a woman's body. This reduction in fat may account for the luteal-phase anomalies in athletic women. The mechanism for the anomalies remains speculative, but results are clear: Subfertile endocrine function is common in muscular female athletes.

Jerilyn Prior, M.D., a sports physician working in British Columbia, published six important studies between 1982 and 1990 evaluating the role of "conditioning" exercise in athletes and its effect on PMS symptoms, ovulation, luteal-phase health, and amenorrhea. Her findings support the conclusion that conditioning exercises can significantly disrupt both ovulation and the luteal-phase normality. Women vary, and some can handle much more conditioning than others without these side effects. For example, in one of her studies marathon runners typically ran about ten miles per training session, or they were in training to build up to that level. Two thirds of the forty-eight menstrual cycles, in fourteen women twenty to forty-five years old, were abnormal. These abnormalities appeared in spite of a pretraining history of normal menstruation and in spite of normal-length menstrual cycles during the training. The abnormalities tended to break down evenly between anovulation and short luteal-phase cycles. Dr. Prior was unable to predict which women would experience these training side effects, but she could discern when they occurred.

Fortunately Dr. Prior has shown that the side effects of excessive exercise can be reversed. If a woman will cut back on her exercise, her fertility will usually return. By reducing the time spent exercising and/or lowering the intensity of training, a woman may regain her fertility.

In younger women cycles seem to be highly vulnerable to disruption from excessive athletics. After age twenty-five a woman is less susceptible, but still not invulnerable, to athletics disrupting her fertility. In athletics as in diet, it is the younger woman whose fertility is most vulnerable to disruption by extremes of behavior. A woman can monitor her own cycle changes through the basal body temperature method in order to determine if her exercise habits are creating such side effects (see pages 14–15 for the BBT method of determining luteal-phase onset).

ALTITUDE AFFECTS FERTILITY

Many of us think nothing of hopping on a plane for vacation or business travel, but studies have shown that exposure to high altitudes inhibits fertility. Studies of migrating Indian tribes in the Andes revealed that a surge of fertility coincided with migrations from high altitudes to lower altitudes. Americans don't migrate up and down the sides of mountains, but the thousands of accumulated frequent-flier miles do attest to the fact that we are moving from lower (ground-level) altitudes to higher cruising altitudes of 30,000-plus feet. And we are doing it far more frequently than the Indian migration. The research cannot conclusively explain if frequent flying disrupts fertility, but FAA regulations do impose restrictions on transoceanic flying for female personnel due to stress-induced menstrual irregularities.

The reasons for these side effects are not known because they have not been systematically studied. I suspect that the reduced oxygen consumption that is a condition of high altitudes may stress the body and lead to a preservation of the person at the expense of the next generation. Such preservation of a woman's own functioning is a characteristic biological response to stress (see page 7). Unfortunately for the legions of infertile women who need this information, systematic investigations of which behaviors correlate to infertility have not been financially supported research areas, even if interested scholars wanted to do the work. There is growing hope that this is now changing. In 1987 the American Fertility Society officially recognized the PSIG (the Psychology Special Interest Group) as one of its legitimate subgroups and has turned to its members for help in developing scientific protocols for future research.

CAREER INFERTILITY

For women who delay conception as they nurture a career, a reduction in fertility can be an unexpected side effect of professional development. The reason? A premature menopausal transition.

On average, menopause, with its last menstrual period, occurs at age fifty. The seven premenopausal transition years, characterized by a wide range of long and short cycles and subfertile menstrual patterns, typically begin in the early forties. Although the average age of entry into the menopausal transition is forty-three, not all women are average. Some enter the seven transition years as early as their midthirties. When this happens, the likelihood of conceiving is severely reduced. How can you know if this premature menopausal transition is happening to you? In two ways:

- By your menstrual pattern: After age thirty-five, if your flow lightens considerably or your cycle length changes, these are signals that you may be entering premenopause.
- By hormone blood tests: If an elevated FSH is found twice in a row during two different tests at least two days apart, premenopausal status can be confirmed.

SEXUAL BEHAVIOR AND FERTILITY

Into this background of general life-style and behaviors and their impact on fertility, we can now project the effects of sexual behavior. I learned about the basics of sexual behavior from rats, the basic laboratory animal.

I was allergic to the smell of rats. I didn't like the way they looked. I didn't want to work with them and I didn't want to get near them, but I did study the work of my professors and their colleagues, because I found the implications of their work fascinating. What they had already found by 1978 suggested a clear picture the role of sexual behavior in promoting the health of the reproductive system in small mammals:

- They found that copulation induces the surge of the hormone from the pituitary that maintains the corpus luteum LH.
- They found that both handling by a human and mating within its own species would trigger the release of the hormone prolactin, which is necessary for milk production from the mammary gland.
- They found that while handling did not trigger an LH surge, mating did in animals that had stopped cycling because of the constant light that was kept on in the lab to inhibit ovulation.
- They found that copulation dramatically increased the incidence of ovulation in species of rodents who normally do not ovulate very often.
- They found that the more times the penis thrusted during mating, the more likely ovulation was to occur. And it wasn't sufficient for a male to mount a female; it took vaginal penetration to produce these physiological effects.
- Finally they found that if the animal copulates, the corpus luteum secretes more progesterone than if she does not copulate.

I looked at these data and wondered: Was female reproductive endocrinology and fertility in rodents being influenced by the kind and the amount of the sexual stimulation the female receives from the male?

Similar relationships have been studied in deer and monkeys. Male deer become sexual a full month before the females begin to ovulate each season. I looked at those data and wondered: Was the male's sexual behavior triggering the ovulation in the female?

In monkeys the findings were similar to those of the deer. There was an annual decrease in male potency two to four months before the annual female loss of menstruation would occur. I looked at those data and wondered: Does the decrease in male potency cause the loss of menstrual cycling in the female?

One other finding in rhesus monkeys intrigued me. All cycles of sexual behavior in the female were abolished after removal of the uterus. I looked at those data and wondered: Why would hysterectomy end a female monkey's sexual life? At that time (and often still today) medical schools were teaching that a women's uterus had nothing to do with her sexuality. The uterus, they said, serves only as the "baby carriage."

These studies in animals—in rats, deer, and monkeys—introduced me to the idea that sexual behavior affects fertility. It takes two healthy beings to procreate. Take away the opportunity for fertility in the female (by removing the womb or the male) and sexuality stops. Alter the cycles, and hormonal and behavioral desynchronization occurs. I wondered whether human beings worked the same way. Did women need men in order to be fertile? Could studies like this be done in the human? To the latter question the professor, who was studying rat sexuality in the psychology department at the University of Pennsylvania, said "No. Studies like this would be impossible to do in the human, and that was why 'animal models' were necessary." However, I was more optimistic and searched for other faculty willing to guide my work as I set out to do the research that led to the discoveries discussed in Chapters 1 and 2. Dr. Jerre Levy, the only woman faculty member in the division of physiological psychology, agreed to be my mentor in my preliminary investigations of menstrual-cycle timing. As positive results began to emerge, she suggested that I find a new faculty advisor with reproductive-system expertise, since her research focus—brain and gender differentiation—was leading in a different direction. She guided me to seek out the most expert university faculty member I could find.

The resulting studies were the springboard for what was to become my central scientific focus for the next fourteen years. Some of these studies have direct relevance to people who want to promote viable pregnancies. Although the details remain incomplete, there are clear relationships between human sexual behavior and human fertility. The knowledge of these relationships can help a couple take direct, private action to increase their chances of successful pregnancy.

INCREASING THE ODDS OF SUCCESSFUL PREGNANCY WITH FERTILE-TYPE SEXUAL BEHAVIOR

Working within the Department of Obstetrics and Gynecology at the University of Pennsylvania with Dr. Celso Ramón Garcia, chief of reproductive endocrinology, I discovered a surprising and still not well-known fact. More than half the couples who seek infertility treatments do not engage in fertile-type sex. During the course of their treatment most do not have weekly sex—the minimum pattern associated with fertile-length menstrual cycles and luteal phases. Perhaps it is the pressure of the treatment itself. Perhaps it is the nature of their lives. But they are unaware of a basic sexual pattern that can enhance their fertility.

Ideally each fertile couple would learn that regular weekly sex enhances their fertility before approaching the infertility specialists. Whether they are tried before or after seeing a medical professional, the behavior patterns I suggest should help. Medically they are harmless. With the sexual-behavior patterns I suggest, a couple can optimize their chances for pregnancy, and they can do it for free.

This knowledge of your own sexual timing is valuable. If you (or someone you know) is infertile the power of this knowledge may help to short-circuit a long, painful, and expensive medical workup. If you are already involved with an infertility evaluation, the material that follows is essential reading. Knowledge of sexual timing is critical to the enhancement of fertility and intimacy. Even if fertility is not your problem, the facts are very interesting. They should lead to a greater sophistication in preventing unwanted pregnancies.

IF YOU WANT TO CONCEIVE AND CAN'T...

There are three basic steps a couple can take—before seeing a fertility specialist. These preliminaries can give a doctor more insight into a particular fertility problem. Basically the couple must raise sex consciousness, engage in fertile-type sexual relations, and seek to reduce stress in their lives.

Start a Sex Diary

Couples who want to conceive but cannot should begin by making no changes in their sex lives. They should investigate their own pattern of sexual behavior by keeping a sex diary.

Appendix C provides a sample diary that can be photocopied. Each partner should put a plus or minus sign on his or her individual calendar each day, without conferring with each other, to indicate if the couple did or did not have intercourse in the last twenty-four hours. This should be built into your morning routine in order to collect accurate information. I recommend against giving one to four stars—like movie reviews—for performance and appraisals. Don't guess a few days later. Trying to fill in a calendar from memory is subject to the Woody Allen effect. Eight weeks of day-by-day record keeping should provide enough data to deter-

FIGURE 5-2: Sample Calendars—Sex and Menses

CYCLE CALENDAR 1983

Code name DEVO		S	M	T	W	T	F	S	S	M	T	W	T	F	S	S	M	T	W	T	F	S	S	M	T	W	T	F	S
Age 20	JANUARY	2	3	4	5	6	7	8	9	10	11	12	13	14	15	16	17	18	19	20	21	22	23	24	25	26	27	28	29
Age at first menses 11	FEBRUARY	30	31	1	2	3	4	5	6	7	8	9	10	11	12	13	14	15	16	17	18	19	20	21	22	23	24	25	26
Gyn. age 9	MARCH	27	28	1	2	3	4	5	6	7	8	9	10	11	12	13	14	15	16	17	18	19	20	21	22	23	24	25	26
Birth control methods RHYTHM	APRIL	27	28	29	30	31	1	2	3	4	5	6	7	8	9	10	11	12	13	14	15	16	17	18	19	20	21	22	23
		24	25	26	27	28	29	30																					

V = masturbation to orgasm **x** = sexual intercourse
O = menstruation \ = genital stimulation by a man

DIRECTIONS

1. Menstruating – Circle each day of menstrual flow or spotting.
2. Sexual activity with a man – Place an X across the date if intercourse occurred
   Place a \ across the date if genital stimulation with a male present occurred (without intercourse)
3. Masturbation – Place a V across the date if only masturbation occurred.

e.g. menstruating on day 16 (16)   menstruating + intercourse on day 17 (17)   genital stimulation with a male on day 18 18 mine the true pattern. Next the couple should look jointly at the data they have generated to see if their records match. If so, it is time to tabulate the results. If not, you must determine if one of the calendars is more accurate than the other. Then the data must be read. If menstruation occurred during these weeks, the marking off of the weeks should begin at the end of each blood flow. Was there at least one sexual intercourse with ejaculation in each seven-day span? From that point each nonbleeding, seven-day span should have at least one plus mark in order for the couple to qualify as having a "weekly" activity pattern. Figure 5-2 shows sample records from my studies. Each calendar is scored to show the weekly versus the sporadic pattern, as well as days of bleeding cycle length and coital total. If you have been having sex during the menstrual period, this should be noted and probably stopped. As I pointed out in Chapter 3, sex during menstruation may interfere with fertility.

Have Sex

If your calendar reveals a lack of weekly sexual contact, you need to start having sex. In addition to frequency, any discussion of sex involves questions of quality: Is orgasm necessary? Is it relevant? My studies have shown female orgasm to be irrelevant to the question of fertility. While certainly relevant to the satisfaction a woman experiences, orgasm does not seem to be relevant to the capacity of the body to become pregnant. The optimal biological time to have a baby is when a woman is young, in her late teens or early twenties, because her eggs are young and at their healthiest. If she's involved with a young man of equivalent age, he may not yet have reached a mature capacity for self-control. The possibility of having to wait for sensual pleasure is an unfortunate fact for her. It makes sense biologically that the evolution of the reproductive system would survive even if sexual skill was absent or if pleasurable sensuality was secondary to youthful vigor. There is time to develop sensuality, but there is not unlimited time for conception. The earlier one gets pregnant, the healthier the eggs are likely to be, increasing the likelihood of a viable pregnancy and a healthy baby.

I cannot overemphasize the importance of consistency in relations. A weekly pattern of sexual behavior is essential. The biological system is designed for optimal fertility in couples who enjoy frequent regular sexual intercourse. Parenting is tough. An intimately bonded couple may have a greater strength for meeting the difficulties. In any event evolution seems to favor their genes.

SEX PROMOTES FERTILITY

One study proves the point that sex is a fertility regulator. It involved healthy, not knowingly infertile women in their twenties.

Table 5-1 shows a lot of information in a small amount of space. It arrays the sexual-behavior pattern of eighty-three different women who

TABLE 5-1: BBT ASSESSMENT BY CYCLE LENGTH AND COITAL FREQUENCY

	COITAL FREQUENCY		
CYCLE LENGTH	WEEKLY	SPORADIC	NEVER
20			⊣
21		- ⊣	
23			+ ⊣⊣
24		⊣	+ - ⊣⊣⊣
25		⊣	
26		⊣	⊣
27	+++	+	++ ⊣⊣⊣
28	+	+++ ⊣	++ ⊣⊣
29		++ ⊣⊣	+ - ⊣
30	+	+++ ⊣	⊣
31	++ ⊣		⊣
32			+++ ⊣
33		+	
34		+	+
35	+	+++ ⊣	++ ⊣
36			+
37	+		⊣
38			⊣
39			++ ⊣
40		+ ⊣	
41		⊣	
50			⊣
54			+
64		++	
71		⊣	⊣
98		-	
99			⊣

kept records in the spring of 1983. It shows that weekly-active women almost always had fertile-type menstrual cycles. After their records were turned in, Belle Erickson, R.N., a graduate student studying statistics, analyzed the data. First she looked at the basal body temperature (BBT) graphs each of the women had kept to determine whether or not the graph revealed an ovulation. (See Chapter 1, Box 1-1, "How to Measure Fertility by Degrees" for a thorough review of what the BBT graph is reflecting.) Before we consider what Table 5-1 illuminates, it helps to see how temperature graphs can reveal so much. Take a look at Figure 5-3, which shows a fertile basal body temperature graph measured across a woman's three successive fertile menstrual cycles. This graph also shows that BBT relates to other events occurring in the body at the same time.

FIGURE 5-3: Fertility Cycles in Women

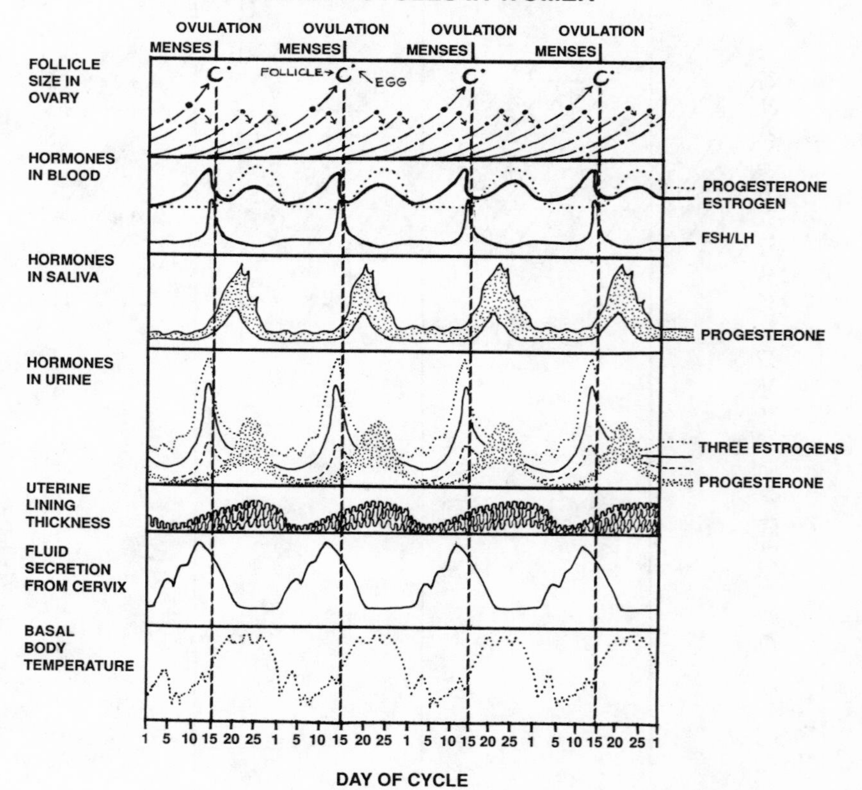

A GRAPHIC LOOK AT FERTILITY

Figure 5-2 shows the remarkable and coordinated series of events that a fertile woman experiences each month. Locate the time line across the bottom of the graph. Read from the bottom up. If, for example, you look at the first Day 15, you see the basal temperature, the cervical fluid, the hormones in the urine, saliva/blood, and thus a cross section of what is happening everywhere in the body with respect to fertility. On Day 15 the basal body temperature is rising. On the same day the quantity of cervical fluid is falling. (It will rise again—the peak of cervical secretions and vaginal moisture may be so copious that a pantyliner is needed.) At the top of the graph is the follicle-size time line. (Those are the structures that contain the eggs and manufacture the estrogen, which is circulated throughout the bloodstream.) On Day 15 you see that ovulation, the opening of the follicle, and the releasing of its egg is occurring.[1] Dr. Erickson evaluated the basal body temperature graphs that the eighty-three healthy women supplied. We found that there were three different patterns of BBT results:

1. *A fertile pattern*—Some women (those denoted in Table 5-1 with a plus sign) showed a completely fertile BBT pattern: a basal body temperature reflective of ovulation, followed by a temperature that stayed elevated for at least twelve days. This elevation reflects an adequately long luteal phase sufficient to generate enough hormones to build the nest.
2. *An infertile pattern*—Five of the eighty-three women showed no elevation in their temperature; that is, they failed to ovulate (these are denoted with a minus sign).
3. *An in-between pattern*—a third group showed evidence of ovulation, but did not show an adequately long period of time in the luteal phase to permit enough nest building for optimal fertility. (These women were denoted by the symbol that looks like the capital letter *T* placed on its side.)

Table 5-1 reveals critical information about each of these eighty-three healthy women. First, it shows menstrual-cycle length. The rows correspond to cycles of different lengths. Each woman is represented by her

....

[1] There is some variation in the normal timing of cycles that are fertile. The numbers arrayed in this graph are conceptual averages.

own symbol. Notice that the cycle lengths range from twenty and ninety-nine days, but those cycles that fall between twenty-six and thirty-three days are highlighted with hatch marks. These "hatched" cycle lengths have been previously shown to be those with the greatest likelihood of fertility. Second, this table groups the eighty-three women by sexual-behavior category. Some women are in the Weekly column, others in the Sporadic column, and others in the Never (celibate) column. In any given column you can see what the cycle length of each woman was as well as what her basal body temperature graph revealed.

Look at the Weekly women: There are no women with short cycles, two with longer than thirty-three-day cycles, and the rest are all within the hatch-marked region of presumed fertile-type cycles. If you study the symbols of each of these women, you will see that all but one had a fertile-type (+) BBT rhythm. The one exception is a thirty-one-day cycle. She did ovulate, but her luteal phase was short.

The next column shows the Sporadics. A number of sporadic women had short cycles, and a number had long cycles. Look closely and you can see that only about half of the Sporadic women showed a fertile BBT graph. Notice, too, that the shortest and the longest sporadic cycler failed to ovulate (as denoted by the minus symbol). Finally a great many BBT symbols, suggestive of luteal-phase defects, was associated with Sporadic women.

A good deal of information comes together here. We see (as before) that Sporadic women have a high incidence of short and long cycles. We see now that sporadically active women also have a high incidence of infertile-type basal body temperature graphs. Even within the hatched region not all cycles are fertile. A Weekly woman with a cycle between

TABLE 5-2
PERCENTAGE OF FERTILE CYCLES VARIES AS A FUNCTION OF SEXUAL ACTIVITY

Analysis: All Cycles
N = 83
Coital Frequency

	Weekly	Sporadic	Never	Totals
Fertile-Type BBT	9	17	17	43
Infertile-Type BBT	1	14	25	40
Totals	10	31	42	83
Percent Fertile Type	90%	54.84%	44.44%	

twenty-six and thirty-three days usually shows a fertile BBT rhythm. A Sporadic woman with the same cycle length may not be fertile.

Finally let's take a look at the Never, or celibate, women's statistics. Here, as in the Sporadics, a wide range of cycle lengths occurs. Abnormal basal body temperature graphs also occur frequently among celibate women. Even celibate women's cycles in the hatched region show deficiencies in their BBT rhythm.

Table 5-2 shows that 90 percent of the weekly-active women had fertile BBT graphs; but only 55 percent of the sporadically active women had fertile graphs. Forty-five percent of the celibate women were fertile in their basal body temperature rhythms.

Table 5-3 goes a little bit farther in this analysis, which considers only the women with 29.5 ± 3–day cycle lengths, those we would normally expect to have the highest likelihood of underlying endocrine harmony. The percentage of fertile cycles varies as a function of sexual activity in fertile-length cycles as well as in aberrantly long or short cycles.

This table examines only cycles in the hatched region—those of 29.5 ± 3 days. Again, the weekly-active women had a higher incidence of fertile cycles than either the sporadically or the never-active women.

THE MEANING OF THE NUMBERS

The point of all the statistics is simple. If you, or someone you know, wants to become pregnant, let her know that it is probably *not* sufficient to follow doctor's orders and have intercourse only around the time of ovulation. Have sex to promote a fertile-type cycle in the first place. A couple should continue to have sex during the postovulatory two weeks of the menstrual cycle. Somehow luteal phase (postovulatory) sex seems

TABLE 5-3
COITAL FREQUENCY AND FERTILITY IN CYCLES 29.5 ± 3 DAYS
N = 40

	Weekly	Sporadic	Never	Totals
Fertile-Type BBT	7	9	8	24
Infertile-Type BBT	1	5	10	16
Totals	8	14	18	40
Percent Fertile Type	87.5%	64.29%	40.48%	

to prolong the length of the luteal phase, a step that may be necessary after fertilization to ensure implantation of the egg in the uterus. Even after ovulation, regular sex seems to keep the system going.

INFERTILE PATIENTS ALSO DEMONSTRATE THAT THE SHORT LUTEAL PHASE AND SPORADIC SEXUAL FREQUENCY ARE ASSOCIATED

Both the healthy young women on a university campus as well as an infertile population who had kept careful records as they attempted to get pregnant showed evidence of luteal defects when their sexual behavior was sporadic. Along with Drs. Garcia and Abba Krieger[2] I published the study "Luteal Phase Defects: A Possible Relationship Between Short Hyperthermic Phase and Sporadic Sexual Behavior in Women." In this report we analyzed the results of sixty infertility patients who had kept careful records before undergoing any infertility treatment: records of their own basal body temperature, when they menstruated, and when they had sexual intercourse.

The result was highly significant and clinically substantial. Short luteal phases are associated with sporadic sexual behavior. In fact 88 percent of the women with short luteal-phase lengths showed sporadic sexual behavior on their calendars. When Dr. Garcia and I published this paper, we suggested that regular sexual behavior might be helpful in promoting the fertility of a cycle, not only at ovulation to allow impregnation but throughout the cycle to prime the reproductive system.

There are many different ways that infertility can manifest and many different causes for it. One thing is clear to me: If you want to be fertile, you ought to be sexual. And if you are being sexual in order to promote your fertility, you ought to be sexual every week, not just around the time of ovulation. At least once each week, and preferably twice. Any specific day? I don't think it matters once ovulation has passed. Keeping up a regular pattern of sex is important even if you have consulted a fertility specialist.

SEX HELPS FERTILITY TECHNOLOGY WORK BETTER

In this era of high-tech baby making, low-tech sex still plays a pivotal role. Working in Brazil and elsewhere, research scientists have added data

[2] An assistant professor of statistics at the Wharton School of the University of Pennsylvania.

to support this conclusion as they investigate the role of human sexual behavior in human fertility.

In 1989 the first study of sexual behavior among couples undergoing reproductive surgery with a new technology—GIFT (gamete intra-fallopian transfer)—confirmed the role of sexual timing. The startling news from Dr. Guillermo Marconi's study was that having sex forty-eight hours before and after the surgical procedure more than doubled the subsequent conception rate. Thirty-six women were studied. Eighteen were told that in addition to the surgical technology they should engage in good old-fashioned sexual behavior right up until they came in for the procedure. The other eighteen were told to abstain forty-eight hours before and after their (similar) GIFT procedure. The difference between the two groups was remarkable. Fifteen of the eighteen (83 percent) who had intercourse became pregnant, whereas only five of the eighteen who abstained became pregnant. The difference is highly substantial. Sex helped make the technology work better. Although at first glance it might seem that having sex might have provided sperm that led to the pregnancy, the authors had medical evidence that led them to conclude that these pregnancies were not resulting from the uniting of the sperm at intercourse with a naturally ovulated egg.

SEX PRODUCES SUPERIOR SEMEN

Another report lent an increased understanding to the role of the man/woman dynamic. A semen sample is usually studied in the laboratory for an infertility evaluation. Normally this sample is produced by masturbating into a cup. Dr. Panayiotis Zavos at the Andrology Institute of Lexington, Central Baptist Hospital in Lexington, Kentucky, working with Dr. Jessie Goodpasture at Syntex Research in Palo Alto, California, evaluated the difference in the quality of semen between men who had masturbated into a cup and those who provided the semen from a condomlike receptacle when the ejaculate had been produced during sexual intercourse. The results were startlingly different. The process of sexual intercourse produced a much higher quality of semen in terms of sperm count, sperm motility, and other usual measures of fertility.

To me these results speak loud and clear. No matter which group has been studied—my healthy young Philadelphia women, Dr. Garcia's infertile patient population in Pennsylvania, Dr. Marconi's infertile population undergoing GIFT technology in Brazil, or Dr. Zavos's male infertility patients in Kentucky—they all show the same thing: Sexual intercourse promotes fertility. This is hardly surprising. What stuns me is that some infertility specialists who know about these data still will not suggest to a couple that they should engage in regular, frequent sexual intercourse. When I have asked about this omission, I have been given the reason that the infertility experience is so stressful that the physician is loathe to increase the stress by prescribing sexual behavior.

SEX AND STRESS

Problems of stress seem to abound in 1991 America. Likewise, problems with achieving a satisfactory sexual life also appear to be very common. Both subjects fill magazines, bookstores, and talk shows because these topics generate public interest. Whether or not these times are more stressful and sexual life is therefore less likely to be successful, the facts are that the issues are troublesome.

I think stress may cause or contribute to infertility. Many modern couples live tremendously stressful lives as a result of what has now become a socially sanctioned attempt to "have it all." In the pursuit of fertility I would suggest reducing stress in both partners. Let's examine how the sources of stress take their toll on fertility.

THE STRESS OF SUCCESS

"Having it all" may be totally unrealistic. During the span of their lives women and men can have many different kinds of experience, but there are only so many hours and so much energy in a day. The needs of the body for fertility, for pregnancy, for parturition, and for caring for infants create profound demands that consume the energy of a woman. It is a woman's body and her life that bear the direct costs of reproduction.

The women's movement has spawned a sense of entitlement to equality among educated young women. I feel compassion for those in their twenties and thirties because, from what I have seen, many of these upwardly mobile, highly educated, hardworking women fully expect to be able to juggle the demands of conceiving, carrying, and raising emotionally healthy children along with career pursuits that demand more than a forty- to forty-five-hour workweek in order to achieve success. The goals seem to be reasonable. Actually playing them out can lead to extraordinary stresses.

Reams of articles have been written about stress in the last thirty-five years. There are a few studies that are directly relevant to the science of intimacy. In 1958 Dr. Leo Tervila reported that individuals who experienced stress ending in death, such as from suicide and automobile accidents, experienced a tremendous and sudden increase in the size of both the ovaries and the adrenal glands. His was an anatomical study to show how stress can affect the output of glands. The hormonal output of both the ovaries and the adrenals directly affects the fertility of women. Stress stimulates the adrenal glands to secrete the hormone cortisol. And elevated cortisol secretion can temporarily turn off ovarian estrogen production.

Thirty-one years later an important study of infertile women showed the direct relationship between levels of anxiety and hormone levels. Anxiety was generated by having the women watch a film showing a successful pregnancy and parturition. All the women in this study found it stressful to watch a film depicting the pleasures and joys of pregnancy and parenthood. This of course is no surprise. What was interesting was that anxiety levels could be objectively measured and compared. This pencil-and-paper test evaluated individual levels of a stable personality characteristic called trait anxiety. Women who scored high in this characteristic showed different hormonal profiles than women who scored low.

Women with high trait anxiety on paper also had higher levels of testosterone, cortisol, and prolactin in their bloodstream than women with low trait anxiety. These three hormones affect the fertile capacity of an individual. Once the high and low groups had been differentiated via personality testing, the study showed that they reacted differently to emotionally stressful experiences. Experimental details are described in the following section.

THE STRESS INFERTILITY EXPERIMENT

The infertile women were brought into the study room, and the experiment was explained to them. Then a catheter was inserted, and blood was taken every fifteen minutes, eleven times during the course of the experiment. The experiment had three stages: a one-hour rest period, a one-hour film, and a one-hour postfilm rest period.

At the start of the experiment high-anxiety women showed higher hormone levels. During the first rest period both groups of women showed a declining level in two of the hormones. Then, when the stress-inducing "happy mothering" film came on, the group with high trait anxiety continued to show a decline, but the group with low trait anxiety (the calmer group) showed a sharp *rise* in all three of the hormone levels.

The authors concluded that highly anxious women tend to anticipate difficulties even when there are none. Consequently their stress-related hormones are always in a state of elevation. In contrast, low-anxious women tend to reflect and assimilate. Their stress hormones are not generally elevated. When a difficulty does emerge, then and only then do low-anxious women's hormone levels shoot up. The authors could not explain why high-anxious women showed a decline in hormone levels when provoked with what all agreed was a highly stressful experience.

STRESS IN NATURE

A related study of wild baboons may help to explain this inverse reaction to stress. The high-anxious women with elevated hormone levels may already be at maximum capacity to secrete hormones before a stress occurs. The decline in hormone levels in high-anxious women exposed to more stress seems to work in the same way that stress affects other primates. Dr. Robert Sapolsky's studies between 1982 and 1986 reviewed the findings of his field research in the wilds of Kenya. Although he was unable to study female hormones and stress (see page 91), the results in the males showed clear relationships between stress and steroid hormones. On annual forays into the field to live among wild baboons, he was able to discern the dominant male and the "pecking order" of the various members of a troop. To measure the hormone levels at any particular time, he would use a dart gun to sedate one animal long enough to take some blood. In the years he worked among these baboons, he observed several events that produced acute stress to the whole troop. Two of the most upsetting were the death of the dominant male and drastic weather changes. He discovered the following:

- *The dominant male circulated lower levels of testosterone than all the other males.* This finding was surprising. I had read other reports that concluded that dominant males circulate higher levels than less-dominant males. That's why they're the top baboon. When I wrote to him with my confusion, he replied that I had read his papers correctly. He explained that in natural conditions, as opposed to the small cages where most animals have previously been studied, dominant males circulate lower levels of sex hormones than all the other males in the troop.
- *When a major stress occurs to the community, the subdominant males show a sudden drop in their circulating levels of testosterone.* This

reaction seems to indicate that they had been circulating maximum levels all the time and had no reserve to draw on when new external stresses were added.

- *The dominant male was hormonally different.* In the face of stress he secreted a huge surge of testosterone, perhaps sufficient to give him the strength to help the troop to fend off the danger.

The discoveries Sapolsky made in baboons resemble those made in infertile women. It seems to me that personality characteristics—the tendency for particular personality traits—do tend to be stable unless a person decides to change and does the work involved. Some women are "hyper," agitated, and live in a state of high trait anxiety. Other women who appear calm probably are in a state of low trait anxiety. Hyper and calm appear to be opposite characteristics along a continuum of personality attributes. I think the data are suggesting that hyper people circulate different patterns of fertility-related hormones than calm people.

In other words a personality characterized by anxiety may be stressed, and the stress may be compromising fertility. When individuals live in a state of low anxiety, they may be conserving their hormonal reserves. They can call on those reserves when they are needed. The individual who has the capacity to stay calm usually emanates power by virtue of this self-containment. Think of the quality of dignity, a personality characteristic that carries with it a sense of calm, of composure. People with great dignity have a stillness to their being. They emanate great power, somewhat like the dominant male in a wild-baboon population. They probably react to crises more competently than hyper individuals. Power and fertility may be more closely connected than is immediately obvious. It would be an interesting area for psychologists to investigate, but for now it is only my speculation.

According to Dr. Sapolsky, the conditions in the wild are generally very peaceful and stable, in large part because the pecking order has already been established. Everyone knows who the boss is, and there isn't a constant vying for top baboon. Only during a major disruption does a new scrambling for power emerge within the community. The animal that has the capacity to draw on reserves of energy is the one who can claim power. The one already working at maximum has no reserve and doesn't get to be top baboon. In other words acute surges of the sex hormone testosterone appear to be necessary for claiming power. I suspect there are parallels in women, since a similar phenomenon was shown for the same hormone. Being able to count on this energy reserve may make the difference in fertility because the physiological demands of pregnancy and parturition may be programmed to "turn off" fertility when the strength is inadequate to the task. Since stress saps the strength, let us focus next on the studies that evaluated how to reduce stress and showed the beneficial changes in hormones that resulted.

LESS STRESS—MORE FERTILITY

In general, stress reduction does promote beneficial physiological changes. The first studies focused on reduction of disease and its correlates; the later ones began to focus on improving infertility.

In 1975 Dr. Herbert Benson introduced stress-management techniques (meditation exercises) that people could effectively use to reduce stress. In his book *The Relaxation Response* Dr. Benson demonstrated how to relax by focusing the mind, step by step, on each muscle system in the body. A person practicing according to his method could learn to relax all the muscles, and in the process the spirit as well. Working at Harvard University, he showed that meditation produced pervasive effects in reducing physiological changes that were known to be stress related. He measured oxygen consumption, cholesterol level, and other cardiovascular risk factors. Dr. Benson called his technique progressive muscle relaxation, and he showed profound benefits that could be achieved by reducing stress in this way. Stress-related diseases (high blood pressure, heart disease, etc.) declined in individuals who took the time to practice his method.

By 1990 he, with his coauthors, had concluded that by regularly eliciting a relaxation response individuals could reduce chronic pain, hypertension, preoperative anxiety, ventricular (heart muscle) arrhythmias, and anxiety.

In 1978 Dr. Julian Davidson's group at Stanford published their evaluation of hormonal response to two to three months of Transcendental Meditation (TM). In the Stanford study individuals who had three months of practice reduced their levels of plasma cortisol by about 8 to 10 percent within the first hour of the meditation exercises. Individuals who were long-term TM practitioners, highly competent in the art of inducing a kind of hypnotic state of relaxation, were even more successful; they reduced their plasma-cortisol levels about 25 percent within about an hour of beginning meditation.

More recently, in 1990 a group of Harvard researchers (which included Dr. Benson on the team) published their study, a preliminary attempt at a behavioral-treatment program based on the elicitation of the relaxation response. Women who were still infertile after ending an expensive infertility treatment workup entered the behavioral-treatment program to learn how to reduce their internally driven stress reactions. The first fifty-four women to complete the program showed statistically significant decreases in anxiety, depression, and fatigue as well as an increased vigor. A little over a third of the group became pregnant within six months of completing the ten-week program.

What they had done was based on the earlier work of Dr. Benson in creating exercises that people engaged in to promote their own relaxation. First the program was carefully explained, because experiments have clearly shown that when an individual knows what to expect, his or her anxiety level is lower.

Then groups met once a week, usually for several hours, to review strategies for stress reduction—social sharing, support time, yoga, physical exercise, behavioral exercises, stress-management training, and so forth.

But the work with infertility patients—conducted by Drs. Alice Domar, Machelle Seibel, and Herbert Benson—broadened the definition of relaxation exercises. The authors commented that while the particular methods they used were good, so were other methods. For some it is spiritual connection during a religious ritual. For others it is taking a quiet walk in the woods. For still others it is repetitive prayer, mental imagery, or absorption in a pleasant task. I think that the work of these Harvard researchers combines with the discoveries about baboons and the high- versus low-anxious women to reveal a profound truth about fertility: Stressful attitudes alter the internal milieu. They cause hormonal overload. And the overload can lead to a burnout—a shutting down of the (expendable) reproductive harmony.

Nature has apparently evolved the fertility mechanism in such a way that the woman who is most able to relax is probably living within a hormonal environment likely to optimize her chances for fertility. The hard-driving demands of modern professional life can render relaxation a difficult art. High-pressure lives can lead to hormonal changes that reduce fertile capacity. It is possible to juggle several demanding activities, but juggling often increases stress. How do you get *out* of the rat race? The solutions are as individual as the people who seek them; and not all problems can be corrected in the present. Some require that we endure.

Short of a full-fledged life change, an intermediate step may be to bring more relaxation into your daily life. This, too, is not easy, because career demands can make it difficult to take the time to achieve the "floating state" that promotes optimal relaxation. Fortunately the experiments in progressive relaxation have shown a way that even busy women can reduce stress. Once an individual makes a commitment to giving regular practice time to relaxation techniques, profound changes in the hormonal environment can result. Consider this: If you cannot find an hour a day to devote to finding peace within your soul, how will you ever find the time to raise a baby? A growing baby or young child needs more than an hour a day of peaceful time with its mother in order to thrive.

THE OTHER SIDE OF THE COIN: REGULATION

If you are involved in either a regular or a sporadic relationship and do not want a child now, the principles that have already been discussed to optimize fertility could be equally well applied to avoid pregnancy.

Understanding the principles of regular weekly sex in promoting a more fertile endocrine system makes the choice of contraceptive method crucial. Since regular sex primes the reproductive hormones, you need birth control that is not only highly effective but conducive to sexual relations. What good is contraception if it kills desire or fails in the heat of passion?

The Downbeat of the Rhythm Method

With the rhythm method, also known as natural family planning, a couple abstains from intercourse when the woman believes herself to be ovulating. The method contains an inherent fallacy: By abstaining for a week, around the time of ovulation, a women might create a luteal-phase defect and reset the cycle to trigger a *new* ovulation. If she does ovulate out of phase, she may not know it. Animal studies in monkeys by Dr. Arnold Goodman done at the National Institutes of Health back me up. Dr. Goodman developed microsurgical techniques by which he could "zap" the dominant follicle in the ovary without hurting the animal. At different times of the month he would interrupt the cycle by zapping the corpus luteum or the dominant follicle. He consistently found that interrupting the cycle would inevitably trigger a new wave of development of follicles, just as I suspect sporadic sexual behavior does. In almost every case the next ovulation occurred 12.6 days later. From other research we know that a potentially ovulable human egg takes between 12 and 13 days to mature once the development of a new wave of follicles is triggered. The complexity of those studies is less relevant here than the consistency of the timing, 12.6 days.

If you do not want to get pregnant, you should know the rhythm. I believe you can trigger an ovulation 12.6 days later whenever the cycle is disrupted. If I am correct, then unprotected intercourse may fertilize that egg. If the egg gets fertilized in this way and you continue to be sexual, the subsequent luteal phase should be sufficiently healthy to build the nest and promote a viable pregnancy—exactly what you are trying to avoid.

If you do not want to conceive, do not practice the rhythm method unless you use a barrier contraceptive (diaphragm, condom, cervical cap) within three days of whenever ovulation may be occurring. This means that if you abstain from sex for a week, you should consider the possibility that a new cycle of ovulation has potentially begun and that 12.6 days later you may be ovulating again. Since sperm can live for three days and await the ovulating egg, unprotected intercourse can trigger a pregnancy as early as nine days after a cycle is disrupted.

The cervical mucus can also provide an important clue, because it changes around the time of ovulation. At this time it becomes stringy and can be drawn between the fingers like a raw egg white, increasing in quantity twenty- to sixtyfold around ovulation.

So, rhythmic timing can work when couples pay attention to the rules:

- Stresses, such as sporadic sexual behavior, can start a new wave of follicle development.
- Ovulation is once again possible 12.6 days after this new wave.
- Sperm can live for up to three days in the fallopian tubes as they await an egg.
- A newly ovulated egg is receptive to sperm for about twenty hours.
- Condoms and diaphragms, when properly used, block access of the sperm to the fallopian tubes—the place where fertilization takes place.

Block That Sperm

Barrier methods of contraception, those that create a barrier between the sperm and the egg, do work as long as the barrier is in place. Condoms, diaphragms, certain contraceptive sponges that are placed at the tip of the cervix, cervical caps, and any other barrier put in place before intercourse can be effective. It is crucial that barrier methods be in place prior to penetration. Men do exude a bit of seminal fluid before ejaculation. The fluid can contain sperm. For women placement of the diaphragm or cervical cap prior to arousal can lead to greater protection. During arousal the cervix shifts slightly. Inserting barrier birth control at this point may not provide complete coverage of the cervix.

This method can also be extremely helpful in reducing the risks of sexually transmitted diseases, which, in itself, safeguards fertility. Whether the spermicides that are used will someday be shown to be harmful to the woman remains an open question. I suspect that there is a potential for such side effects but that this potential needs to be weighed against the other risks to life from *not* using contraception.

The Pill

The oral contraceptive almost always works because of the underlying effects that it produces in the pituitary gland, ovaries, and uterus. Women who take oral contraceptives circulate about three times higher levels of estrogen than the body normally produces. These high levels of estrogen produce changes in the pituitary secretion of FSH and LH. These changes inhibit fertile cycles, both in the ovary and in the womb. Going off the pill may delay conception until hormone levels return to prepill levels. Keep this timing in mind for pregnancy planning.

Oral contraceptives are highly effective and, for most women, appear to be not only safe but beneficial—to their bones, cardiovascular health, and general well-being. More than fifty years of research on and use of the oral contraceptive has suggested that the relative balance of competing influences favors their use in individuals who can take them. The newer lower-dose hormonal supplements have not had as long a test, although they seem to work well.

Risky Sex

The method called coitus interruptus; in it the man withdraws before he ejaculates in order to interrupt the ejaculatory stream from proceeding to its target, the egg. It is extremely dangerous. Men usually leak a little bit of seminal fluid long before they ejaculate. And this little bit of fluid can contain sperm, sperm that can swim and fertilize an egg. Coitus interruptus is risky sex. It should be avoided.

HOW TO DECIDE

Everything we do as we move through the days of our lives adds risks to the duration of our lives. When we get in a car, we risk an accident. When we walk down steps, we risk a fall. When we use contraception, we risk an accident related to side effects. The critical relevant question is not whether or not there are side effects. What needs to be considered is how the potential side effects of one choice balance against the potential side effects of alternative choices. These decisions are individual ones. There are no formulas that can be applied, because individual needs and skills are too variable.

For me as a biologist the broad view is clear. The design of nature favors the proliferation of its species. When I consider the millions of seeds produced from one plant, it seems akin to me to the tremendous array of sperm produced from one ejaculate. The design of the system is for reproduction; and human beings interfere with reproduction at their own peril. There is a risk associated with whatever choice we make in preventing conception. Likewise, pregnancy itself represents an enormous physical, emotional, and financial burden to an individual or a couple. It, too, has its risks. I believe we should make choices based on solid knowledge. We each decide which risks and which benefits we select. Life is better when we know what we are doing and why.

CONCLUSION

The science of intimacy has important messages to teach about fertility —first and foremost that you are in control of your own fertility, more often than you think. Fertility is not something that medical science determines, although physicians can help when physical problems are present. When and if fertility becomes problematic, a couple can first take action themselves, without spending the time and money or risking the stress and embarrassment of a medical workup for infertility. Couples should be aware that even with the knowledge provided in this chapter and the skills of the best infertility specialists, results may not always be as they would wish. Still, it helps to improve the odds by making use of the facts.

In sum, fertility is often of your (you and your partner's) own making. The reasons why include the following:

- Regular life-style choices and behaviors can improve or imperil your ability to conceive.
- Stress—from any source—can disrupt fertility.
- Relaxation can positively affect the hormones that regulate fertility.
- The timing of sexual relations with the aim of producing pregnancy can affect fertility.

- Ironically the factors that promote fertility (relaxation, stress reduction, regular sexual relations) may need to be learned, but sensual skills and expertise do not have to be learned to promote fertility.

For many the true sensual skills do not bloom until the children are grown. Meanwhile, the chemistry between people is affecting us unconsciously. Men, and women, transmit essences, and these are discussed in the next chapter.

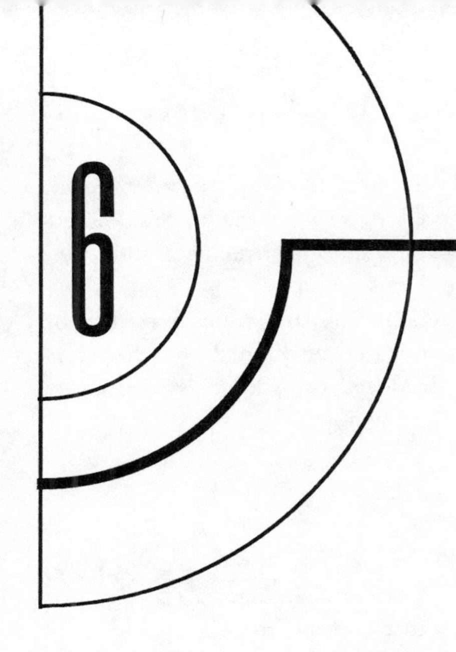

PHEROMONES: MALE AND FEMALE ESSENCES

Time past and time future
Allow but a little consciousness.
To be conscious is not to be in time
But only in time can the moment in the rose-garden,
The moment in the draughty church at smokefall
Be remembered; involved with past and future
Only through time time is conquered

—T. S. ELIOT, "Burnt Norton,"
The Four Quartets

Intimate partners affect each others' reproductive biology, for better or for worse. Whether intimacy is likely to enhance a woman's endocrine system or disturb it depends on the timing of her sexual behavior. For men the discovery (pages 108–9) that monogamous men show testosterone peaks in phase with the estrogen peaks of their partner suggests that they, too, might be vulnerable to the timing of intimate behavior.

As Chapters 1 and 5 showed, regular weekly heterosexual contact seems to promote fertile-type reproductive cycles in women. Moreover, sperm viability after sexual intercourse is significantly better than after masturbation (see page 137 in Chapter 5). What remained unresolved was how this was working. Self-stimulation appeared ineffective, whereas either intercourse or genital stimulation by a man (provided it was regular) did seem to promote fertile-type endocrine cycles. The same goes for men. Self-stimulation does not alter testosterone levels; sexual activity with a woman can.

The data seemed to suggest that sexual activity in the intimate presence of a man was different from sexual activity when alone and left to one's own devices. A man seemed to add something.

The search for that "something" is the story of this chapter. It led me to a preliminary answer, which turned out to be the male's "pheromones." This subject kept me busy designing experiments, conducting research, and analyzing data for more than six years.

Although my own research focus centered on the influence of men on the endrocrine system of women, earlier writers on women's lore had tried to study women's cycles and had introduced the idea of synchrony of cycles much earlier.

THE SYNCHRONY OF CYCLES

Women's lore has a rich history of allusions to a synchrony among women's menstrual cycles. One example is the Yurok Indians of California, a group that continues to follow tribal customs and to this day practices mysterious rituals involving menstruation. The aristocratic women of the tribe leave their family tent to live together within a menstrual hut for ten days during each monthly bleeding cycle.

The first anthropologists to record this menstrual isolation were men. These male investigators drew false conclusions about the female reaction to their menstrual experience. They missed the meaning of what they observed because they interviewed the Yurok men but did not interview the women. When they learned of the monthly sequestering, they presumed that the time apart meant that the women were being segregated during menses.

In 1982 anthropologist Dr. Thomas Buckley published his study after interviewing the women. He learned that the women did not consider themselves banished; they felt privileged. The men were required to take over the household chores during the ten days each cycle when the women were in seclusion. Relieved of work, the menstruating women used the time for reflection, prayer, camaraderie, and rest. The women also said they were menstruating in synchrony—at the same time of the month as their friends. Just imagine how women of the 1990s might enjoy a monthly cycle of seclusion and recreation among women friends.

Anthropological investigations are limited because they are based on hearsay, not objective evidence. However, perceptions like these are scientifically valuable because they can trigger studies that prospectively gather data.[1] The actual publication of prospectively gathered data has confirmed the fact that women do synchronize their fertility cycles.

---

[1] In a simplistic sense prospective data represent data that a scientist went prospecting for. In prospective research gathering, the events are recorded as they are happening, and the subsequent

The statistically valid *science* of menstrual synchrony began in 1971 with a then startling publication in *Nature,* a prestigious British scientific journal. Martha McClintock, the author, was still a graduate student at Radcliffe College when she published the first proof that women who live together start to cycle together; that is, the timing of their menstrual onsets tends to synchronize. Her discoveries of menstrual synchrony set in motion a generation of research studies attempting to repeat her study and to figure out how synchrony might work. By 1986 my colleagues and I were able to confirm that human pheromones provided an underlying mechanism that could alter menstrual timing. In addition we suggested some of their relevant chemical constituents.

Dr. McClintock first told me about her work when we were both graduate students at the University of Pennsylvania. She had almost finished her doctoral dissertation. I had just begun graduate school. Her discovery of menstrual synchrony was well known to the scientific community but had not yet been replicated by other investigators. She told me the story and explained how she had made the discovery.

Box 6-1: Martha McClintock's Discovery of Menstrual Synchrony

As a young graduate student living in an all-female dormitory in the late 1960s, Martha McClintock thought she had noticed a cyclic pattern in the dorm bathrooms. About once a month the trash cans overflowed with women's sanitary products. She decided to test scientifically her perceptions that women were menstruating in synchrony. After obtaining faculty permission, she enrolled student participants who would be willing to record the date each time they menstruated. What she found *statistically* confirmed her intuitions.

In the beginning of a semester, when dorm assignments would first trigger a new set of living and housing relationships, women tended to have random onsets of menstruation. As the weeks of the term progressed, those women who lived together or were close friends started their menstrual flow at increasingly closer times. She documented the phenomena scientifically and published her findings. (Women's lore was now scientifically validated.)

Her conclusion was that women who are close together tend to menstruate at or close to the same time.

charting and analysis permit statistical tests. These tests predict the probability that future (similar) tests would confirm the finding. The major value the statistical analyses can render is in their predictive capacity, for example, Will there be a full moon on the twelfth?

Since 1974 a number of other investigators have replicated her discovery. The question of how the phenomenon worked remained unanswered. Dr. McClintock didn't know for sure, but in 1971 she guessed that *pheromones* were involved.

WHAT ARE PHEROMONES?

Pheromones are defined as chemical substances that are secreted externally by some animals. These chemical substances convey information to other animals of the same species and in the process produce a specific response in their behavior.

The word *pheromone* is something like the word *hormone.* A hormone is a substance produced within the body, usually in a gland, that is secreted into the bloodstream, travels through the body, and exerts its effects on some other gland or tissue of the body. Chapters 3 and 4, on the hormonal symphony of women and of men, describe the specific relationships between five such sex hormones. A pheromone is similar. It, too, is a substance that is produced in the glands of the body. It, too, is a substance that travels some distance from its origin to act in the glands within the body. The difference between hormones and pheromones is this: Whereas hormones exert their effects *within* one body, pheromones exert their effects *between two or more different bodies.* With pheromones one person or animal produces the substance, and another person or animal is acted on by the substance. We are just beginning to learn how the process works.

PHEROMONES IN NATURE

During the time that my colleagues and I were investigating menstrual cycles in women, many other scientists were studying pheromonal effects in ants, birds, bees, rats, and other animals. Research had shown that the glands of a variety of animals secrete substances through urine, sweat, and external openings that attract other animals of the opposite sex within the same species. Consider, for example, dogs.

When a female dog is in heat, she emits pheromones as airborne chemicals that produce an odor that attracts male dogs to her for mating. A built-in "maintenance of the species" system broadcasts her availability for a fertile mating whenever her body is ready. And the males get the message. Somehow though the nose, the brain is activated, and the males come running.

By the time I got to graduate school in 1974, pheromones as sex attractants were well understood to exist in nature. The work at the University of Pennsylvania in the experimental psychology division was focusing on the role of pheromones in modulating the endocrine system of female rats. Pheromones were serving not as sex attractants but rather as sex primers, priming the reproductive system to be ready for fertile mating. Some of Dr. McClintock's extraordinary predoctoral work showed that if female rats were housed in a separate room from males and a connecting air duct was opened or closed to permit or prohibit the odors of the males to pass to the female room, different but predictable effects on their fertility would occur. Male odors stimulated fertility in females, but only in females of the same species.[2]

PHEROMONES IN HUMANS

Until 1979 no studies had been published on the possibility of human pheromones, either as sex attractants or as primers of the endocrine system. Then, in 1980, a study was published and was immediately criticized by members of the scientific community. Dr. M. J. Russell, working with two colleagues, published an extraordinary report—"Olfactory Influences on the Human Menstrual Cycle."

In this study the underarm secretions from a single woman had been collected and rubbed on the upper lip of other women three times a week for a period of four months. These substances were reported to cause the women who received them to show the menstrual synchrony that Dr. McClintock had earlier reported. At the start of the experiment the recipients on average had a menstrual cycle starting nine days after the donor's. After four months of this regular upper-lip application, the average onset difference between the menstrual cycle of the donor and that of the recipient had whittled down to three and a half days (i.e., less than half the original difference). The finding that the underarm secretions from one woman could cause menstrual synchrony in other women was startling.

Members of the scientific community criticized the study because it was not conducted with appropriate double-blind procedures. The technician who applied the odors was the same person who supplied the secretion. Therefore scientific critics argued that this study had not proven the cause of menstrual synchrony because the technician could

[2] The definition of *species* is "a reproductively isolated group," i.e., it is defined by the genetic pool from which new offspring are created.

have been producing synchrony just by being close to the subjects. It seemed that pheromones might account for menstrual synchrony, but a more rigorous study needed to be done in order to prove it.

I was fortunate to be in the right place at the right time to contribute to that proof.

THE DISCOVERY THAT SHOOK THE WORLD PRESS

In November 1986 science reporter Boyce Rensberger of *The Washington Post* broke the story on the front page. Within a month it had been heralded in *Newsweek, Time* magazine, *U.S. News & World Report,* and throughout the international newswriting community.

My colleagues and I had demonstrated that it was possible to bottle the essences of men and women, freeze them for a year, then thaw them, apply them to the skin under the nose somewhat like perfume, and change the endocrine milieu of the women who received them. This intrigued many people. It excited the pharmaceutical industries and led the Monell Chemical Senses Center, where we made the discovery, to begin the arduous and expensive process of attempting to get the U.S. Patent Office to approve patent applications for five related discoveries. It led to signing a licensing agreement with the Tagasako Company of Japan for the development of the Japanese rights to what I named female and male essences. Since these were discoveries I coauthored with colleagues, and since they came out of the scholarly work of many different scientists who had preceded us, I can offer you, the reader, a set of uniquely personal perspectives.

HUMAN SEXUAL PHEROMONES IN A PHOTOCOPY LINE

The photocopy line in the Obstetrics and Gynecology Department where I was a research associate was where it began for me, in 1980. I was standing in a photocopy line more and more frequently in order to fill requests from scientists from around the world who were writing and asking for a reprint of the seven studies on women's sexual behavior and its relationship to menstrual cycles that Dr. Garcia and I had coauthored. (These were discussed in Chapters 1 and 2.) While I held a copy of "Sexual Behavior Frequency and Menstrual Cycle Length," a man standing behind me looked over my shoulder and asked if he could read the paper while waiting in line. When he asked if I knew the author, I introduced myself. Within two years George Preti and I had forged a partnership for the series of research studies that led to the confirmation of human sexual pheromones. Other colleagues worked with us as well, and their help was critical. Dr. Celso Garcia was involved in the design and interpretation of the experiments. Dr. George Huggins examined most of the patients to determine their general state of health. Henry Lawley helped us in innumerable ways with the research analysis and chemical monitoring.

THE MONELL CHEMICAL SENSES CENTER

George Preti was an organic chemist on the staff of the Monell Chemical Senses Center, a place where seventy scientists were investigating sensation and perception. Monell was affiliated with the University of Pennsylvania Department of Obstetrics and Gynecology, where I worked with Dr. Garcia as a research associate. Dr. Preti had spent at least fifteen years or more contributing to scientific knowledge about human odors: which parts of the body produced them, what their chemical constituency was, and how the odor could be chemically analyzed to predict the underlying function of the body. His special interest was human malodors, those smells that people find offensive. He told me that his work concerning mouth odor was revealing that the nature of the odor could predict the potential for certain kinds of cancer. As a chemist he understood the molecular configuration of these odorous materials and he had the sophisticated equipment to analyze them. Dr. Preti had an abiding interest in the possibility of human pheromones. A number of our colleagues at Monell and at the University of Pennsylvania were actively studying different aspects of pheromone research in small mammals, and occasional seminars to discuss the research were held on campus. He was interested to learn about my discoveries relating sex frequency to menstrual-cycle length because he saw the possibility of synergism in our separate interests.

He invited me to give a seminar to the scientists at the Monell Chemical Senses Center about my work. A few months later he invited me to work with him. He wanted to figure out whether the effects that Dr. McClintock and I had found—that women who had close contact with women or regular sex with men had a change in their endocrinology—were due to a testable pheonomenon. If pheromones did in fact produce these effects, we might be able to locate and discover them.

He suspected we might be able to identify human pheromones as the cause for the previous findings. I had discovered that self-stimulation, or masturbation, did not seem to affect cycle length, whereas genital stimulation in the presence of a man did. Dr. Preti and I considered the possibility that the presence of a partner might add something. We believed the difference was probably less related to sexual completion (i.e., orgasm) than to arousing the nervous system. Genital stimulation by a man, even in the absence of orgasm, had a more significant endocrine-system effect than self-stimulation. The presence of a man provided some added biological value. Perhaps some airborne substance transmitted from the man through his body was responsible. Dr. Preti's hypothesis was that the substance was found either in mouth odors or mouth gases or in the underarm. I thought that underarm odors were the source. Whatever the origin, these odors entered the woman's nose, fired a sensory nerve to the brain, and altered her endocrine response.

Originally Dr. Preti worked with Dr. Lynette Geyer, a psychologist who subsequently moved to California. Dr. Preti asked if I would help him continue the work as partners, or coprincipal investigators. I would direct the behavioral biology (nonchemistry) elements; he would direct the chemistry elements. We set to work for about a year refining the proposal. In late 1982, a year later, our project was funded.

Unfortunately the amount of money offered was half of what we had requested and we were forced to redesign our experimental plan. We had planned to test both mouth and underarm odors as a potential source of pheromones. Now Dr. Preti and I had a difference of opinion. He thought that the pheromonal substance was transferred from the man to the woman during kissing. I thought the essence was in the underarm secretion and argued that we should study it rather than the kissing gases since we did not have enough money for both. I convinced him by reasoning that

1. Romantic relationships usually place men taller than women.
2. Sleeping and hugging do not occur near the partner's mouth but in the crook of his arm.
3. Dr. Russell's study using underarm secretions *had* seemed to work on menstrual synchrony.

He agreed.

THE FIRST DOUBLE-BLIND PHEROMONE EXPERIMENTS

We set up our experimental design to cover three distinctly separate research phases:

For Phase 1, we would collect the male and female essences, extract the critical elements, freeze them, and put them into the deep freezer for subsequent thawing in Phase 3.

For Phase 2, we would enroll a large number of women to start tracking their menstrual-cycle patterns, their basal body temperature records, and their sexual behavior.

For Phase 3, women who had successfully kept accurate records in Phase 2 would be invited to enter into the essences experiment. Here they would continue recording menstrual, temperature, and sexual behavior and would now receive the essences placed under their nose (on the upper lip) three times a week. In addition in Phase 3 the subject would give blood three different times during the luteal phase of the last cycle of the study in order to test the sex-hormone levels circulating in her blood.

PHASE 1: THE COLLECTION OF MALE AND FEMALE ESSENCES

Since this was the first prospective study ever done on human sexual pheromones, we knew that there were many variables that would inevitably be discovered to affect the production of human pheromones, if they even existed. We decided to control for variables by carefully preselecting donors who met a very stringent set of criteria; They had to

- Use no drugs
- Take no hormones
- Have hairy underarms that they did not shave, because the hairs were thought to serve to collect and disperse the odors
- Be engaged in a sexually active committed relationship, to ensure the stability of their own endocrine pattern
- Be younger than thirty-five, the optimal reproductively fertile years (and more than seven years menstruating if they were women)

Previous scientific work had already shown some aspects of the nature of underarm odor. Three different kinds of glands—apocrine, sebaceous, and eccrine—exist in the underarm. The secretions that emerge from the apocrine glands have no odor when collected at the skin surface. If those secretions are placed in a dish along with bacteria of the same type that normally live on the hairs of the underarm, the bacteria manage to produce the characteristic odor of the sweaty human underarm. Two different kinds of bacteria had, by 1980, been identified on the underarm. One kind, the micrococci, produce an acidic odor. The other bacteria, the diptheroid, produce a more pungent odor. Sweat seems to enhance production of the odor by the bacteria. Chemists had analyzed the particular constituents of these odors, and some of their elements had been identified.

When Dr. Preti analyzed the freshly collected secretions from the apocrine glands, he was able to show that sex hormones, namely androgen sulfates, were present in the underarms. When a hairy underarm produces moisture in the presence of these weak androgens and the bacteria, the characteristic odors of sweat are produced. In August 1990 Dr. Preti's focus on the secretions of the underarm again made news. *The New York Times* reported his discovery of the specific culprit in the malodor—and the promise of a potentially simple solution through chemistry. His interest in malodor took years to bear fruit.

The Female-Essence Donors

Our donors were selected for their ability to produce all of the underarm odorants that had been characterized so far. Some individuals produce more odor than others when they sweat, and this may be the result of having higher hormone levels or more bacteria.

The requirements for using the essences of women donors included very rigorous additional criteria in order to "control for" possible variables that might influence menstrual-cycle length.

- Each of our five women donors of female essence had a regular twenty-nine-day menstrual cycle with a clearly fertile-type basal body temperature graph.
- Each was gynecologically mature; she had been menstruating for at least seven years. Each had a menstruation onset that occured within seven days of a full moon (the reasons why this was desirable are described in the Chapter 8).

Collecting the Essence

To collect sweat, each woman wore an underarm patch that was pinned to the inside of a T-shirt that she put on after washing with Ivory soap. As a precaution, deodorants were prohibited for the duration of Phase 1, even though we had no reason to believe they would interfere with our results. The collection pads were worn for six to nine hours a day, three days a week.

Each day the pads from each donor were placed in an individual glass jar (that had been carefully cleaned chemically) and then frozen in the deep freezer. The jars were marked to identify what day of the cycle each pad had been collected.

The women kept records of their own basal body temperature graphs and menses onsets. After fourteen weeks of collection we were able to go back and determine which of the fifteen cycles of pads had:

- the optimally fertile basal body temperature rhythm; and
- which menstrual periods began near the full moon.

Only the pads of these five "best" cycles were selected for the Phase 3 experiment, that is, the ones that fit the vigorous criteria of cycle length, lunar phase, and so forth.

Batching the Pads

In order to prepare the sample to be used on the research subjects the next year, pads were grouped into three-day segments according to time of cycle. All pads from each three-day group were placed in a five-foot-high glass column and allowed to soak in alcohol to extract the essences into vials. These vials were labeled and frozen.

When Dr. Preti combined batches of female essence, he created the equivalent of a "Day 2" essence, a "Day 5" essence, a "Day 8" essence, and so forth. In doing so, he created essences that are characteristic of the underarm secretions of fertile women at particular "timed" cycle stages. Each batch (such as Day 2 or Day 5) had a pooled mixture of essences from different women.

The Male Essence

In a similar manner male essences were obtained, extracted, pooled, and frozen. The male samples were batched consecutively according to the date received. We suspected there might be a seasonality to the concentration of substance produced if male essences contained the pheromones that worked to alter women's endocrinology. Since the cyclic variations in sex-hormone levels of men showed an annual cycle (see Chapter 4), and since underarm secretions contained male sex hormone, timing might be relevant to pheromone research. Testosterone levels in men are highest in autumn each year. For that reason we collected essences in autumn, believing that they might be at their strongest concentration then. Phase 1 of the experiment, the collection, extraction, and preparation of the male and female essences, required almost a year. Essences were collected in the autumn of 1982 for administration to subjects in the autumn of 1983.

PHASE 2: FREE CYCLING

By the end of 1982 the female and male essences were in the deep freeze, and I was ready to recruit research assistants. Since I already knew what I expected to find—that women with regular weekly sex behavior with men would be likely to have a more fertile-type menstrual-cycle pattern than those who had sporadic encounters or a celibate life-style—I set up a process for *double-blind collection* of data with a double-blind experiment. Neither the research subjects nor the research assistants knew our hypotheses.

I recruited ten research assistants—college students who would be willing to work with me for fourteen weeks to learn how to conduct research correctly. They would have the job of recruiting the research subjects who would be willing to keep records of their sexual behavior, menstrual-cycle onsets, and basal body temperature record graphs.

These extraordinary young women enrolled a total of 210 research subjects. The 210 women research subjects began recording their own menstrual, sexual, basal body temperature, and other daily records throughout the spring semester at the University of Pennsylvania. Each of these research subjects was contacted regularly by her own research assistant.

Toward the end of the fourteen weeks I sent a letter to the 210 participants inviting each woman to consider continuing on into Phase 3 of the experiment when the fall semester, 1983, resumed. From the closing survey of Phase 2, which was anonymously administered, we learned that 96 women qualified for entry into Phase 3. For example, women who during Phase 2 reported that they did use drugs could not be accepted into Phase 3 because we knew that drugs might affect menstrual-cycle length. Likewise those who had failed to record their basal body temperature on at least 80 percent of the days of Phase 2 could not qualify for a more exacting phase. It was tricky to serve privacy while accurately following each woman's continuing involvement, even though all data were collected in an anonymous fashion. We managed to convince women we would assure their anonymity, and we did.

Protecting Privacy

Each of the 210 women of Phase 2 had selected a code name for herself, which she wrote on all her documents. I never knowingly met any research subjects, and the original ten research assistants did not see the code names on the documents. There was no way that I could crack the code to match code name to research subject. We did know which code name went with which questionnaire, BBT graph, and calendar, so we were able to determine who qualified when we invited Phase 2 women to go forward. We used these data to learn whether a particular coded name did or didn't qualify. For Phase 3, 49 of the eligible women agreed to continue and were able to complete participation.

The results of Phase 2 of the pheromone studies in 1983 confirmed and expanded on what my colleagues and I had discovered in the mid-1970s at the University of Pennsylvania and in perimenopausal women in the late 1970s at Stanford University (see Chapters 1 and 5).

PHASE 3: THE IMPACT OF ESSENCES

Our first task was to assign women into one of two experiments:

- the male essence study or
- the female essence study

We began the experiment with the assumption that men would have a stronger influence on women than women would have on women. Since male odors were stronger, we reasoned that they would dominate if they were effective at all. We decided to assign women with aberrant-length menstrual cycles into the "male essence" experiment, because we were testing to see if male essence would mimic the effect of regular sex with a man. Since regular sex with a man associated with normal length 29 (±3)–day menstrual cycles, we wanted to see if regular application of male essence would change their cycles from aberrant to normal when regular sex was not occurring.

The female essence study was designed to test whether the menstrual cycle of women who already had a regular 29 (±3)–day cycle would be synchronized to the female-essence samples. Therefore women who reported regular 29-day cycles were to be assigned to the female-essence experiment.

On the first day of Phase 3 we asked women to say whether they had regular menstrual cycles averaging 29.5 (±3)–day length or aberrant-length cycles, those shorter than 26.5 or longer than 32.5 days.

Since Phase 3 participants had been keeping records of their menstrual pattern the previous spring, we assumed they would be aware of what kind of pattern they had. We also realized that they might make mistakes or that their cycle length might have changed over the summer or would change once they entered the experiment.

The two Phase 3 experiments began in September 1983. Each subject reported to a laboratory to meet a female technician who would see her three times a week from September until mid-December 1983. At each visit the technician applied a substance to the skin of the upper lip.

The Female-Essence Experiment

Each subject was given a basal body temperature chart, thermometer, and calendar for recording sexual behavior. She was reminded how to use these. As part of the "blind" experiment design, subjects were told that we were testing to see if a natural fragrance affected menstrual-cycle timing.

At random half of these subjects were assigned to the placebo and half to the female essence. Both fluids smelled lightly of the alcohol on the pad, so subjects and technicians could not discover what substance was being applied.

Due to individual variations in menstrual patterns a woman could enter the study at any point during her cycle. Whenever the essence recipient entered the study, she was given the "Day 2" essence. Each time she came to the lab, she got the next essence timing sample in sequence (Days 5, 8, 11, and so on). The placebo group received a similar alcohol swab each time.

At the end of the experiment the data were analyzed to see whether menstrual synchrony had occurred. We wanted to know:

- Did the women who received female essence show a synchrony in the timing of their menstrual cycles?
- Did their menstruation cycles move into sync with the batched donor samples?
- Was this result absent in the women who received placebo?

We asked the question in a similar way as Martha McClintock had studied her original menstrual-synchrony data. Our results matched hers. Women who received female essences had a statistically significant likelihood of menstruating in synchrony with pooled female essence. Placebo recipients did not show synchronicity.

THE PHEROMONE FINDINGS

At the end of the first menstrual cycle we calculated Cycle Day 2 of each woman and counted how many days away it was from the Essence Batch Day 2 (see above). The average cycle difference between women and essence onset was what would be expected by chance, about 8 days.[3] After two more months of female-extract application, the number of days' difference in menses onset relative to the donor cycle decreased significantly. It halved—from 8.3 days to about 3.9 days. In contrast the placebo group, those who received just alcohol, showed no change from the start of the experiment to the end.

In other words the women who received female essence showed a timing change in the onset of their menstrual cycles. When we looked at the hormone levels of the women who received female essence and compared them to the hormone levels of the women who received placebo, no difference emerged in estrogens, testosterone, or progesterone between the two groups. The time of the cycle when the blood was taken was the same for both groups. Since hormone levels hadn't changed, something else was acting on endocrine functions. Although these cycles were all affected, the explanation of this phenomenon has not yet been reported.

This study was the first systematically designed, prospectively conducted, double-blind research in humans that attempted to manipulate the human menstrual cycle with female essence. Our results rather neatly paralleled previous studies:

- Menstrual synchrony was shown.
- Female essence from underarm secretions appeared to be effective in producing the synchrony.

We were excited, but not really surprised. Considering the menstrual-hut tribal custom among the Yurok Indians, the menstrual synchrony that Dr. McClintock found in the Radcliffe College dormitory, and the

[3] By chance, a difference between any two women, or between any woman entering the study and the female essence Day 2 would be about 7½ days. We calculate this number as follows: since a criterion for inclusion in the study was a cycle length of 29.5 ± 3 days, any two menses onsets will average somewhere between 0 and 14.75 days apart (half of the 29.5). As a consequence, the theoretical average difference is 7.38 days (half of 14.75).

underarm essence applied in Dr. Russell's study, we were on a trail that was continuing a chain of research discovery.

Women affect one another, and menstrual synchrony appears to be caused, at least in part, by the nearness that allows female essences from the underarm to stimulate the endocrine systems of recipients. Since menstrual synchrony has now been shown both by direct application of female essences on the upper lip and by friends whose natural, nonsexual behavior places them at the normal social distance (of more than twelve inches), if the same mechanism is at work, then the essence must normally be airborne.

THE POWER OF MALE ESSENCE

The male-essence experiment provided more answers. We wanted to know how the cycle lengths would be affected in the women who received male extract versus those who received placebo. At the end of the experiment we expected to discover what proportion were aberrant in length and what proportion were normal (29.5 ± 3 days) in length with women when they did not have a regular partner but received experimental essence or placebo three times a week for close to fourteen weeks. Our results showed that:

- 67 percent of male extract recipients showed a normal-length cycle, and
- 17 percent of the placebo recipients showed a normal-length cycle.

In other words women who started with aberrant-length menstrual cycles and who did not have a regular weekly sex partner showed a tendency for their cycles to shift to normal length. When they received male essences, cycles improved in the direction of improved fertile capacity, mimicking regular weekly sexual behavior. The male essence seemed to substitute for regular weekly sex. Is male essence better than having the male around? Not on your life! (See Chapter 7.) If he is not around, perhaps the essence can serve as a substitute for keeping the endocrine system going. Although we cannot be sure, the likelihood is high. Studies will need to be further developed first.

THE CHEMICAL BASIS OF INTIMACY

Our preliminary experiments in human pheromones suggest a number of mechanisms that help to explain the intimate effects of being close to another person. Human pheromones seem to work as endocrine-system primers rather than sexual attractants. The length of time it takes to show these effects, the twelve to fourteen weeks of experimental discovery, very closely matches the length of time it takes for ovarian follicles to develop and mature—about eighty-five days. The parallel between the timing of follicular development and the woman's response to male and female essences suggests that these essences are working in the ovary at the level of the follicle.

MALE ESSENCE REQUIRES PHYSICAL INTIMACY

Our data also suggested to us that physical intimacy counts. The usual social distance between men and women is not adequate to produce male-essence effects. Women work with men, live with men (brothers, fathers, and so forth), and often have aberrant-length cycles (< 26 or >33 days) when they don't have regular weekly sex. In other words if male essence is effective, it appears necessary to get it close to the woman. In our experiments, that meant placing it on the skin just under the nose. Our experiments also suggest the power of these substances, because they work effectively in the absence of social contact. The women who received essence did not have regular contact with the donors of the essence. In fact the essence had been collected a full year before and frozen.

FEMALE ESSENCE IS MORE SUBTLE

Powerful but subtle female essences in our experiments were only applied to women who already had a regular, normal (29.5-day) cycle lengths. We discovered that women do affect other women at the normal social distance close friends maintain.

PUTTING MALE AND FEMALE ESSENCE TOGETHER

Many questions remain unanswered, particularly whether the essence can substitute for the person and what the precise relationships are between sexuality, synchrony, and male and female influences. I have some ideas, but tests need to be done to evaluate them. Even so, it is intriguing to speculate. I suspect that there is a continuum where both female and male essences influence a woman in a predictable way.

Perhaps regular weekly sex with a man normalizes the cycle length to the fertile pattern of 29.5 ± 3 days. Once the cycle is normalized, then the more subtle menstrual synchrony has a basis on which to work.

POSSIBILITIES FOR FUTURE INVESTIGATION

Further tests will need to be conducted to determine whether the essence works through the nose by smell or through the skin—perhaps it is airborne and then enters the skin by absorption and subsequently enters into the bloodstream.

In order to explore this research, we needed more money for subsequent testing. If we had the funding, our tests would have repeated the experiments for longer periods of time and we would have placed the essences on some women under the nose and on others behind the knee, where absorption is known to occur rapidly. Answers of this sort will have to await future experimental studies.

Until April 1991, I thought that it would be the Japanese who would make the discoveries and glean the major profits. When George Preti and I published our findings and attempted to acquire more grant money from the National Institutes of Health and the National Science Foundation, we were unsuccessful. Repeatedly our proposals were approved but not funded. The approvals meant that our scientific colleagues agreed that the work was worthy. The severe restriction on research money beginning in the early 1980s affected our potential for developing this work through government-funded research. The priority structure favored other types of research. We next turned to industry and met with executives from a number of pharmaceutical and cosmetic companies. Unfortunately we kept hearing the same thing. The American executives liked out work, but they were looking for a product that could be developed rapidly to improve their "bottom line" within the next few business cycles, or quarters. Pheromone research would take years.

In contrast, Japanese executives had no problem with the idea of investing their resources to develop a product that might yield rich rewards but take many years. They were slow to negotiate; slow to conclude; and were the only investors to purchase license rights. As of 1990 only the Japanese had licensed the rights to develop our pheromonal research into marketable products (in Japan only). It is my ardent hope that an American pharmaceutical company will choose to work on this project, because cultural differences inhibited my ability to influence Japanese businessmen and contribute to this project's development.

Although the discoveries were coauthored by chemist Dr. Preti, the experiment's biology and behavior came from my own expertise and creative style and from the studies described in earlier chapters. The successful collaboration between biology and chemistry enhanced our ability to make the discoveries. After a formal introduction and hand shaking, the Japanese businessmen never again communicated with me. They seemed to have had no interest in learning my opinion about how to design experiments to develop the product.

Constrained by their own culture, these Japanese businessmen could not find any value in talking to me, the female member of the pheromone team. In April 1991 I learned through Monell lawyers that the Japanese had dropped the project.

PHEROMONES AND DEODORANTS

In a culture where deodorants are so common, it is reasonable to wonder whether they would inhibit the influence of pheromones.

I suspect that deodorants are irrelevant to the power of pheromones. The critical element in the male and female essence is probably effective independent of the sweat odor, similar to carbon monoxide gases, which are potentially lethal but themselves odorless. Gas companies actually add a separate odorous component to carbon monoxide, "tagging it" in order to enable people to identify leaking gas by odor. We smell the "tag" rather than the carbon monoxide when we notice a gas leak.

Pheromones may work the same way. In our experiments and those of Dr. McClintock, many of the participants did wear deodorants and thus were odor free. They were still affected by the weekly rhythm of the intimacy.

One more potential discovery may well emerge. It is likely that female essence may serve to increase the amount of sexual intercourse women engage in. For now this is speculation, but it is based upon my evaluation of the research results.

THE APHRODISIAC EFFECT

When I was reviewing the data arrays my research assistants had constructed for me, I thought I noticed something startling. During the fourteen weeks of the female-essence study, a substantially higher incidence of weekly sexual behavior occurred in the group receiving female essence compared with the control group.

During the first three weeks of the experiment the sexual activity of subjects in the two groups was not significantly different, although a trend may have already been starting.

- Thirty-six percent of female-essence recipients had weekly sex.
- Eleven percent of placebo recipients had weekly sex.

In the second period, weeks 4 to 14 (the end of the study), this difference increased and became statistically significant. Differences this large would occur by chance less that 1 percent of the time.

- Seventy-three percent of female-essence recipients had weekly sex.
- Eleven percent of placebo recipients had weekly sex.

REGULARITY IN SEXUAL BEHAVIOR

The total number of sexual encounters per week did not relate to any aspect of the female-essence treatment. What the treatment seemed to cause was not simply more sex but an increased incidence of regular stable weekly sexual behavior. The regular application of female essence to women with normal cycles led to a substantial number of them (73 percent) engaging in more stable and regular heterosexual contact—about six and a half times the rate of the placebo group! How interesting it would be if the test replicated.

When the data were analyzed by the average count per week, or likelihood per day, no significant association emerged.

In other words the theme repeats. There is something critically different about the love-cycle timing that requires or produces stable timing of behavior from that of feast-or-famine behavior.

THE CHEMISTRY OF LOVE

The discovery of male sexual pheromones was published in late 1986. Two other discoveries, one that occurred around the same time in Dr. Preti's laboratory and the other presented a year later by investigators at the Kinsey Institute, gave me the missing pieces of the puzzle to suggest what might be going on. What the Kinsey Institute scientists found in their studies of lesbian women helped me to make the link to what Dr. Preti and I published about the chemistry of male and female essences.

Seventeen years passed after the discovery of menstrual synchrony in women before systematic investigation of the effects lesbians might have on each other were tested and reported. In 1987 I was fortunate to learn what investigators at the Kinsey Institute, Drs. Stephanie Sanders, Marie Ziemba-Davis, and June Reinisch had found when we all attend the same scientific meeting. Dr. Stephanie Sanders presented the work that she and her colleagues at the Kinsey Institute had just completed. The Kinsey people had had an interest in my work and had decided to ask questions about sexual behavior similar to the ones I had been asking. They wanted to know if regular sexual behavior among lesbian women would also associate with normal-type menstrual-cycle lengths. They patterned their experimental data collection after my own. They found statistically significant effects similar to ours.

Women who had regular sex with women had menstrual cycles of about 29.5 days. Women who had sporadic sexual behavior with other women had aberrant-length cycles. They did not define or describe the nature of this sexual behavior, but they did report on the pattern of its timing. Although their finding echoed my own discoveries of stable sexual relations with men, there was one difference.

In order to achieve the fertile-type menstrual-cycle length, the stability of behavior needed to occur three times as often in lesbian women than I had found was necessary among heterosexual women. Sexual relations three times a week among lesbian women were equivalent to one time a week among heterosexual women.

As I listened to their work and considered the discoveries that Dr. Preti and I had made on the chemistry of the essences, I was intrigued. I realized that there was one substance in male essence that is three times as concentrated in men as it is in women. It was one of those moments that scientists love to experience, one where my own work perfectly dovetailed with somebody else's to lead to a conclusion that might render a new discovery. Consider the following:

> If lesbians needed three times as much activity as heterosexual women, and
>
> If the essences were the cause of fertile-type menstrual cycles, and
>
> If one of the essences had one third the concentration in women as in men,

Then it seemed that this one substance might have produced the pheromonal effect.

THE ACTIVE INGREDIENT

DHEAS, dehydroepiandrosterone sulfate, is found in female essence in about one third the concentration it is in the male, according to the 1987 paper that Dr. Preti, Dr. Garcia, and our colleagues were about to publish.

I think that it is reasonable to suggest, therefore, that DHEAS may be the responsible agent in these sexual-behavior physiological relationships.

If the study were replicated, we would have significant implications for a biomedical treatment in the promotion of fertility for a subset of infertile patients. This would seem possible because we already know that regular weekly sex with a man seems to increase the fertility of a woman. For those women whose libido renders them uninterested in adequate sexual frequency, female essence might be used to stimulate the propensity to engage in intercourse.

Male essence would also have its uses. Consider the times when couples are forced to be separated, by business travel and other events. Infertile couples who are undergoing fertility treatment often do not want to be sexual during the luteal phase of their cycle. The intense intrusion of a fertility workup into their intimate lives can inhibit the desire for lovemaking with one's partner. If the essences were to be substituted during this time, one of two positive benefits might result: (a) Women might be more inclined to engage in sexual intercourse, or (b) they might not need to, because their cycles might become more fertile.

CONCLUSION

Isolating the "sex primer" compound responsible for our intimate behavior was exciting, but the true excitement is in its application. Pheromones offer the possibility for breakthrough research in the areas of fertility and fertility regulation. They also suggest the possibility of a far more profound benefit than simply improving the fertile capacity of women.

I look forward to the time when research money will again be available in America, through either public or private grants, in order to develop these extraordinary essences as products that may be used for

- Birth control
- Reducing the devastation of aging
- Possibly helping to strengthen bones
- Enhancing cardiovascular health (as do hormone-replacement regimens)

These substances may also have other profound effects that we have not even begun to understand.

We live in a thoroughly interconnected world. Each person's emanations affect those around them. Although we may not be able to see these innerconnections, science does and in its systematic way is beginning to unravel the relationships. With that knowledge comes the power to live more richly.

With these basic scientific studies reviewed, let us turn next to the experiences of the senses: our sensuality cycles.

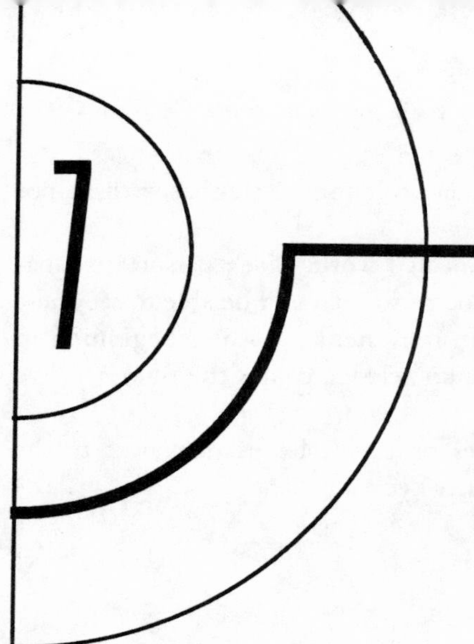

SENSUALITY CYCLES

The softest thing in the universe
Overcomes the hardest thing in the universe
That without substance can enter where there is no room

—LAO-TZU, #43

When I was about four years old and did not yet know how to read, I used to watch my parents absorbed in a mysterious process that I struggled to understand. When I would try to talk with them, they would shush me, saying they were busy. What were they doing, I wondered? I pondered this with my four-year-old mind and was confused. Eventually I did figure it out, and I still remember the special moment of understanding that came suddenly more than forty years ago. In a flash of clarity I experienced a remarkable joy—and a certainty that I *knew* what they were doing! They were "reading." The marks on the newsprint stood for some kind of a code—a code that had meaning enough to engross their attention. I knew I wanted to learn it. I begged them to teach me, but they smiled gently and told me to wait. They said that I would be taught the code in the normal course of things when I got to first grade.

By the time first grade rolled around (two years, and half a lifetime, later) and I was permitted to learn my letters, the excitement was tremendous. I quickly took to learning the letters, combining them together, learning words, and eventually—the joy of reading.

I tell this story because in large measure it reflects my understanding about the nature of sensuality. Reading is a sensual experience making use of vision to evoke creativity. We read and worlds of internal experience expand. We read and the workings of our silent imagination evoke images and memories, perceptions and sensations. When the writing works, we are engrossed in it. Sensuality in love cycles is akin to the endless variation with which words can be placed in relation to each other to give ever-varying meanings. In the intimate exchange of sensual experience the players act and are acted upon by each other. In the process of sensual experience the forms change and so do their meanings. The permutations are infinite. The ordinary and familiar have the potential of being rearranged to become novel and exciting.

The senses—touch, taste, vision, smell, temperature, and a number[1] of others—are biological realities. To the scientist sensual experience represents a complex field of inquiry. I want to explore sensuality first in its positive, life-enhancing perspective; that is, in terms of its biological definition, without any moral overtones. (In the concluding chapter, "Monogamy and Restraint," I'll add my view on the appropriate constraints.)

YOUR INTERNAL SENSUAL WIRING

For researchers in experimental psychology the systematic study of sensation and perception provides an important focus for understanding the way the nervous system is hard-wired to be sensitive to specific kinds of stimuli. The nerve cells, like the wiring in a computer, are actually laid down physically and form interconnections between various parts of the body. The discoveries of sensory physiologists have value in discerning the facts about sensuality cycles. Our bodies contain specific kinds of nerve cells that are sensitive to different kinds of stimulation. Sensory nerves are receptive to different kinds of sensation, in contrast to motor nerves, which are stimulated by sensory nerves to fire on muscles and glands to activate movement. Sensory nerves serve to sense and later promote our interpretation of the meaning of external stimuli. Consider some of the following examples:

- Nerve cells in the eyes, called rods, are particularly sensitive to shades of gray. They fire electrical impulses into the brain in response to stimulation from true colors.

[1] In the strictest definition, "senses" refer to the processing of sensory nerves. The list of sensory capacities includes pressure, heat, cold, itch, tickle, color, pitch, tone, intuition, logic, and more.

- The cones, also optical nerve cells, are sensitive to different colors and fire impulses into the brain in proportion to the color stimulation received by them.
- "Temperature" nerve cells just beneath the skin have endings particularly sensitive to heat or cold. These nerve cells fire impulses into the central nervous system when they are stimulated by heat or cold. Although a cold nerve will also fire if it receives enough heat stimulation, it will fire much more rapidly and easily if it receives cold stimulation, to which it is sensitive.
- There are also "pressure sensitive" nerve cells. The tip of the cervix (see Figure 7-1) contains pressure-sensitive nerve endings. These nerves fire impulses into the nervous system when pressure is applied. For example, the tapping of the penis during sexual intercourse stimulates electrical impulses of nerve energy in the brain in small mammals—and apparently in humans too.

Human sensuality has to do with the way we perceive the sensation of nerve impulses. There are multiple levels that can engage our attention. Our sensuality is based as much on the particularly sensitive nerve cells that respond to touch and pressure as it is on our perception of experience. Hormones also contribute to sensual experience. They affect the capacity of nerves to fire during stimulation and they also affect the way we interpret what we experience. For sensuality the intellect is as active as the sensory nerve endings are. Ultimately it is the splendid interaction between sensation and perception—the body and the brain—that produces the extraordinary range of human sensual experience.

Sensuality is more than a nerve phenomenon. It is a complex interlocking mosaic of nerves, hormones, thoughts, and memories.

THE LIFE CYCLE OF LIBIDO

The desire to have sex is an almost universal drive. The biologist in me sees desire as the basis of the survival of the species. Since we are genetically programmed to want to reproduce, desire is a biological reality, a force that serves to promote the existence of our kind.

Sexual desire can be documented in most people at least as early as the onset of puberty and then with varying degrees of intensity as the years continue. The desire for sex usually precedes any knowledge of its elements. Before young people understand sexual intercourse, their fantasy life is expectant. They do not know what sex feels like; its dynamic

FIGURE 7-1: The Woman's Cervix—Showing Secretions

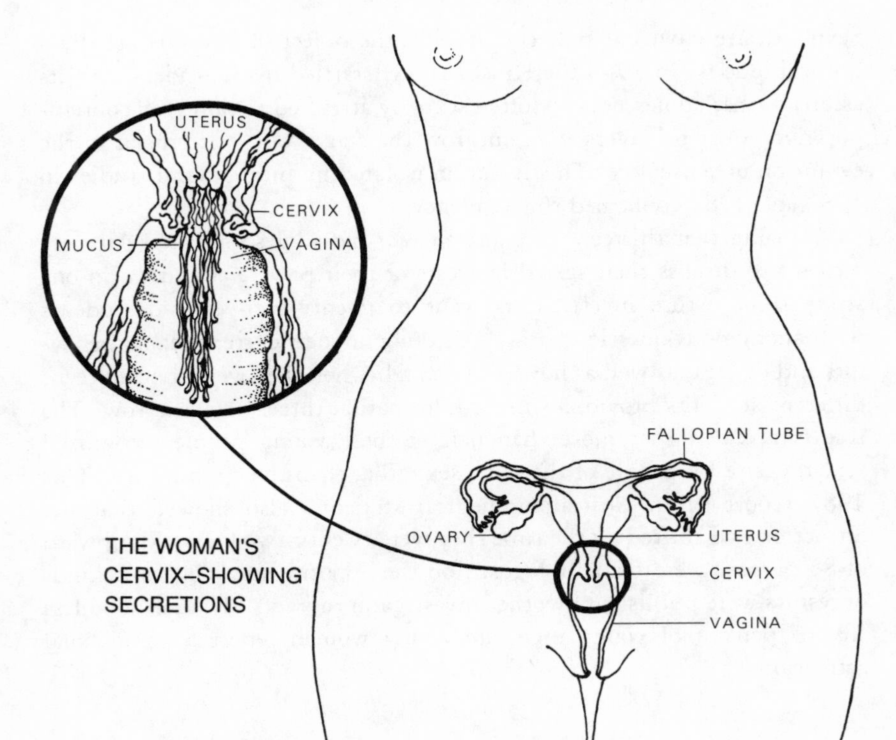

elegance eludes them. Still, desires exist within the pubertal child. Although desire is often attached to the attraction for another person, if no one is available, the creative mind will fill in the gaps with fantasy.

Sexual desire is difficult for scientists to investigate because the subject is so personal and so wrought with emotional overtones. But scientists still try. Between 1985 and 1990 a wonderful series of research studies broke through some of these barriers.

THE DAWN OF DESIRE

Sexual desire dawns at puberty. At first the object of desire is usually a bit ambiguous. In 1948 the Kinsey report startled many readers with its assertion that adolescents are often sexually attracted to same-sex contemporaries. Most teenagers soon outgrow the stage, but it is crucial in the evolution of sensuality. This is not an isolated finding: Several studies in 1987 and 1989 confirmed this tendency.

Knowing that thirteen- to eighteen-year-old adolescents are extremely reticent to discuss their sexual interests or their private activities, in one study some astute investigators went to twenty-two-year-old students with anonymous questionnaires. Asked about their current sexual behaviors and desires, as well as how they had felt when they were about fifteen, different students provided similar information three years in a row. The results revealed that more than half of these young people recognized sexual desire for people of the same sex rather than the opposite sex. This 1987 report (from Australian medical students) also showed that the attraction had shifted by the time they were twenty-two years old; upward of 85 percent were sexually focused on the opposite sex. The same kinds of results were published by other investigators as well. By adulthood less than 5 percent of young men and young women report a homosexual attraction.

THE URGE TO MERGE

Sexual urges are extremely strong in young people. In 1990, a study documented how often college students around eighteen and a half years old had urges for sex, and it examined differences in reactions between these young men and women. The scientists defined sexual urges as being distinctly different from sexual fantasies, although both were elements of desire. Sexual urges that were provoked by an external stimulus (seeing a "sex object") occurred more often in the young men than in the young women, four and a half times a day versus twice a day. But sexual fantasies, defined as internally derived spontaneous thoughts about sex, occurred just as often in the women as in the men: about twice a day. In other words, four to seven times a day clear thoughts of sex and a desire for it were reliably reported among eighteen-year-old college students. Men were different from women; the sex objects they saw triggered urges twice as often. The mind remains an active participant in desire, but the frequency declines. By the age of seventy desire has dropped to about once a week.

HORMONALLY DRIVEN DESIRE

The surge of testosterone in pubertal boys may account for their omnipresent desire. The monthly ebb and flow of hormones in women affects their desire as well. Women who have a fertile menstrual cycle usually show a monthly variation in their libido. In fact, those who show a monthly cycle of moods are the ones most likely to show a monthly variation in desire. This effect can be marked. Oral contraceptives usually obliterate both mood and libidinal swings. The libido may remain, but at a steady state (as in men)—except for the pill-free days, when it often increases.

About 50 percent of cycling and fertile women report a peak interest in sexual activity just around the time they are ovulating. A smaller group (12 percent) show a peak interest in their desire just before they menstruate. The rest report peaks of desire at other times. Not all women are the same in the time of their peak of desire.

Since both estrogen and testosterone levels in women are usually rising together (twice each month), desire may derive from either or both. I suspect that adequate levels of estrogen and elevations in levels of testosterone work together to provoke this increase in desire.

A few studies have shown that the testosterone levels in blood are related to the levels of sexual desire in women. Women with higher levels of testosterone reported higher levels of desire and, in some of the studies, more sexual behavior in their married lives, just as men with high testosterone levels. Equivalent studies of unmarried women have not been published. One other element was discovered in Philadelphia in the mid-1970s by Drs. Harold Persky, Harold Lief, and colleagues. In monogamous couples the testosterone levels of the men tended to reach two peaks each month, in phase with the two monthly sex-hormone peaks of their wives—at ovulation and then seven days later. As her sex hormones rise, his follow. The sensual dance between the couples seems to be directed by her fertility cycles. And these can be measured by her hormone secretions. Love cycles each month in monogamous young couples, and the woman's ovaries can serve the role of pacesetter.

DESIRE CAN DIMINISH WHEN SATIETY IS REACHED

Although no systematic scientific research has been published on the question of satiety, common sense suggests the answer anyway. In a monogamous relationship, if sexual appetites are mismatched, someone will be left wanting. The couple will need to negotiate. No study has yet been published to factor in the nature of the appetite disparity when measuring desire or its change, which seriously limits our ability to provide meaningful answers about desire. Still, we can get some answers from the published research.

There is evidence that desire decreases with age. The studies are consistent. Women often report a decline in desire as they pass from their forties into their fifties and sixties—as their sex hormones decline—unless they take hormone-replacement therapy.

Between forty-four and sixty years of age 25 to 30 percent of women report a decrease in their desire for sex. Equivalent studies to verify the cultural myth that the peak desire for women occurs in the thirties have not yet been reported. Yet the timing of the life phases of the hormonal symphony probably tells the true story pretty well. There seems to be a bloom in women, so often described in literature, well before the decline in desire starts. That sensual peak happens at a later life stage than it does in men. Although the studies have not yet provided complete answers, significant research about the decline is available.

Here's what has been established about the decline through research:

- In 1980 a Danish physician, Dr. Karen Garde, published her study of two hundred forty-year-old women. She reported that 33 percent never experienced libido—desire—spontaneously; yet most were sexually active, and 96 percent could attain orgasm. The absence of libido can occur because hormones are missing, but an absence of libido is also common when romance is missing. When I look at data like those, I wonder about several different things. Are these women who never experience libido having sex with partners who never permit them to build up their appetite because the partner's libido is higher? In other words, do they lack the hunger but enjoy the meal? Or are the women who never experience libido the ones who do not have romance in their relationships? The studies do not answer these questions, but the composite body of data is clear about midlife changes: desire diminishes.
- In the 1987 report of the early stages of the Stanford Menopause Study, the women studied were around forty-nine years old and premenopausal. All were still menstruating, although irregularly, had not yet entered menopause, and most were suffering from hot flashes. The desire for sex was monitored in one hundred of them. Thoughts about sex occurred:

once a day or more in 16 percent;
once a week or less in 55 percent; and
somewhere between these two extremes in 29 percent

Overall, desire was much lower than among college women.

- Similarly, studies that collect data on "absence of desire" show that it increases with age. In one large study of Swedish women, absence of desire was reported in:

  3 percent of thirty-eight- to forty-six-year-olds;
  9 percent of fifty-year-olds; and
  29 percent of sixty-year-olds

  Hormones are not the only cause.

  Two investigators evaluated women seeking medical help for a loss of sexual desire and compared them with women of the same age who showed a regular sexual desire. There were no differences in the sex-hormone levels of these two groups. The conclusion of this study is that a loss of desire is sometimes unrelated to sex-hormone levels.

If desire declines, does contentment increase? I suspect it may—for some. It depends on one's attitude.

Women aren't the only ones with declining desire. It happens in men too. A young man might have thoughts of sex seven times a day, an old man four to seven times a month. With this typical age-related change the desire remains, but the frequency declines. Physical fitness seems to play an important role in the sexual function of men.

One 1990 report of forty-eight-year-old men bore this out. The researchers divided the men into two groups. One group agreed to participate in a rigorous physical-fitness program meeting three times a week for a one-hour exercise regimen. The other group exercised as often but took a long walk instead. The results showed an increase in desire and sexual behavior in the men whose physical-fitness program was rigorous and whose program produced the usual cardiovascular health changes expected for men who exercise rigorously. But no sexuality changes emerged in those who took walks. Apparently for men the exercise must be intense if the libido is to change. Middle-age men who exercise vigorously experience sexual changes that return them to the sexuality patterns of younger men as they improve their cardiovascular fitness.

WOMEN MAY BE DIFFERENT

Although physically fit women might also show sexuality changes as their hormones change, too much exercise might work either way. For women too much exercise might lead to decreased sexual interest, since fertility and the steroid sex hormones tend to be diminished in such women (see pages 123–124).

Alternatively, since estrogen levels are inhibited after excessive exercise, desire might increase if the hormonal balance is altered with a higher relative level of testosterone. There is no clear answer because studies of sexual changes after fitness exercise have not yet been reported in women.

WHEN LIBIDO IS LACKING

Among couples who seek professional help for sexual dysfunctions, a lack of interest is one of the most frequent reasons for approaching the professional. The exact figures are unknown because broadly based data have not been gathered in women who do not seek help. Among those who do seek professional sex therapy, men outnumber women, perhaps because virility in our society is essential to a "real man's" self-worth. Those who lack sexual desire usually have other problems as well. Among the men with inhibited sexual desire, more than half showed a lifetime history of depressive disorder. Even more women who came for professional help—71 percent—showed a lifetime history of depressive disorder. Other problems are also common among women who have low sexual interest. PMS was reported by 71 percent of the women with inhibited sexual desire, an incidence three to four times the rate found in a healthy control group and fourteen times the rate found in the general population (5 percent). Since regular, rigorous exercise reduces depression, it seems likely that exercise might increase sexual desire as well. Studies remain to be done to tell us if women respond as men do.

THE IMPACT OF PELVIC SURGERY

Women who undergo a hysterectomy (with or without the removal of their ovaries) often lose desire. Even when ovaries are retained, they tend to age more rapidly once the uterus has been removed. Since one in two American women is prescribed a hysterectomy, the problem is pervasive.

The first studies designed to determine whether estrogen-replacement therapy overcame desire problems in women whose ovaries had been removed provided some conflicting evidence. Women who had no hormone-replacement therapy showed lower levels of sexual interest than those who received estrogen. These early studies could not reveal whether the sexual interest represented the return of a lost libido or a recovery from the lubrication deficits during arousal. Vaginal dryness is a pervasive condition, and painful intercourse common when estrogen is lacking. If a woman cannot lubricate, if she cannot get aroused because her physiology isn't working, her libido may shut off and the shutdown may be unrelated to testosterone. More recent studies from Dr. Barbara Sherwin's laboratory at McGill University in Montreal have provided some answers. Testosterone combined with estrogen therapy restored libido in double-blind studies of women who lost libido after having their ovaries removed. Estrogen alone usually does not. Therefore testosterone can clearly be connected to libido in women. Estrogen plays a role as a more or less priming factor.

AGE AND LIBIDO

To summarize, sometime between the age of eighteen and the age of forty-nine women experience major declines in the frequency of their desire for sex if they do not take hormone therapies. The onset of the decline is pretty variable from one woman to the next. At about the age of eighteen sexual thoughts four times a day are pretty common. In the thirties there may even be an increase. By the age of forty perhaps one third of women do not have a spontaneous desire for sex, although they may have some interest if someone approaches them. By the age of fifty most women are experiencing a sexual thought once a week or less. By sixty close to one third have *no* desire. This pattern of decline can probably be changed if hormone-replacement therapy is taken. Physiology is only one part of desire. Romance changes the picture. A new romance, or the reblooming of an old one, can dramatically increase desire—at any age.

For men the story is similar, from sexual thoughts seven or more times a day in the young man to a precipitous decline to a few times per month in the older man. And for men the studies have shown (pages 99–107) that while the amount of desire perception declines, the perception of pleasure tends to remain strong. Here, too, a blooming romance often overrides these age-related declines.

Whether or not desire is present, the will of the individual plays a powerful role in the cycles of sensuality. Some remarkable insights have recently been provided by researchers.

THE WILLINGNESS TO RESPOND

The desire for sexual behavior is powerful in young people. One study, published in 1989 by Drs. Russell Clark and Elaine Hatfield, illustrates the power of the drive. The experiment was conducted in two different years on two different American college campuses. The researchers enrolled confederate experimenters[2] who were twenty-two-year-old students. The investigators tested these confederates among their contemporaries to find that the students ranged from "slightly unattractive" to "moderately attractive." The purpose of the study was to find out how willing the students were when invited to one of three options for relationship and to discern whether attractiveness affected the outcome.

Experimenters stood on a college campus quadrangle and repeatedly approached strangers of the opposite sex, other college students, to test their willingness to explore intimacies. The experimenter told the student that she or he had been noticing this person around campus and found this student attractive. Then the experimenting student asked, in random preassigned order, one of three questions:

- Would you go to bed with me tonight?
- Would you come over to my apartment tonight?
- Would you go out with me tonight?

The results were different depending on gender. In response to the question "Would you go to bed with me tonight?" none of the women said yes; but 69 percent of the men said yes in 1982 and 75 percent of the men said yes in 1975—yes, to a young woman, regardless of how attractive she was. However, men were *less likely* to say yes to the invitation to go to the woman's apartment and even less likely to accept a date for that evening.

In contrast, while no woman said she would go to bed with a man that night, and less than 6 percent would go to his apartment, half said they would go out with a young man that night.

This study showed that young men were usually willing to agree to sex with a young woman provided they didn't have to get involved with a date. Women were just the opposite; they were willing to explore opportunities to get to know a strange young man but unlikely to accept an

---

[2] The term "confederate experimenters" is used to denote the status of a trained accomplice, one carrying out the secret purpose of the experiment, without telling the subject that he or she is being used as a subject when the lure is offered.

offer for sex. Whether AIDS has affected this potential is doubtful, yet not fully tested. Media reports of sex surveys tend to report very little fear of catching diseases on the college campus—disturbing possibility if true.

The reason for the sex difference may be that young women fear for their physical safety even if they feel as promiscuous as young men. Alternatively, young women may simply be more cautious or less promiscuous. Other studies have provided data to suggest that it is women who tend to be more cautious than men and it is women who serve the role of social brake setter in the dynamic sexual dance. Young women have their first coital experience later than young men and have fewer partners, according to several studies. Figure 7-2 shows the average age when intercourse starts, in men and women, blacks and whites. The data were published in 1989 and represent a large population sample from North and South Carolina. By age sixteen 50 percent of the white males and 75 percent of the black males have begun. The girls start later. If women start later and have fewer partners, with whom are the men having sex? Apparently a relatively small proportion of women are extremely active and have sex with a great many young men. Most young women tend to be more conservative than most young men.

Although the women's liberation movement, in its early days of proclamations for sexual freedom, tended to promote the right of young women to engage as freely in sex as men always had, some sexual patterns, such as who sets the limits, have not changed very much. But some have. The double standard appears to be evaporating throughout the educated world. In 1989 Dr. Ulrich Clement published a study reviewing data of large groups in the United States and West Germany from 1938 through 1981. He showed that the sexual behavior of young men had hardly changed, but the sexual behavior of young women had come closer and closer to that of the young men in the age of onset and the frequency of sexual behavior. Young women were changing to a pattern that permitted more sexual expression.

Another study in 1988, by Dr. Gail Wyatt, found the same pattern of change in the sexual behavior of both black and white women living in southern California. Intercourse was beginning earlier, and first coitus was less likely to be with the husband or fiancé. Moreover, in recent years there was an increased number of sexual partners, and young women participated in a greater range of sexual behaviors than they did thirty, forty, or fifty years ago.

Similar patterns have been reported in 1989 in South America among Colombian university students. These twenty-two-year-olds completed

FIGURE 7-2: When Sexual Intercourse Starts

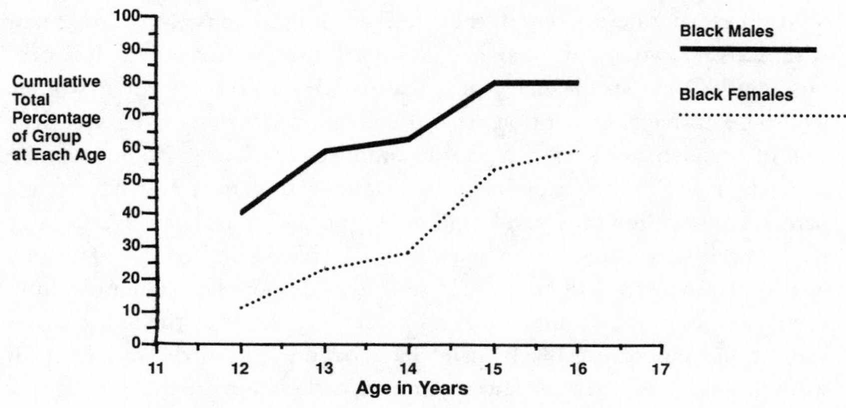

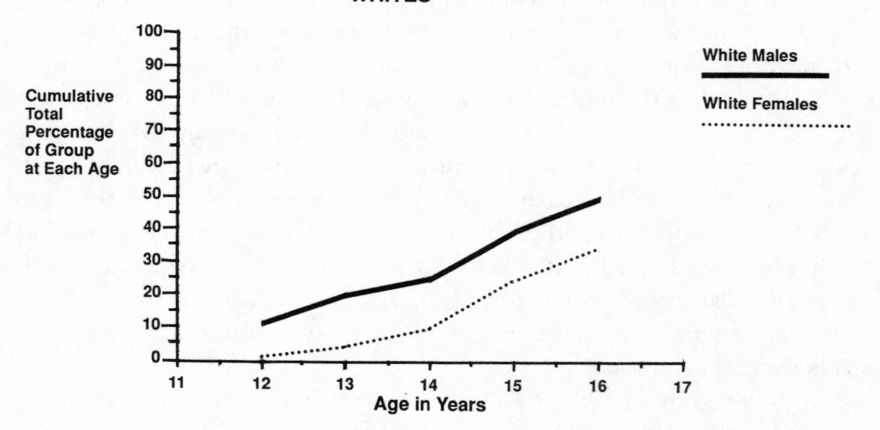

anonymous questionnaires in 1980, and so did another crop of twenty-two-year-olds for each of the next five years. Sexual behavior was changing during the course of study, moving in the same direction as in North America and Europe; that is, there was increasing liberation of the sexual behavior of young women. In Colombia there were still substantial differences between the sex behavior of young women and that of young men, and in part this may be attributed to the strict laws against abortion.

For about 80 percent of the young South American women, the first coital partner was the sweetheart; but only 16 percent of young men had their first sexual experience with their sweetheart. As might be expected in a culture where young men more often have their sexual experience with prostitutes, by the age of twenty-two, 49 percent of the men reported that they had had a sexually transmitted disease. Only 14 percent of the young women had had a sexually transmitted disease.

Other studies have shown that once young people experience their first sexual intercourse, they tend to continue to have regular intercourse. One further note that could be attributed to the presence of desire combined with the limitations posed by laws against abortion: Heterosexual experience with anal coitus was reported by one third of these young women and by almost as many of the men. These results in South America are shocking in light of the mode of transmission of AIDS. Unfortunately, there is confirmation of this same pattern in the United States. In 1988 Dr. June Reinisch of the Kinsey Institute reported to the International Academy of Sex Research professionals a similar conclusion: Anal coitus had been experienced by about a third of American women.

WOMEN ARE BECOMING MORE WILLING

As the 1990s approached, willingness to have sex appeared to have increased in women but stayed about the same in men. Starting at earlier ages, and in unorthodox ways, such as oral or anal intercourse, young women were engaging in sex with more partners as well.

Reports from West Germany as well as the United States, South America, and Australia are showing the dissolution of the double standard and the consequent increase in sexual behavior of young women. The women are becoming more willing.

While willingness has been increasing, sexual problems abound. Although different studies show varying degrees of difficulties, they all show a pervasiveness of them. In 1990 investigators reported the results of 500 people who answered questionnaires about their sexual lives. Respondents ranged in age from eighteen to fifty-nine years old, and most (91 percent) were women. The data came from questionnaires distributed at church meetings, college classes, and Tupperware-like parties that sold sexual aids. The data sample was probably skewed with an abundance of sexually open people and not really representative of a general population. Even so, problems were common. Close to half acknowledged that they had sexual problems in their relationships. Similar data came from a more representative population in Denmark. Among 225 healthy forty-year-old Danish women, at routine physical exams and subsequent interview, fully 35 percent acknowledged sexual problems. Close to half of these said the main problem was they had too little motivation to engage in sexual behavior—in other words, no will. Other common complaints were that they derived nothing from intercourse or felt it to be an obligation.

THE LIFE CYCLE OF WILLINGNESS

The only scientific measures of willingness trends available are negative —that is, when willingness goes away. Studies consistently show that sexual problems are common in every age and stage of life; but the nature of the problems seems to be different at different life stages. Probably with young people the biggest problem is a difficulty in matching libidos and sensitivities to each other. Lacking experience but feeling powerful desire, young men may be unable to offer a good, competent sensual experience to their partners. In older people the problems tend more to be related to blood flow, arousal, and declining function that is disproportionate between partners and leaves one unable to satisfy the other fully.

In between these extremes one of the major challenges to lovers in a stable relationship is negotiating differences in willingness for sensual expression. The needs for privacy and space compete with those for regular sexual activity and satisfying the "hungrier" partner. When people care about each other, they come to understand the importance of balancing these sometimes conflicting personal requirements. With some sensitivity and the knowledge of sensuality cycles, the person with the greater need for sexual contact may be able to offer sensual gifts that stimulate the arousal of the less needy one. A kind of bartering system operating on the sensual level can develop where needs are acknowledged and solutions tested.

The scientific data are sparse for those who do not have a stable relationship. In women my studies have shown that instability can be disruptive to the endocrine system (see Chapters 1 and 5). For men the data are lacking.

Desire and willingness may set the wheels into motion, but a third factor, arousal, is necessary for a complete sexual contact.

THE ABCs OF AROUSAL

Sexual arousal may start in the brain, but it plays itself out in the genitals. In both women and men the basic physiology is the same. During arousal, and as long as it is sustained, the blood moves, sending its pulsations into the pelvis and into the genital region. For men the result is obvious —an erect penis, a blood-filled organ held turgid by fluid. For women the evidence is more subtle, but the physiology is similar. Arousal is reflected in engorgement of the genital tissue, labia, vaginal walls, and entire pelvic region. And vaginal lubrication follows as the engorged tissue transudes fluid from the capillaries in response to the pressure. A delicious series of pelvic sensations grabs one's attention and directs the focus onto the senses.

The desire for sex can lead to arousal, and arousal can enhance desire. They are difficult to separate. Still, it helps to try to tease apart the components in order to look at human sensuality cycles. In young people arousal seems to come pretty easily. In older people arousal may be repressed or enhanced by uniquely individual reactions. What triggers one person may not trigger another, and for this reason the study of arousal is particularly difficult.

SOUND BODY, SOUND MIND

Arousal in men seems to be simpler to acknowledge. For one thing, the fact of their arousal or its absence is obvious: the presence or absence of an erection. This obvious physical manifestation has made men somewhat easier to study in the laboratory. Several recent studies have added to our knowledge of how arousal works for men and their sexuality. First, the man's state of health counts. Men who exercise vigorously perform better. They are more arousable. They engage in sex more often. Other habits affect arousability; for example, smokers who stop smoking and simultaneously build up their vigor by an exercise regimen profoundly increase both their sexual arousal and their sexual behavior.

Physical fitness counts, but so do psychological stress and cognitive style, the way a person perceives. In 1986 an Italian investigator showed that the way a man thinks affects his sex-hormone responses. To begin this study, the men took tests to see whether they were deniers or nondeniers. Deniers are men who tend to deny feelings and emotional reactions when they are treated badly, the "macho" males. Nondeniers are men who allow themselves to express what they feel when they are treated badly. This experiment reminds me of the trait-anxiety tests in women described in the last chapter. The tendency to deny or not deny the reality of unpleasant experience tends to be a stable character trait, an attitude of being. The results of this study support the work described in Chapters 4 and 5.

Sex-hormone responses are different in men with different character traits. Deniers showed a significant increase in two of their sex hormones, LH and testosterone, minutes after a sexually arousing stimulus (a series of photograph slides) was presented to them. The nondeniers did not have a significant change in their hormones. In other words, the "macho" males experience a surge in sex hormones when they see sexy photos; the other men don't. More research is necessary before a complete and proper interpretation can be accomplished. For now we can conclude that the cognitive style, the general pattern of a man's way of being, interacts with his erotic style. Sexuality and personality characteristics clearly do affect one another.

THE PULSE OF AROUSAL

Arousal is measured in the lab by the movement and pooling of blood. A number of experiments that measure blood flow in the pelvic region have been published about both women and men. Blood flow in men is measured as erection. Science quantifies. To put the measurement into numbers, investigators measure the change in circumference of the penis in comparison to some baseline, such as nonerect circumference. Put a man in front of an erotic film with a penile-strain gauge around the base of his penis and chances are the record will show that his penis responds to the erotic stimulation on the videotape. Somehow visual images enter his eyes, fire nerves, alter perception, and provoke an erection. He may be able to use an act of will to inhibit the arousal or he can just let it happen.

Blood flow in women is measured by the vaginal pulse amplitude or light reflectance, which quantifies the force and change in amount as blood moves into the genitals during arousal. Both tell a similar story. When an external erotic stimulus is presented to a willing experimental subject in private, the results are consistent. The blood flows into the genitals as the person looks at or listens to erotic stimuli. Research scientists published the proof between 1976 and 1990.

In 1976 Dr. Peter Hoon, working with Dr. John Wincze and Emily Hoon, published their study called "Psychophysiologic Assessment of Sexual Arousal in Women." They developed techniques for assessing the vaginal blood volume using a method they called photoplethysmography. One end of the instrument was inserted vaginally like a tampon. An external wire leading to a computer hookup could record and then compute changes in vaginal blood volume. The instrument proved to be a highly reliable indicator of sexual arousal. The computer-recorded local blood volume was shown to increase as the woman experienced sexual arousal. The blood flowed. A number of studies followed using variations of this testing method.

In a synthesis of the research results, five physiological principles seem most relevant to sexual arousal in the sexes.

A Woman's Physical Capacity to Be Aroused Is Related to Whether She Has an Adequate Level of Estrogen

In 1982 gynecologists Drs. J. P. Semmens and G. Wagner published a study in the *Journal of the American Medical Association.* They showed that vaginal blood flow increased significantly after estrogen therapy in women who were estrogen deficient. This deficiency is a characteristic of menopause and is easily remedied with appropriately dosed hormone-replacement therapy (see Box 3-2, 80). In the Stanford Menopause Study, women who were still menstruating but approaching their menopause with declining levels of estrogen showed variations in their ability to experience sexual arousal. Eighty percent said that they do experience arousal "usually" to "almost every time" they begin heterosexual activity with their partner. But 20 percent said that they often "are not aroused" during sexual connection. The blood doesn't flow as well. Starting in their early forties women may begin encountering estrogen deficiencies during the week before and after menstruation each month. By age fifty estrogen is often low throughout the entire month (see page 73, and Figure 3-8, page 79). Likewise, an adolescent girl or one in her early twenties may also circulate relatively low levels of estrogen. This may be why girls begin sexual activity later than boys.

Other vaginal changes occurred with estrogen replacement:

- Fluid (lubrication) increased.
- pH dropped (making the vaginal chemistry more acidic, as in younger women).
- The electrical potential measured as transvaginal potential difference increased. These electrical changes reflect the capacity of current to flow. Although the human body creates changing electrical fields—as seen when an EKG is taken to evaluate cardiovascular health—the meaning of this for sexuality is not known. Age-related vaginal changes are now documented. Estrogen creates a younger-type electrical physiology.

Dr. Semmens's very important discovery is that a woman's physical capacity to be aroused requires some minimum level of estrogen. A younger fertile woman's ability to be aroused may be higher when her estrogen levels peak each month, but research remains to be done to test the possibility.

Women who menstruate do show a regular rise and fall in estrogen. The studies published through 1990 show graphs that seem to array a cyclically repeating increase in certain vaginal blood-flow changes in response to erotic audiotapes, but these apparent variations are not statistically significant. This does not mean that fertile women have no variation in their ability to be aroused. Rather it means that scientists have not yet devised a way of proving such a possibility, as they have in other animals. The scientific method tests experimental data in order to elucidate principles. The failure to find a phenomenon *cannot* in itself prove that the phenomenon searched for does not exist. Like a needle in a haystack, science cannot disprove what it cannot find.

In rabbits vaginal blood flow increased when estrogen levels were artificially increased by injection after ovariectomy. Estrogens are specific in effect. The kidneys and the bladder, for example, do not show any change in blood flow when estrogen therapy is given. Only the vaginal blood flow increases after estrogen therapy. Progesterone works the opposite way. When combined with the estrogen, it lowers the estrogen benefit.

Physiological Arousal Is Different from Sensation and Perception

Compared with menstruating thirty-year-old women, premenopausal, fiftyish women do not perceive any difference in their level of arousal. However, the tests show that they are different. Older women produce a significantly lower vaginal blood flow when they are sexually aroused. When estrogen levels get low enough, as is inevitable after menopause for about 80 percent of women, painful changes in intercourse may follow. When intercourse is painful, it is often due to a deficiency in blood flow to the region. Over time the tissues thin out and do not lubricate well. In 1983 Dr. Semmens reported that painful intercourse tends to occur before there is any cellular evidence of atrophy of the tissue. This painful intercourse can be a warning sign of too-low estrogen. His study was important because it confirmed what women were reporting but what their doctors were unable to see when they studied the tissue under a microscope. For these women estrogen therapy restores the tissue and restores the potential for a comfortable, enjoyable sexual life.

Women Have Basic Maintenance Requirements

The adage "Use it or lose it" has poignant relevance in terms of a woman's sexuality and sensuality. Vaginal atrophy (visually apparent deterioration of tissue) is a common problem in menopausal women. In 1983 Dr. Sandra Leiblum and colleagues reported that there was significantly less vaginal atrophy in menopausal women who had more active sex lives. For those women who did not have a partner, self-stimulation—masturbation—helped to reduce atrophy. When an area of the body, such as vaginal tissue, is gently massaged, blood tends to flow into the region, the cells tend to multiply and thicken, and atrophy is reduced. Keeping the blood flowing helps to delay the hormonally induced declines.

The Arousal Pattern in Men Mirrors That of Women

Arousal in men is as subject to loss of youthful function as it is in women. In the Stanford Menopause Study 109 women answered the following question about their partner: Does he have a problem with erection? And 24 percent of the men partnered to these perimenopausal (average age forty-nine) healthy women "usually" had a problem with erection. The blood doesn't flow so easily. Other investigators have provided similar kinds of data.

The studies of men and women tell the same story. As people grow older, their physiological capacity declines, but their perception of pleasure tends to be maintained. We learn better to appreciate what we have.

Instructing Your Partner Can Work Either For or Against You

Explicit instructions—telling your partner what you want—can kill desire, willingness, or arousal, or they can enhance the outcome. Although the tone of the request counts, the outcome seems to depend more on whether or not the man is sexually dysfunctional. Drs. Julia Heiman and David Rowland published evidence in 1983 showing that sexually dysfunctional men reacted differently to instructions than sexually functional men did. Their experiment focused on clinic patients who had sought help with various sexual dysfunctions. Most common were erectile failure, premature ejaculation, and retarded (delayed) ejaculation. A control group of volunteers of similar age who claimed to be sexually functional was enrolled for comparison. Men came into a laboratory, were hooked up with the penile-strain gauge described earlier, and listened to erotic tapes of a female voice that was describing a scene involving various heterosexual activities. The results are fascinating for what they suggest about when and whether a woman should tell a man what she needs in order to enjoy the sexual dynamic.

THE SEXUAL PERFORMANCE—DEMAND EXPERIMENT

Each man, in private, was presented with two conditions at two different times. The first was a high-performance demand. The man was instructed to listen to the tape and to try to get and maintain as hard an erection for as long as possible. Dysfunctional men did poorly when they were given instructions to maintain a strong erection on demand. Men who had claimed to be sexually functional did well, even in the potentially stressful setting of a laboratory. Dysfunctional men had great difficulty with the "high demand" instructions and tended to get very small, half-limp, short-duration erections.

The second condition was a low-performance demand in which the men were told to relax and enjoy the experience. They were told that it was not important to respond sexually and that they should not make any effort one way or the other as they listened to the erotic tapes. Here the results were reversed. The dysfunctional men produced stronger erections than they did in the high-demand condition. The dysfunctional men produced stronger erections than the functional men in this low-demand condition.

When functional men are told just to relax, to enjoy it, and that it is not important to respond sexually, they follow directions and tend not to respond sexually. Their erections were less than half their normal strength. In contrast, sexually dysfunctional men responded sexually only when there was a sexual stimulus and no pressure to perform. Their penile response at their best was only half as good as the erection response of the functional men when they were told to get an erection.

If a man is fully functional, a woman should be able to tell him what she wants and what she needs. He in turn can give it to her if she has the self-knowledge, the erotic capacity, and the timing to ask for it. Of course he must have the willingness, the patience to learn, and the desire to give her great sexual pleasure. If a man is dysfunctional, he probably cannot provide satisfactory sexual conclusion with his penis. With such a man it may be wiser to seek less in order to get the most that he can give, which will not be as much, in terms of penile-vaginal sexual performance, as that which a sexually functional man can provide. Whether a man with erectile difficulties can be cured is unclear. The causes vary, and the studies are inconclusive. A man with erectile dysfunction may still provide tremendous sensual gifts if he wants to give them. Studies that test women's ability to respond to erotic requests have not been published.

The need, then, is for a woman and a man to get the self-knowledge, the erotic capacity, and the timing. No single fail-safe manual serves as a guide. Books may purport to teach, yet it is in the dynamic of relationship that the individual variations bloom. A sense of openness to discovery, honesty, taking the time to listen and learn through the offerings of the culture and one's lover—each of these blends with the individuals' experience and reflection. To receive these sensual gifts, we must seek them out.

SUBTLE SIGNS OF SEXUAL AROUSAL

Sexual arousal is also expressed in a number of other ways—ways that are every bit as obvious as erection to the observant lover. A particular set of muscles in the face, the zygomatic muscle, contracts when women are sexually aroused. This is the muscle sheet that raises the mouth at the side—as in a smile or a grimace. According to a 1986 study of sixty young women, sexual arousal was shown to occur in these muscles during a variety of sexually stimulating audiotapes.

Nipple erection is another common sign of arousal. More rapid breathing also occurs. When all goes well, the sensation and perception of desire can be exciting, triggering arousal that in turn can trigger an orgasm, followed by satiety and peace.

The most obvious sign of sexual arousal in a woman is vaginal lubrication. The lubrication is derived from two sources. One form of lubrication is a transudate, that is, a "sweating" along the vaginal walls, which results from the pressure of engorged tissue that pushes some of the liquid of the blood through the walls to form a fluid lubricant. Other lubrication in the vagina derives from cervical mucus, which varies in quantity depending on how much estrogen is circulating in the body. When there is more estrogen, as at the time of ovulation, there can be as much as sixty times more cervical fluid copiously lubricating the vaginal walls. At orgasm there is a female version of ejaculation, which is one further source of lubrication. In some women a propelled glandular fluid is squirted vaginally at orgasm.

Clearly, sexual arousal plays a critical role for men who want to engage in sexual intercourse. Arousal plays an equally critical role for women. An unaroused woman tends not to lubricate. Forcing intercourse when a woman is not yet lubricated is the sensual equivalent of having sex with a man who does not yet have an erection. A sensitive man will help his partner to lubricate before connecting coitally because he will understand that otherwise he is likely to cause pain and tissue damage. Although the use of lubricants is widely touted, I'm not so sure they shouldn't be used only as a last resort. They do solve the abrasiveness, but I wonder if it wouldn't serve the couple better to honor her biology, learn how to manually (and gently) stroke and stimulate the labia and clitoris, and take the time her body needs to promote her own arousal. The use of lubricants does not solve her absence of arousal.

One of the many valuable contributions that the pioneers Dr. William Masters and Virginia Johnson made was the fact that arousal takes longer as people get older. Young people can become fully aroused in fifteen to thirty seconds when the dynamic is going well. A woman and a man in their midlife years may require five minutes or more of nondemanding stroking and petting to get the blood to flow. Sensuality cycles more slowly as people mature. The urge to come and go in a heated rush should give way to a slower, more sensuous pace.

A close relationship between a woman's arousability and her likelihood of experiencing orgasm exists. Figure 7-3 arrays data I collected at Stanford from women who were thirty-seven to fifty-five and still menstruating, though irregularly. Most of these were women who were approaching menopause were both orgasmic and arousable, as the figure shows.

Each dot represents one woman's report on both her capacity for being sexually aroused during sexual activity with her partner and her likelihood of experiencing orgasm. As the figure shows, most of the women who are always aroused are always orgasmic. Likewise, the one woman who is never aroused is never orgasmic, and the few who are seldom aroused tend not to experience orgasm. Although this finding supports common sense, it is good to see real data that support intuitive perceptions. More than one hundred healthy women completed this part of the questionnaire, and this healthy population is probably representative of this age group. To get from a state of arousal to a state of orgasm, the final component of sensuality, a sensual journey must be traversed. Before examining the sensual journey to orgasm, let us first consider orgasm in its most simple condition, as a reflex conclusion to a triggered impulse.

FIGURE 7-3

RELATIONSHIP BETWEEN THE CAPACITY FOR ORGASM AND AROUSAL

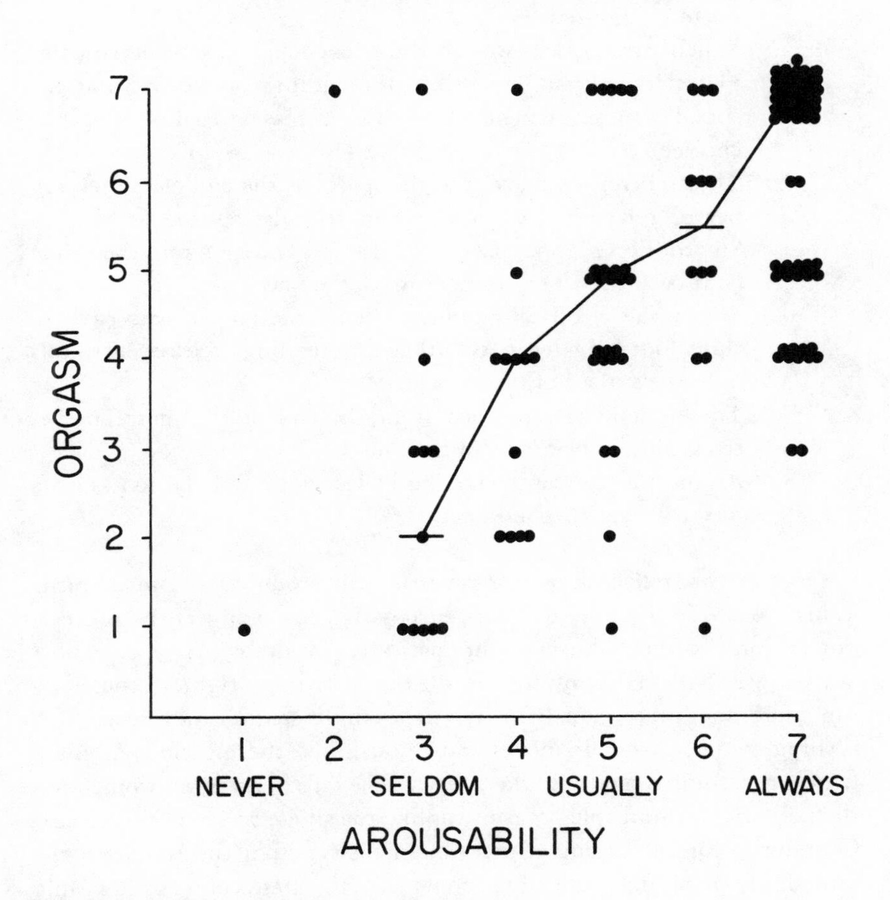

THE SIMPLE FACTS ABOUT ORGASM

In the physiological sense orgasm is a muscle reflex. In reflex physiology a predictable sensory and motor nerve path is followed into and back out of the spinal cord. A classic example is the knee-jerk reflex, outlined below:

1. The physician uses a rubber mallet to tap the tendon just below the knee.
2. Pressure-sensitive nerve endings that are receptive to this tapping are activated.

3. Once triggered, they fire electrical impulses from the nerve ending near the mallet down the long shaft of the nerve, and then onward into the spinal cord. It is somewhat like the flow of electrical current from the switch to the bell when a doorbell button is pressed.
4. When the single, two- to three-foot-long sensory nerve completes its split-second firing, the endings of the nerve at the spinal cord release neurochemicals, such as adrenaline or acetylcholine.
5. These chemicals move into the spaces at the end of the sensory nerve, relaying the impulse and stimulating the firing of a "motor nerve" (one that is attached to a muscle) from the spinal cord back out to the muscle above the knee.
6. When the electrical impulse reaches the end of this two- to three-foot-long leg nerve, the nerve endings release chemicals onto the surface of the muscle.
7. These chemicals, released from the end of the motor nerve, trigger the contraction of the muscle.
8. As the muscle contracts, the bones move and the foot shoots forward a split second later.

Orgasm is a reflex akin to a yawn in women and a sneeze in men. When you have the urge to yawn, a sensation first triggers the process, but the process then continues automatically, inevitably, independent of any further control. The muscles in the throat begin to tighten, constricting more and more, until finally, in a single spasm, they release. A rushing sound as the air moves adds "music" to the muscles. A single female orgasm is something like a yawn. The reflex orgasm in women has the potential for multiple orgasms, unlike the single spasm of the sneeze. Orgasms occur in varying degrees of intensity, from the trivial to the profoundly moving. Twenty-four hours later the uterus of a postorgasmic woman may still be contracting, a muscular process that can help send sperm on their way into the fallopian tubes.

Orgasm in men is like a sneeze, characterized by the tightening of pressure followed by an emission, a bursting forth of fluid. Again, the intensity varies. For both men and women orgasms are muscular reflexes. In both cases nerves play a role, serving to trigger the reflex.

What is important to remember for love cycles is that orgasm is a reflex, a biological inevitability. You can help yourself and your partner when you know what reflexes are and how to elicit them.

Once sufficient stimulation has been applied, with due respect for what kinds of rhythm and pressure trigger the nerves to fire and the muscles to contract, the reflex response of muscle contraction is usually inevitable. However, reflexes can be inhibited. An act of will, nonrhythmic thrusting, or some mental function that uses higher brain centers to repress can be capable of overriding and inhibiting reflex responses.

Orgasm can be delayed or totally inhibited not only through physical processes but through emotion, tension, or other mental processes. When orgasm is not successfully completed, tensions and stresses are created and not released. If the body holds on to these stresses, I suspect that they have the potential to affect pelvic pathology. Imagine the blood rushing to the genital region and then failing to be appropriately disgorged. I think of fibroid tumors in the uterus as the twisted growth of benign tissue. I suspect these relationships could be profitably tested, although no studies have yet evaluated these ideas.

ORGASMS ARE EASY...FOR MEN

Usually a boy discovers orgasm in his sleep in the early stages of puberty. The urge to complete the arousal process with a concluding orgasm can be as urgent as the need to sneeze once that need is triggered. The way to proceed from arousal (erection) to do what is necessary to produce the orgasm involves penile massage. From manual self-stimulation, and later to shared stimulation via entry into a woman, the man first learns to move to the rhythm that provides his own release. But his rhythm isn't hers, and he may not realize that his approach to orgasm may be inhibiting rather than stimulating hers.

Meanwhile the women waits—for his rhythm to learn to accommodate to hers. The sophisticated man understands that moving to her rhythm first, postponing his own ejaculation and loss of rigidity, creates more profound and shared experience. Such an aware man reaps the rewards of his sensual gifts throughout life. For the more naïve man it may take many years—if ever—before he discovers that her needs are not the same as his and that any woman can easily feign orgasm out of a misguided wish to enhance his sense of manhood.

Do your partner a favor, women: Don't fake an orgasm. As with many lies, this ultimately backfires, because it limits his ability to learn how to satisfy you. In addition, it limits your ability to recognize and develop the rhythms you need to experience. If the man flits from woman to woman, he may not be around long enough to develop the orgasmic rhythm with any one of them. Likewise, if women permit too easy access to their sexuality, they will promote this promiscuity in men and might consequently limit their own erotic fulfillment. Men who never get to know the extraordinary sexuality of which women are capable—a sexuality that can enhance their own—ultimately lose out in the cycling of their own love lives.

WHAT'S INSTRUMENTAL FOR FEMALE ORGASM

An enduring image first captured by the avant-garde artist Man Ray pictures Woman as a violin. The analogy works beyond the physical similarity between a women's shape assessed from the rear and a violin's curves.

A sexually lush woman, like a violin, is a delicate and a powerful instrument. As a man, in order to play her, to make music with her, you must learn the instrument so that you can produce the full range of music of which she is capable. If you succeed, you will create consummate harmony. Play her badly, ignore the way she is designed, and you will reap a whole set of discordant notes. She is the same instrument either way. What differs is the skill that the musician brings to the instrument. The more finely tuned the instrument, the greater its potential for harmonic or discordant music.

The analogy can go only so far. Both the man and the woman play and are played upon. If you, as a man, can tap into the rhythm of a woman's biology and sensuality cycle, you can enjoy the pleasure that comes from erotic fulfillment. When you do learn your lover's unique biology and sensuality cycle, you will be amazed to discover how easily, and how often, you can fulfill your own needs. She will rarely use a headache as an excuse to say no. Rather she is likely to ask you to cure her headache with your loving skills.

FREUD MISUNDERSTOOD THE INSTRUMENT

To play the "violin," one needs to know where and how to stroke the strings. That subject has long been mired in confusion. By the time Freudian theory had permeated the culture, there was widespread conclusions, without any scientific evidence, that the "highest" place that genital stimulation could be applied to lead to orgasm was by the penis in the vagina. The underlying assumption—that a women should be able to accommodate her rhythm to whatever rhythm led to a man's ejaculation —had its counterpart in the style of dancing in the same era. Her job was to follow. According to Freudian theory, as women matured, they became orgasmic through vaginal stimulation. According to this theory, women who required clitoral stimulation in order to achieve an orgasm were somehow less mature, lacking, in need of therapy. To this day, there are no scientific data to support this premise, which has probably caused a great deal of damage, because the theory was couched with such authority.

What Really Is Known

Here are some of the principles that can be suggested. They are my theories and I base them on my scientific study of anatomy and physiology as well as on my perceptions of what women and men have been describing as I reviewed more than five hundred questionnaires and three hundred letters received in response to my earlier writings.

There appear to be three different genital locations in which the reflex of orgasm can be triggered if appropriate pressure and rhythmic stimulation are applied. These are: (a) the clitoris; (b) the G-spot and other vaginal regions; (c) the cervix.

The key in all three places has to do with the rhythmicity and pressure of the stroking. If you consider the coital thrusting that any male mammal engages in, you can observe a predictable pattern to the rhythm. Penile thrusting begins at one speed and continues with a regular, repeated deep rhythmic thrusting, and a smoothly accelerating speed follows.

Perhaps this need for a smooth rhythmic nerve stimulation helps to explain why vibrators have been so effective for women in producing clitorally stimulated orgasms. The nature of a vibrator is in its regular rhythmic pulsing stimulation. The rhythm is so regular that orgasm becomes inevitable when the vibrator is placed gently on the clitoris. Physiologists label the inevitable approach toward any reflex response, from an orgasm to a sneeze, nerve recruitment. Stimulation fires sensory nerves. As nerves fire, other nearby nerves are recruited to fire in parallel. It is something like getting a bonfire started: One piece of wood fires the next, and so on. When enough nerve energy stimulates the trigger point, the resolution of the response is inevitable. Genital stimulation during coitus seems to work just the same way. Figure 7-4 shows the location of the clitoris, as well as the genital anatomy, when viewed from the surface.

In 1981, in the *Journal of Sex Research,* Drs. John Perry and Beverly Whipple published a now-classic study in which they demonstrated that there are two forms of sexual stimulation, carried on two different nerve networks. Either can produce an orgasm. In a way one could think of this as the solution to the question of whether there are two different kinds of orgasm, clitoral and vaginal, and whether one is better than the other. In fact, that old argument misses the point, because both orgasms described by Perry and Whipple involve vaginal stimulation. What distinguishes them is that one (triggered by clitoral or vaginal stimulation) fires the pudendal nerve while the other (triggered by either vaginal or cervical stimulation) fires the pelvic nerve.

When women research subjects took Valium, it interfered with the firing of the nerve in the clitoral/vaginal area but not with the firing of the vaginal/cervix area, the pelvic nerve. These data show that the two different sensory nerves influence orgasmic responses in different regions of the genitals. Perhaps just as relevant, these results, collected in women who had their uterus intact, demonstrated there is a sensory-nerve path from the cervix to the uterus that affects orgasm. Preserving the cervix can preserve a certain quality of life.

Vaginal stimulation was studied not only by Perry and Whipple but

FIGURE 7-4

THE SURFACE GENITAL ANATOMY OF WOMEN

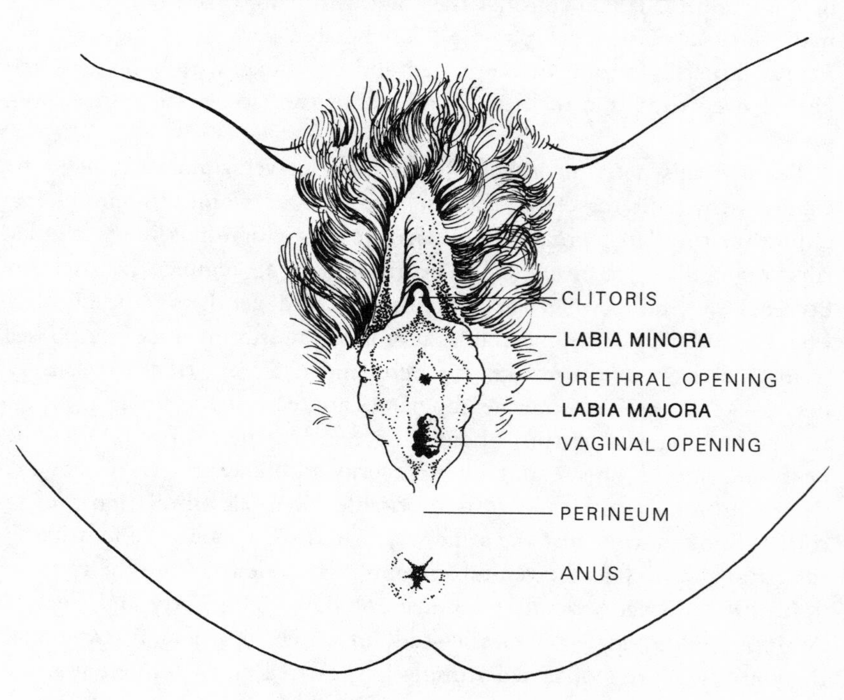

also in Colombia by Dr. Heli Alzate, whose series of papers provides support for the idea that it is the rhythm of the stimulation, rather than the exact location of it, that may most profoundly influence the capacity for orgasm.

THE CHANGING CLITORIS

Another principle sophisticated lovers should know was taught by Masters and Johnson. When a woman is not yet aroused, the clitoris is at its most accessible. Once arousal has begun, the clitoris begins to retract. Clitoral stimulation in an unaroused woman should be extremely gentle, because the nerves are easily accessed and easily bruised. As arousal progresses, greater pressure can and will probably need to be applied in order to achieve stimulation that is both comfortable and orgasmic.

THE THRUST OF THE MATTER

Men are different from women. When a man is thrusting into a woman's vagina, he does not have to think what he needs to produce his ejaculatory inevitability. He moves to the rhythm his penis requires. Although the female-superior position might seem adequate for the woman to manipulate the motion, the mechanics here produce extraordinary muscular stress, limiting fine-tuned control. Consider the problem this way. With the male superior, the angle of the penis allows the man to rest his weight on his knees and pivot back and forth the way children do on a seesaw. He achieves penile/vaginal massage without having to support his weight at his ankles. In so doing, the erect penis is appropriately aimed for entry —upward toward both of their waistlines. Female-superior generally requires the woman (with shorter legs and no thruster) to perch on her heels and awkwardly attempt deep knee bends. Even with extraordinary strength she cannot afford to lose concentration or she will fall. If women were able to manipulate the thrusting, they would achieve orgasm as frequently as men. But it seems that in most mammalian species the coital pattern holds the female motionless while the male does the thrusting.

As a biologist I think of the biological requirements for reproduction and suspect that the male-superior position evolved as the fallback position because it worked best for face-to-face contact. First, the male has the thrusting equipment and can adjust the woman's hips because she is not supporting her weight when she bends to receive him. Second, a biologically successful coition requires that the sperm be kept inside the woman so that it can make its way up the vagina, through the cervix, into the uterus, and out into the fallopian tubes in order to await the ovulating egg. The longer the female can stay still, the less sperm will spill out.

This biological design for reproduction has led to the evolution of a series of stimuli and responses that differentiate the male from the female. Put another way, females that kept still were the ones whose genes survived to reproduce. The offspring with those genes have a natural tendency to keep still. The more sperm that gets to the fertility site in the fallopian tubes, the greater the chance of conception. Thus the biological imperative is to keep women still.

What works for the rest of the animal kingdom does not necessarily follow for man and woman. When we look at what it means to be human —the humor, adaptability, cooperation, choice, and creativity—we can transcend the basic biology. For both people to achieve arousal and satisfaction, each must have his or her own rhythm met. Many people do. Different sexual positions, to take one example, can enhance the fun and exchange the power.

If a woman needs a particular kind of thrusting, one might naturally wonder why she does not just tell the man what she needs. The answer is simple: It is too hard to do. I think it is something like asking a man to define what kind of thrusting pattern he needs. He may be able to do the thrusting just right. But it would be almost impossible for him to verbalize and articulate: how often, how fast, how deep, how hard, and how many thrusts to make before ejaculation will begin. As his moods change, so do his needs. Add to that the extraordinary complexity of a woman's cycle, the varying hormone levels throughout the month and throughout her life. At certain times in her symphony certain kinds of stimulation will be appropriate. The very same form that may be exciting in one phase might be aversive at some other phase. The nonverbal processes involved in the sexual dance of movement and speed do not always lend themselves to verbal communication.

When a woman may be realizing what she needs, she may be unable to speak about it. During the height of erotic pleasure the act of discussing, giving orders, and directing the thrusting pattern of her partner may kill her own desire—and his.

If the essential communication is beyond words, the need for long-term commitment on the part of the couple becomes significant. Each partner must learn to recognize and act upon the other's nonverbal arousal cues.

FIT FOR ORGASM

Just as a physically fit man who engages in vigorous exercise promotes his sensuality and sexuality, so it is with a woman. For a woman the fitness of her internal genital musculature counts too. These muscles will have a powerful influence on her capacity to experience *and* to promote sensory pleasures in her partner.

The pubococcygeal (PC) muscles form a muscular band that encompasses the bladder, the vagina, and the anal region (see Figure 7-5). They connect the pubic bone at one end and the spinal coccyx bone at the other and wrap around the three openings of the urethra, vagina, and anus, much like the figure eights of an ice skater. The greater the PC muscle strength, the easier it may be for a woman to achieve orgasm through clitoral and/or vaginal stimulation.

When the PC muscles are very weak, a problem characteristic of a majority of women after forty-five, vaginal orgasm is probably unlikely, and urinary incontinence is very likely. Just take a look at the shelf space

FIGURE 7-5

THE PUBOCOCCYGEAL MUSCLES OF WOMEN

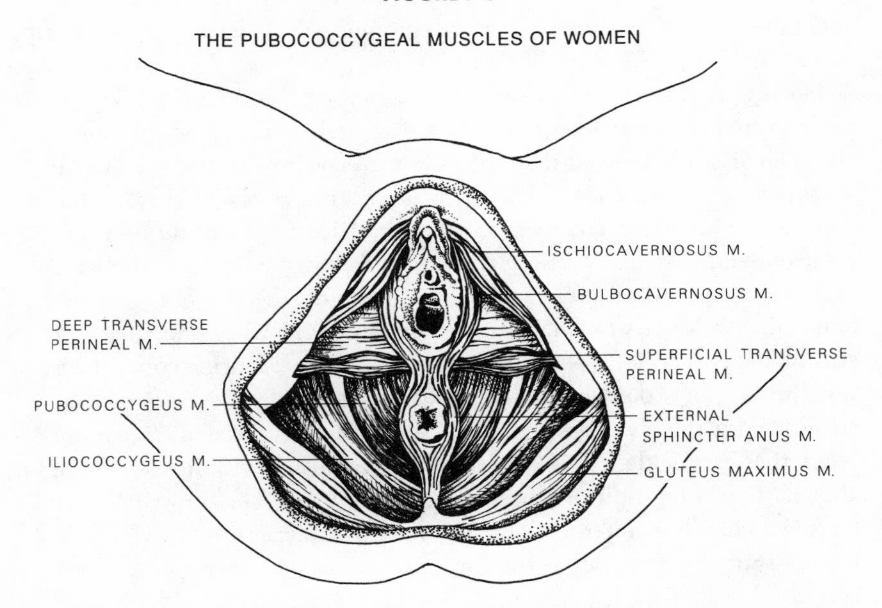

now given to adult "diapers" in supermarkets to get an idea of the pervasiveness and the market value of this problem. As early as 1951 Dr. Arnold Kegel published his first series of studies showing that dysfunction of the PC muscles was present in all cases of urinary stress incontinence.

He developed the perineometer, so named because it measures the strength of the muscles of the perineum, the region between the pubic bone and the coccyx bone. He also developed a set of exercises and a way of teaching the patient to do them that produced cures for approximately 85 percent of the women who had the problem. The good news is that PC muscles can be strengthened through exercise. As the muscles get stronger, urinary incontinence recedes and clitoral stimulation can produce an orgasm. As the muscles get even stronger, either vaginal or clitoral stimulation adequately produces the response, according to some studies. In other words, a woman who has difficulty controlling her urine may have difficulty achieving a vaginally stimulated orgasm.

More recently other researchers have shown that the lower the PC muscle strength, the higher the likelihood of urinary stress incontinence as well as the higher the likelihood of an inability to achieve orgasm through either vaginal or clitoral means. Drs. John Perry and Beverly Whipple measured the strength of the PC muscles through an improved electronic perineometer that Perry developed. They also developed techniques by which a woman could be trained to develop her muscle strength and improve the conditions that are common when the muscle strength is poor.

During their studies of PC-muscle functioning Drs. Perry and Whipple rediscovered the Grafenberg spot as a potential source of sexual stimulation. Their study showed that about an inch and a half inside the vagina, at about twelve o'clock high, there is an area of tissue that becomes engorged with blood and swells, somewhat like the swelling of a man's erection, and that this swelling can be elicited by regular, repeated stroking of that spot. This area forms a mound about the size of a quarter or half-dollar when a woman is sexually aroused and when she is properly stimulated by gentle, repeated stroking of that spot. During coital thrusting the G-spot would normally be massaged and would be particularly stimulated if the penis has a ridged head, from circumcision, providing a ledge that can tickle the G-spot each time the penis withdraws. Presumably if there were no ridge present, the stimulation of thrusting would be somewhat less intense. The PC muscle contracts reflexively when the G-spot is stimulated rhythmically, much like any other reflex action.

And as in the recruitment of nerves for a sneeze or yawn, the particular rhythm of the thrusts is important in producing the effect of reflex contraction, the vaginal orgasm. When the thrusting pattern stops and starts, starts and stops, the nerves will be recruited in a kind of herky-jerky effect, probably producing more irritation than recruitment toward orgasm. Since it is the inherent rhythmic harmony of nerve firing that allows the nervous system, like a harp, to produce resonance, the stimulation of the G-spot, or other vaginally sensitive region, is the highest when the couple have learned the rhythm the woman needs. In 1985 Dr. Whipple, working with Dr. Barry Komisaruk, published their paper "Elevation of Pain Threshold by Vaginal Stimulation of Women" in the journal *Pain*. They showed that direct physical stimulation of the vagina works to decrease the sensitivity to pain. In their study the regular, repeated massage of the vagina did not diminish sensitivity to touch, only to pain. Their results suggest that the stimulation works to trigger the release of the endogenous opiates, but research must still be done in order to test for this effect. Although there has been a great deal of scientific controversy over whether a G-spot actually exists, most women who have answered questionnaires on the subject have confirmed an awareness that they have a region within their vagina especially sensitive to erotic stimulation. Perry and Whipple showed that the stronger the PC muscles, the better this reflex process works.

If Freud and his followers had focused on helping a woman to build her PC-muscle strength rather than overcoming her childhood toilet training, vaginal orgasms might be the birthright of every woman. Freud was close to the truth, but he missed the meaning. Rebuilding the muscle strength rather than the memories can lead to a recovery of function.

THE BENEFIT OF CHILDBIRTH

After women give birth, the passage of the fetus through the birth canal alters the shape of the vagina. Imagine passing a large head and shoulders through what was a tight and narrow muscular barrel. The muscles will be expanded and stretched. In the 1960 *Annals of the New York Academy of Sciences*, Dr. Kermit Krantz documented the change in vaginal anatomy after childbirth. Different is not necessarily worse. Women who have strong PC muscles will have a better muscle tone and better vaginal function than those whose muscles are weak. For the woman with well-toned internal musculature, the vagina will now offer a series of undulating folds that caress, massage, and enhance the sensual pleasure of the penis. The environment becomes lushly sensual after vaginal childbirth.

This is a gift that is often overlooked. Popular wisdom would have us believe that a woman is "stretched out" after vaginal childbirth. This is not necessarily true. With age, declining sex hormones, and multiple pregnancies unexercised PC muscles do tend to weaken and sag. I strongly recommend that women learn the following exercises and do them regularly because of the pleasure they provide to their partner as well as to themselves. Short of senility, it is never too late.

THE KEGEL EXERCISES FOR PC MUSCLES

The first step in the Kegel exercises (originally developed by Dr. Arnold Kegel) involves locating the muscles that must be exercised. If you (man or woman) can perfectly stop and start, stop and start a stream of urine in midflow, you have located the muscles and probably have good, strong muscle tone.

It is not possible to detect when a person is doing Kegels. If you contract your PC muscles correctly, neither your stomach nor your buttocks should simultaneously tense up. An electronic perineometer, which measures the strength of the PC muscles, can be used very successfully to help the individual work with biofeedback as she builds her muscle strength. Some women can improve their muscle strength by exercising without the perineometer, but the feedback with the instrument usually produces substantially better results, and much faster. In either case only women who begin muscle building with a baseline of reasonably good muscle strength can expect to solve their problem without biofeedback.

For women whose initial muscle strength was weak, having a doctor telling or showing them what to do in the office was not sufficient, according to studies by Dr. Kathryn Burgio and her colleagues at the National Institutes of Health. The women needed the instantaneous feedback information that a perineometer provides on the computer screen. This equipment can be rented.[3]

An excerpt from Dr. Kegel's 1956 paper, published in the *Journal of the International College of Surgeons,* might be a helpful starting point.

> I give my patients the following detailed instructions for exercises without the additional apparatus:
>
> 1. Before arising in the morning, contract the muscles five times, as demonstrated with the perineometer; then contract them again five times on first standing up; try to hold the muscles in a contracted position while walking to the toilet.

[3] Contact Leslie Talcott, R.N., 242 Old Eagle School Rd., Stratford, PA 19087.

> 2. Interrupt the urinary stream several times during each voiding.
> 3. Repeat contractions of the same muscles five times every half hour throughout the day. It is brought to the patient's attention that permitting the muscles to sag for long periods retards progress and that to "bear down" during exercises may aggravate her complaints.
>
> The early morning exercises are especially important, for the patient begins the day with the perineum in a high position, whereas previously she started out with a sagging pelvic floor and considered the associated tired feeling normal. Because so many of my patients complain of losing urine late in the day, I have come to regard stress incontinence as part of a pelvic fatigue syndrome that must be relieved if one is to obtain permanent results. With adequate exercises the woman learns to maintain the perineum in a higher position and enjoys a sense of pelvic strength, which, by many, is appreciated as much as the relief of incontinence. Those who attain this new pelvic comfort continue the exercises with confidence and enthusiasm.
>
> When the patient is reexamined after three to six weeks of diligent exercises with and without the perineometer, it will be observed that much of the slack in the supportive muscles has been taken up; the perineum, bladder, and uterus have assumed a higher position, and the vagina is longer and tighter; at the same time the strength of contractions has increased.[4]

A regular, repeated practice of contracting these muscles will ultimately overcome the urinary stress incontinence. Many women begin to notice improvement quickly—within a few weeks. Once you learn the muscles that are to be worked, they will become stronger and the work will become easier. Once you have formed the habit of contracting these muscles, you will be able to do it anytime, anywhere, knowing that your privacy is maintained even though you may be in public.

Women who have stronger pelvic muscles may be more orgasmic. If you develop muscles sufficient to overcome urinary stress incontinence, you may find orgasm achieved more easily and more often. It should be noted that many women begin to experience orgasm while they are doing the exercises. This is perfectly normal.

DOES PENIS SIZE MATTER?

Just as size in a woman should not detract from her—or her partner's—pleasure, the same holds true for a man. It is not that the size of a man's

---

[4] Originally in Cutler, Winnifred: *Hysterectomy: Before and After* (1990)

penis or a woman's vagina does not count, rather it is the realization that such differences can be accommodated. The *Kama Sutra (The Hindu Art of Love),* postulates that the size of the genitals of men and women varies. In the process of selecting suitable candidates for their children's marriages, parents were remiss unless they considered which genital category the young man would grow into: the hare man, the bull man, or the horse man. It was acknowledged that when both lovers are of equal size, there is likely to be satisfaction to both. When there is a mismatch, one or the other is going to be unhappy.

The Biology of the Cervix, a scholarly book edited by Richard J. Blandau and Kamram Moghissi, provides anatomical facts that support the *Kama Sutra*'s premise:

In Chapter 5, "The Intrinsic Innervation of the Human Cervix, A Preliminary Study," the authors show that the nerve endings found in the human cervix are the most abundant in the estrogen-rich stages of women's lives, ages twenty-three to forty-eight. After the menopause (with its characteristic drastic reduction in estrogen) the loss of these nerve endings is common.

Chapter 3, "The Comparative Anatomy of the Mammalian Cervix," provides anatomical drawings of the penis and vaginal-cervical ending, which reveal the almost perfect lock-and-key fit between the male and female of different species. Figure 7-6 reproduces two of these sketches.

In looking at these sketches, you can't fail to notice the different endings of the penis and how beautifully they are shaped to tap against the species-specific cervix. In the regular, rhythmic thrusting, the anatomy is designed to stimulate what we now know to be the location of the G-spot. Considering the biology of mammalian reproductive organs, the importance of anatomic fit seems obvious.

WORKING WITH WHAT YOU'VE GOT

The idea that "the size of the penis does not count" seems to me to miss the point. What counts is that the sizes of the particular vagina and penis being fitted to each other work well together. Among couples who do not have a perfect fit, knowledge of the adjustments they need to make can help. Fortunately people are flexible.

A man with a small penis who finds himself bonded to a woman much larger than he can still give regular, repeated deep rhythmic thrusting. Such a pattern taps the end of the cervix, and the ridge of the penis

FIGURE 7-6

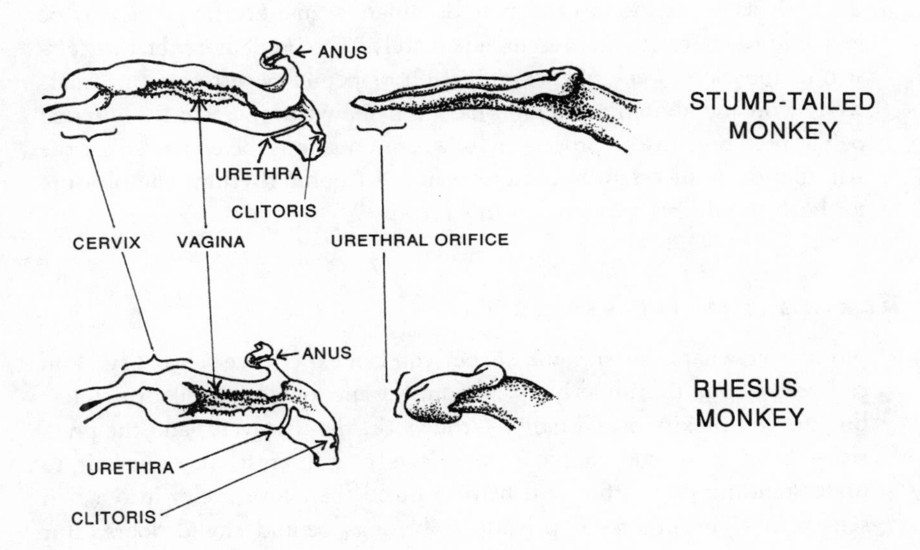

strokes past the G-spot, stimulating his partner until she reaches orgasmic inevitability. If his penis has a pronounced coronal ridge, each stroke will offer pleasurable stimulation if the positioning is right. On the vaginal "clock" the G-spot region is twelve o'clock high! If the rhythm is right, the nerves will be recruited. Such a couple will probably learn the importance of position. When her legs are high, the vagina gets shorter, and he can more easily reach the G-spot region en route to the cervix. Unfortunately, when this happens, the vagina will also get wider, and he will receive less penile stimulation. According to the studies of Alzate, women vary in the vaginal region that they find particularly erotically sensitive. Couples need to become aware of their individual differences, not the average tendencies. Good sex does not involve just one position. Improving the fit by varying positions, strengthening her PC muscle, and the exercise of patience, humor, and sensitivity will all help to enhance their potential for good love cycles.

Likewise, when the man has a very large penis and the woman is small, he will need to learn what to do in order to avoid damaging her. Awareness of anatomy and awareness of the need for rhythmic stimulation should help. The base of the penis is not heavily wired with sensory nerves. The tip of the penis and its coronal ridge are where the sensory nerves have been placed. Even if they do not fully penetrate, men who outsize their partners can probably experience equal satisfaction.

If a man is paired with a woman whose PC muscles are weak, sexual positions are especially relevant. If her legs are spread with her feet above the body as is common in the male-dominant sexual position, she may be less able to contract the muscles adequately. To compensate for this less-strong muscle system, the combination of penile penetration and male weight on the abdominal region has been shown to increase the internal pelvic pressure. In the process muscle tension should be enhanced. If the stimulation is successfully patterned in a harmonic rhythm, the pleasure for both should be increased by this pressure.

MAKING THE FIT PERFECT

No matter what your size and shape, you can experience sensuality. You can read about the underlying anatomical and physiological principles, but the complexity of sensuality—the work, the discovery, and the practice—belongs to the couple. The efforts and interest they commit to understanding each other will help to build their love cycles into a rich sensuality. The process is private. It belongs behind closed doors. But some of the principles are scientific. They can be studied. Still, there is art.

CONCLUSION

These are highly individual principles that should be understood not only in terms of sensuality cycles but also in terms of monogamy and restraint. The scientific principles of nerve recruitment combined with acts of monogamy and restraint allow a couple the time they need for building and enhancing a lush sensuality. Cultural rules vary. Where women are "educated" by misinformation, the possibilities for sensuality cycles are severely limited. To take just one example, if women are trained to endure the sexual demands of a man, and not expect pleasure themselves, inhibition of their sensory arousal can be expected. The culture can teach frigidity. In the context of an intellectually open society the woman's enjoyment of sensual pleasure should enhance the experience of the man and vice versa. Within the private spaces of one's relationship, being open to discovering the sensitivities of each person enhances sensuality cycles in both.

Let us next consider the cosmic cycles, where the external control over the affairs of life on earth is mediated. All life on earth moves to the rhythm of cosmic motion.

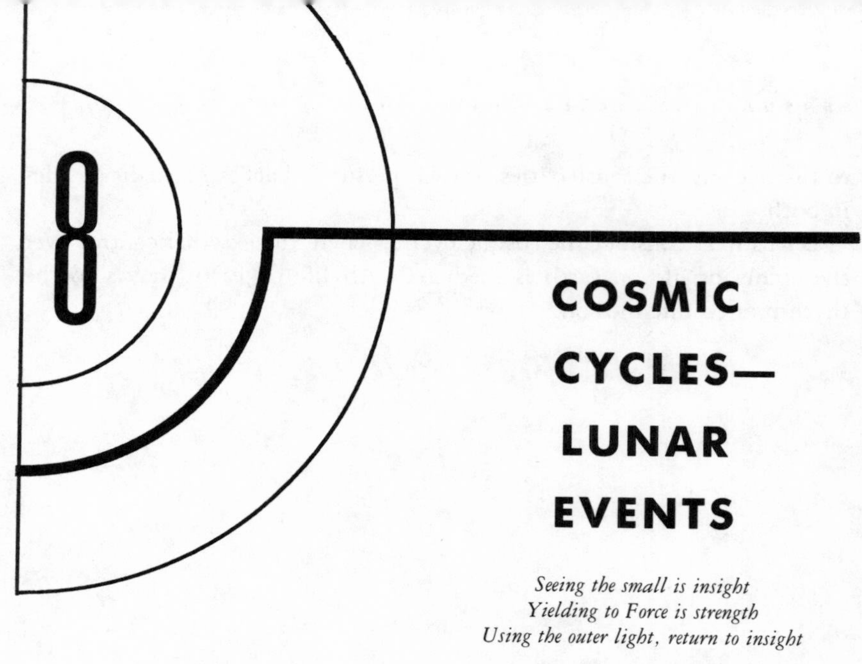

8 COSMIC CYCLES— LUNAR EVENTS

Seeing the small is insight
Yielding to Force is strength
Using the outer light, return to insight

—LAO-TZU, #52

Even a shallow consideration of the cycles of the moon and of men and women triggers the idea that human cycles and lunar cycles have something in common and that both men and women are involved in the process. Even trouble comes in cycles. The numbers of suicides, hospital emergencies, and other crises rise at the full moon. Lunacy is a real phenomenon, and aptly named. And recently published studies have begun to prove it.

THE MAGIC OF 29.5-DAY CYCLES

Rigorous research by physicists has shown that the moon becomes full with predictable regularity just about every 29.5 days. Similarly, rigorous research by reproductive biologists reveals that the average menstrual cycle among fertile women is usually 29.5 days long. The moon cycle coincides in length with the average menstrual cycle. This coincidence has biological significance and is nonrandom.

The hormonal cycle of a fertile woman is an exquisitely complex symphonic interrelation between events in her own body and interactions with her lover. The pulse beat of this symphony is further controlled by larger, cosmic forces. Many of us sense the inevitability of such a reality. The astrologer, the philosopher, the biologist, and the religious person each in his or her own way appreciates that there are heavens and heavenly bodies whose forces exert controlling influences on the affairs of men and women.

Most of us recognize that these cyclic forces are there even as we appreciate that an unsolved mystery remains. Just how it all comes together forms the subject matter of the scientist, the poet, and the priest. Some things can be explained by the scientist. It helps to begin with two elements of cosmic timing: the lunar month and the lunar day.

EARTH TIME VIA MOON TIME

The "lunar month" of 29.5 days and the "lunar day" of 24.87 hours set the tone for much of human life.

The lunar month refers to the fact that every 29.5 days the moon becomes full once again because of the timing of the movement of heavenly bodies in space.

Once each lunar month our moon travels in its orbit around the earth. During the monthly cycle we can tell that the moon is traveling by looking up each night and noticing how its appearance keeps changing. These changes are known as its phases.

The lunar day, of 24.87 hours, also reflects changing moon location. While the earth rotates on its axis to line up with the sun every 24 hours —a solar day—the earth is also moving within its annual orbit around the sun. It takes .87 of an hour longer than 24 hours for the earth to move into the same position in relation to the moon that it was in the day before. The reason? The earth and the moon have to catch up to each other because the earth moves, each day, 1/365 of its annual trip around the sun.

Once every 24.87 hours the sun, the place where we are on earth, and the moon form three consecutive points on a straight line. The moving earth and the moving moon again line up with the sun to mark a new complete cycle every 24.87 hours, which is the daily moon cycle or lunar day.

Life on earth moves to the rhythm of this cosmic dance. And a variety of animals—from crabs, to earthworms, to monkeys—demonstrate a harmonious choreography within their reproductive cycle. The "sexual dance" of each species, including our own, is set to the motion of the cosmic cycles. For some animals the sun appears dominant, for others the moon. The timing of the music comes from the cosmos, what the Bible calls the heavens.

EXPERIMENTING WITH MOON TIME

In the last twenty-five years a number of scientific experiments conducted in Germany, the United States, and elsewhere have helped to provide the factual underpinnings. Some experiments have removed humans and other animals from their normal exposure to sunlight to show how important the sunlight was in keeping their rhythms going. Other experiments have placed people in underground apartments to show the role of moon time in maintaining human biological rhythms.

Experimenters have removed certain nerves from animals' brain tissue in order to identify which part of the nervous system controls the biological rhythms.

Taken together, these and other experiments have provided a great deal of information. They all reveal that the cosmic journey of the earth sets the pulse of our neuroendocrine systems. Biological forces are harmonic. They form overlapping cosmic rhythms. In a methaphorical sense we are traveling in an ark on a sea that subjugates us to its tidal rhythms. The same goes for all life on earth.

BECOMING AWARE OF COSMIC CYCLES

The sense of a "harmony of natural law" compels us to behold the moon when we focus on menstrual rhythms of women and their effects on men. If you look up at the moon, you probably notice when it is becoming full (waxing) and when it begins to wane. Have you noticed the direction of the filling? Have you noticed that the moon always become full from the right side of the circle to the left? The day after a new moon the very first crescent of light that shows is curved to the right, and it's black on the left, the way a capital letter *C* would look written backward. The left part of the circle gradually fills in night by night until 14.75 days after that first right-arched crescent a fully lit moon completes the circle.

Waxing to its fullest, the moon then starts waning—until, 14.75 days after full moon, it's nothing more than a narrow left crescent, a thin sliver of a letter *C* before disappearing on the night of new moon. If you look up at the moon, you can tell whether the crescent is at the beginning or the end of the month by whether the curve's outer arch is on the right (waxing) or the left (waning). There is reason to care about this. The moon can affect the harmony of our lives. When we learn to check the moon time, we elevate our awareness of the relevance of cosmic cycles and their impact on our lives.

Although most of us appreciate the sun's importance in our lives, we may not realize that the moon may provide an even more powerful influence on the rhythms of our internal secretions and that its influence is far more subtle. Grasping the meaning of the moon in our lives is useful because an appreciation of the cosmic forces can help us to live more lovingly. Only since 1980 has the scientific community "discovered" that there is a relationship, a harmonic synchrony, between the menstrual cycles of fertile women and the lunar cycle of the earth's moon.

The scientific discovery of these coincidental forces is recent, but the cultural myths of many peoples have long reflected their awareness of a symmetry between the moon and the menses. The science is ten years new, and I happened upon its revelation.

DISCOVERING THE MOON AT THE UNIVERSITY OF PENNSYLVANIA

Back in the late 1970s, when I was pursuing the research studies leading to my doctoral degree in biology, an enormous amount of potentially accurate menstrual-cycle data was being collected. I have earlier described the thousands of months of women's menstrual records I collected in 1976, 1977, and 1983.

The 1977 study had originally been designed to gather reliable data from several hundred women on sexual and menstrual behavior: when women had intercourse, how often, and what, if any, relationship their sexual behavior had to the length of their menstrual cycle. My pilot study, a year earlier, collected similar types of data that sixty other women had recorded.

When these data were being collected, I didn't know that the average menstrual-cycle length in each of these two years was going to turn out to be exactly 29.5 days. I didn't know that when cycles approximated this average length, women had the highest likelihood of being fertile. And I didn't know that the moon revolved around the orbiting earth every 29.5 days. I was looking at a different issue: the relationship between sexual behavior and menstrual cycles.

Meanwhile a suggestion was made to me by the only woman faculty member in the Department of Biology. Dr. Ingrid Waldron had an intellectual interest in biological cycles.

Knowing that a large data base of menstrual records would soon be available, she suggested that I investigate what the recorded scientific literature had already published on relationships between moon cycles and menstrual cycles. Perhaps my own data could further contribute to knowledge. I liked the idea of finding out what had been discovered about menses/moon relationships in recorded science. She also suggested that I read one of the studies she had published seven years earlier.

WHEN INSECTS LEARN TO DANCE

Dr. Waldron had conducted an experiment on biological rhythms when she was a graduate student in zoology at the University of California at Berkeley. Those were the days of the strobe lights in the dance parlors. Fast flickering light was common in the nightclubs of the sixties and seventies. Somehow the flickering lights combined with either the loud rock music or the sultry slow-dance tempos to add glamour and romance. I knew the feeling of strobes because I used to go dancing in such places. And she had used strobe lights for her experiments.

Dr. Waldron had used strobe lights in order to study the oscillatory rhythm of the wing beat of locusts (who couldn't dance). These insects could be impaled on a stick, somewhat the way a candied apple is, and placed near a strobe light. Dr. Waldron would change the frequency of the strobe light to make the light flicker either faster or slower. She was able to measure and then to prove that the wing flapping began to beat in synchrony with the strobe light. When the strobe light flickered faster, the wings of the insect beat faster; when the strobe light slowed down, so did the flapping of the wings. (Maybe they *could* dance, after all.)

LOOKING FOR THE STATE OF THE ART

As I searched the library stacks for previously reported relationships between the moon cycle and the menstrual cycle, the strobe light flickered in my thoughts. I searched to find a report of a synchronized relationship between women's menstrual cycles and moon cycles similar to Dr. Waldron's wing-beat discoveries. I expected that the thousands of volumes in the university's biomedical library would contain studies that reported on moon-menses relationships. After all, it was one of the largest such libraries in the world.

I was wrong.

Repeatedly throughout the history of science one or another scholarly investigator had set out to study that relationship, fully expecting to find it. Each of the scholars was a man. Each had concluded, publishing his data along with a careful analysis, that there was no relationship between the natural timing of women's menstrual cycles as revealed by prospectively gathered data and the timing of the moon cycles.

This didn't make any sense to me. I respected science and the scientific method, but I had also been taught a fundamental principle. The failure to prove something does not necessarily mean that it doesn't exist. It might only mean that because of the way the potential relationship was sought, someone failed to find it. If you cannot find a needle in a haystack, you cannot rationally conclude that there is no needle there. You can only know that you failed to find it. I studied the way other investigators had searched for the needle and then designed a different way of looking at the issue.

SETTING UP FOR SUCCESSFUL REVELATION

I created a circular graph, shown in Figure 8-1. On this graph I created thirty-two sectors to allow a different sector for each day of the lunar month as the almanac showed would be required. Variations in days between the four quarters would emerge, sometimes 7, sometimes 7½ days from one quarter to the next. Using an almanac, I filled in the dates when the moon would be *full* (at the top), *new* (at the bottom) at *first* quarter, and at *last* quarter for the four months in 1977 the data would be accumulating. Once this had been completed, I had the beginning of a plan. Here was a way for graphing the menstrual onsets.

It was October 1977, and the data from the women had not yet been turned in. I looked at this finished graph and started to think about it.

What would happen, I asked, if I began to chart the menstrual data of one woman whose cycle was regular every thirty-two days? Or what would happen if I charted the data across these same few months of a woman whose cycle was regular every twenty-six days, or thirty-five days? Imagining these hypothetical women, I selected, at random, August 28 for her first period onset and put a dot on the sector of August 28. Then I started to play with the graph. If she had a thirty-two-day cycle, her next period would occur on September 29, so I put my next dot on that date sector. If she continued to have her regular normal thirty-two-day cycle, her next period would come three places over on October 31, and I put another dot there. Then December 2.

FIGURE 8-1: The Circular Lunar Graph

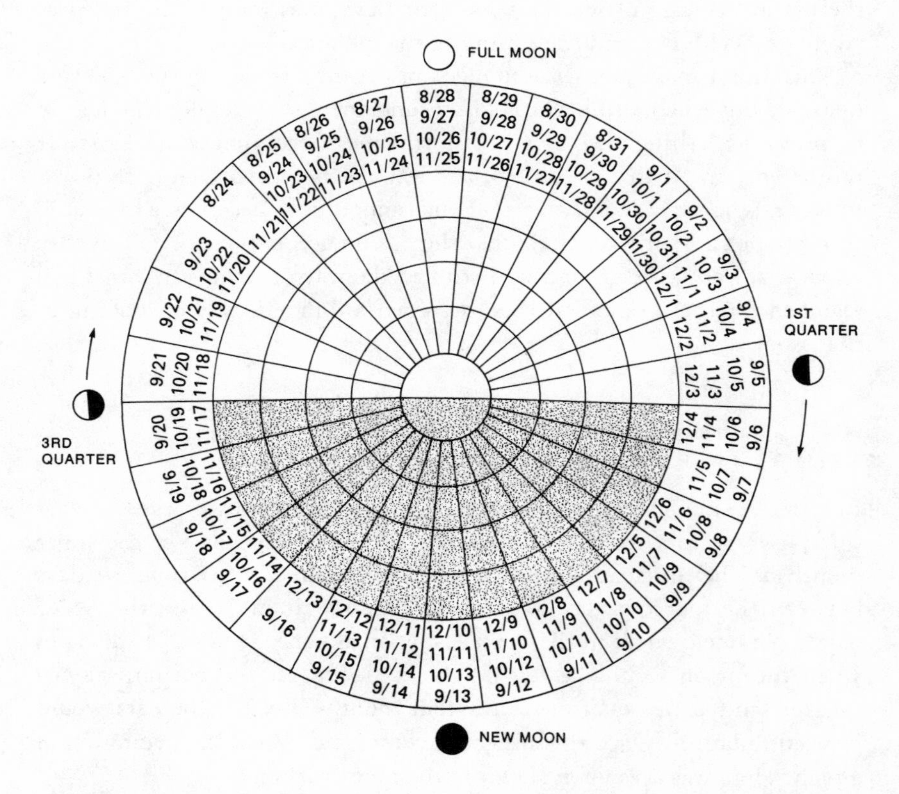

Figure 8-2 has four dots placed within these four sectors to show the idea. Can you imagine what I realized as I placed these four dots? I had just figured out why the other scientists who had reported on menstrual/lunar events failed to find any relationship.

THE LOGIC OF THE MOON GRAPH

Here is how I reasoned. In the course of a year our hypothetical thirty-two-day cycler would have eleven dots placed on this circle. The eleven dots would distribute around the circle pretty evenly. Anyone looking at those dots would conclude that the woman was not having her period at any particular phase of the moon. She wasn't more likely to menstruate in the dark half or the light half or the first quarter or the last quarter. A

FIGURE 8-2: Charting Four Months of a 32-Day Cycler

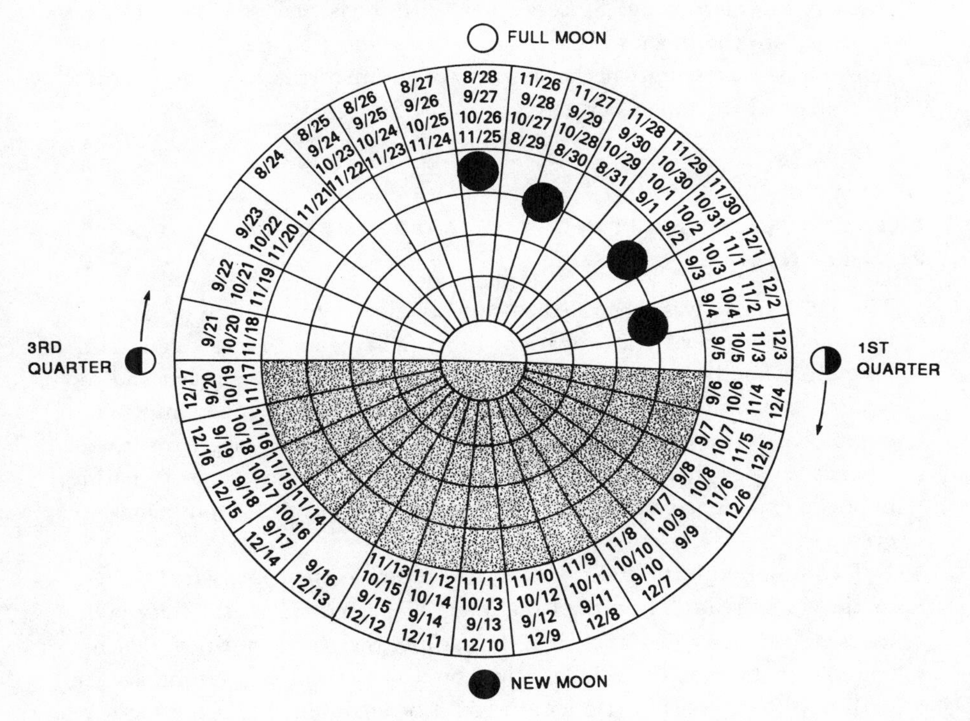

similar result would have to happen if a large group of women with different menstrual cycles each reported one or more menses onsets and these data were plotted on this circular chart.

But women who got their period as often as the moon did, every 29 days or so, would be different. A woman who had a regular 29.5-day cycle would end up having her dot placed in the same sector each month. Why? To understand, just consider any woman with a 29.5-day cycle and look at what happens when you graph her data. For example, if she menstruates on August 28, her next menses 29.5 days later on September 27 places her dot in the same sector of the circle. Whatever her first menses onset is, all of her subsequent ones will appear in the same sector if she has a 29.5-day cycle.

The only question that the moon graph could meaningfully address was whether a group of women who cycled as often as the moon cycled arrayed themselves in any particular phase of the moon cycle. If the moon did not affect the menses of these women, we would expect their onsets to occur all over the circle, in a random, nonorderly way. If the 29.5-day cyclers turned out to be nonrandom, a biological event would be shown and the moon graph would reveal in what way the moon affects the menstrual-cycle timing. Since women with menstrual-cycle lengths that approximate the moon's cycle length are the most fertile of women, this research is addressing whether fertile women menstruate during a particular phase of the moon's cycle.

THE RESULTS—CHARTING THE MOON/MENSES RELATIONSHIP

The 1977 study was scheduled to close as Christmas vacation began, and by mid-November I was impatient to see what would be revealed. Mid-December finally approached. The data arrived. The students left and the campus was empty. In the old stone biology building, now empty and quiet, I began to plot the data. And, lo and behold, the magic was there! I began to see what every scientist who explores the mysteries of nature hopes to experience: A biological event unfolded on the graph before my eyes.

Two hundred forty-eight women had recorded their menstrual pattern in the 1977 study. It turned out that 27 percent of these (sixty-eight women) had an average cycle length that met the criterion—being within one day of the moon's cycle length. The data of these sixty-eight women with a 29.5 ± 1-day cycle length could be analyzed. I selected October as the month to plot. Only one cycle onset was plotted for each of these women because, by definition, each additional cycle onset would fall in the same sector. By plotting only one datum for each person, her October onset, we avoid the problem of artificially multiplying the occurrence of a real result and making it look bigger than it really is.

Since all of the students tended to be on campus for that entire month, October seemed likely to be the most accurately recorded month of data.

Of the sixty-eight women who had a lunar-period cycle length (29.5 ± 1 day), forty-seven had their dots placed in the light half of the month —forty-seven compared with twenty-one, a ratio of more than 2 to 1. This was nonrandom. Twice as often fertile-length cyclers were starting their flow in the light half of the month as opposed to the dark half. It was a statistically significant difference. Next I looked at the pilot data collected the year before from sixty women whose menses were recorded for three months. The same thing showed up.

In 1976 the ratio was 3 to 1. In other words, in 1976 three times as

FIGURE 8-3: Charting the 29.5 ± 1-Day Cyclers

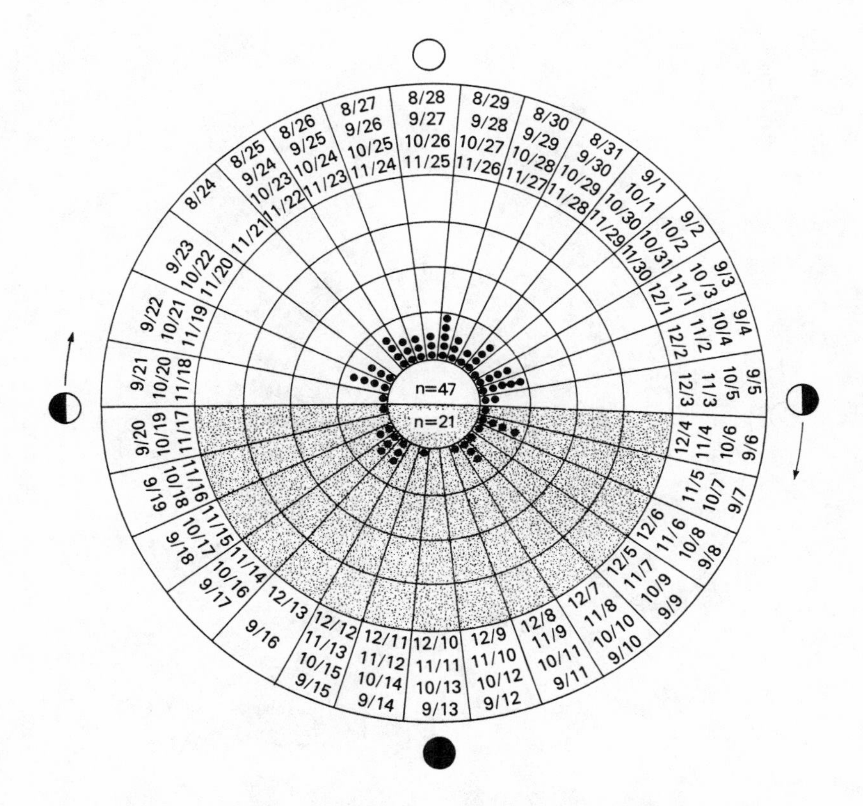

many menstrual periods began in the light half of the month as in the dark half of the month among women with 29.5 ± 1-day cycles.[1] Returning to the larger body of data on the 248 women and their 772 menses onsets, I wondered if I would find the same inconclusive results as previous investigators if I plotted the data in the same way that previous investigators had done.

My results matched those of previous investigators. When I plotted the data of all of the women and all of their cycles from the 1977 autumn data gathering, no statistically significant difference in onsets was revealed. Figure 8-4 shows the data array of what I call the entire haystack. No significant phenomenon was revealed when I obscured the 29.5-day cycles within all of the cycle lengths.

[1] The pilot-study data were also tested statistically. They *were* significantly different.

FIGURE 8-4: Charting the 772 Menses Onsets of 248 Women

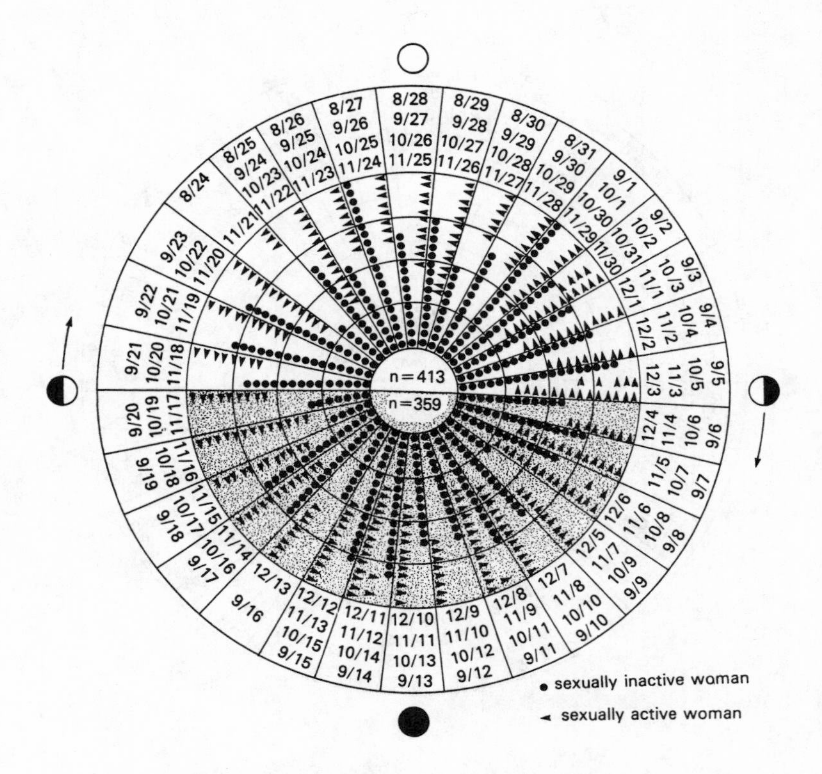

If you look at the center of this circle, you can see N = 413, the number that represents the total number of menses onsets in the light half of the month. The other number, N = 359, reflects the number of menses onsets in the dark half of the month. Although it looks like a little bit more onsets in the light half than in the dark half, a difference this small (53 versus 47 percent) was not statistically significant. In other words, if I had only analyzed my data the way others before me had analyzed theirs, I, too, would have had to conclude that the moon and the menses do not appear to be related. But now I knew better. When I looked at women whose cycle was approximately the same length as the moon cycle, those women had a much higher likelihood—2 to 1 in 1977, 3 to 1 in 1976—of beginning their menstrual cycles in the light half of the month.

REPLICATION IN BROOKLYN

A year later Dr. Erika Friedmann, working in Brooklyn, confirmed the discovery in another sample of women. Dr. Friedmann had been a graduate student at the University of Pennsylvania, and we became friends. Her Ph.D. work, collaborating with professors at the University's veterinary school, led to her discovery of how important love is to life (see page 47).

As fellow biologists Dr. Friedmann and I shared a profound interest in how human behavior affects health and well-being. Although her focus at Penn had been on human/pet relationships, she had been interested in my work also.

When she left for Brooklyn to begin her new job as an assistant professor in the SUNY system, she took with her the memory of my discovery. One day she called to ask me to show her how I had designed the menstrual calendars. She had decided to conduct a similar study in Brooklyn—the opportunity for replication.

In her study 305 Brooklyn College undergraduate students and their associates kept records on similarly preprinted calendars. She found that 32 percent of the women had what was now to be called lunar-period cycles. In her sample she also charted the October menses onset and she also used a circular lunar graph. She found what I had found—a statistically significant increase in menstrual onsets in the light half cycle of the month. In her sample about a *three-to-one* division occurred, as had emerged in my 1976 study. She published those results about a year after I had published mine, in the same journal.

One exciting aspect of being a scientist and publishing discoveries is the rich field of friendships that scholarly work brings. After my first paper phase-locking lunar phases to menstrual cycles was published in 1980, I received letters from several physicists and biologists whom I had never met. They were particularly intrigued with the idea that a human biological phenomenon responded to the moon, and their letters helped to educate me about lunar timing and certain aspects of tides, electromagnetic radiation, and the force of gravity. We all assumed that the menstrual cycle of women was responding to cosmic cycles rather than controlling them. We all wondered how it was working. The cosmic influence was probably not working through the eyes via light, as it does in small mammals. Although some competent scientists had shown that when the visual pathway was deprived of light, small mammals became infertile, this was untrue of the 29.5-day cyclers—women and certain monkeys. In the large mammals, the primates, similar disruption of visual input, altering the light or being born blind, did not disrupt fertility.

In animals that depend on light for maintenance of their fertility, disrupting the visual pathway disrupts fertility. In animals that depend on the moon, disrupting the visual pathway does not seem to affect fertility.

ANALYSIS FROM BALTIMORE

A letter came from Dr. Wolfgang Schleidt at the University of Maryland. It was 1984, some months after I had returned from Stanford to the University of Pennsylvania. I was working to establish the Women's Wellness Program in the hospital at the University with Dr. Garcia. Dr. Schleidt, a professor of zoology at the University of Maryland, wrote describing his reaction to my 1980 lunar menstrual phase-locking report. He had just come upon it and was interested in cycling phenomena. He told me that he had taken the points from my sixty-eight women (Figure 10-3) and replotted them on his own graph somewhat like a stock market moving average is plotted.

His graph is reproduced as Figure 8-5.

It certainly looked harmonic—a smoothly flowing cycle somewhat like the monthly estrogen cycle of Figure 1-1 (page 9). What an intriguing way of looking at the moon and menses.

THE MOVING MENSES ONSETS AVERAGE

Dr. Schleidt's method showed much more information. Each of the points he plotted showed the number of women in the specific half-circle of the moon's cycle that was centered on one particular day.

Centering at the full-moon day revealed forty-seven women in the half-circle, as I had already shown. His graph shows this event as one point, the highest point on the graph right near the day of the full moon. Rather than simply dividing the circle in half one time as I had done, his method had divided the circle in half thirty-two times. So he showed the events that were happening on the half-circle for every single day of the month. In other words you could look at his graph and discover how many women started their period in all thirty-two half-circles which I might have selected. I had divided the circle into light half and dark half of the month. One half circle.

His method was better. It plotted *all* the half-circles that one

FIGURE 8-5: Dr. Schleidt's Graph of 1977 Lunar Cyclers

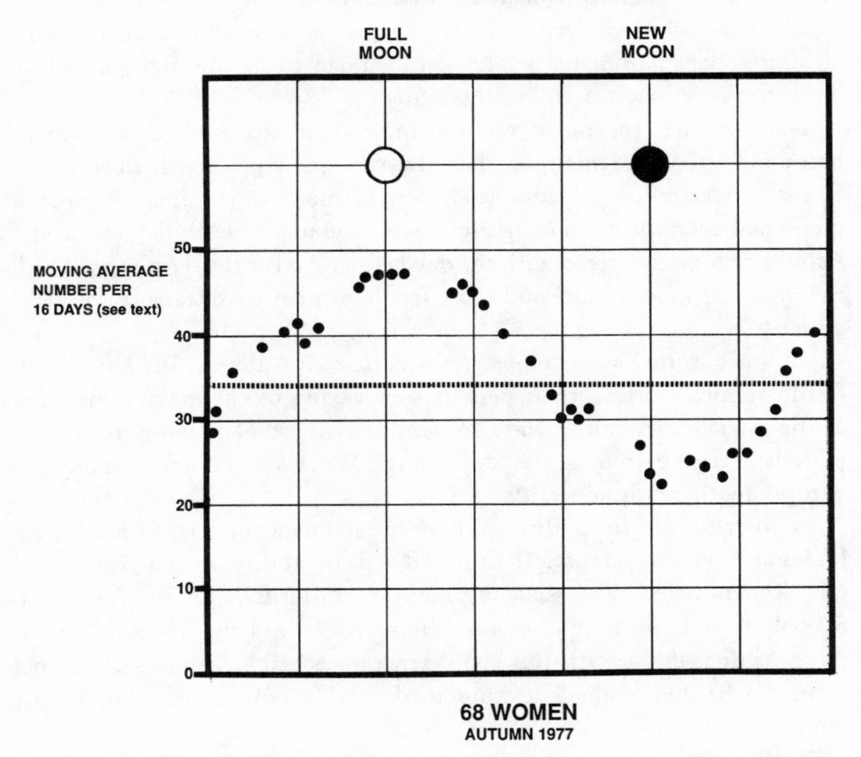

could have chosen and then created a graph to show what the results would be for each of these bisections, at new moon or full moon or any other day. You could look up along the vertical axis of his graphs to see how many women were in the half-circle that any particular moon-day was the center of. So Dr. Schleidt had created a moving half-circle with the number forty-seven (the number I had shown when full moon was at the center of a half-circle) being just one number along a continuum of many other numbers.

ENHANCING THE ANALYSIS: MOON ORCHESTRATING MENSES

Dr. Schleidt wanted to know if I had any more data that might be graphed according to his more sophisticated moving-average method. I was happy to write back to say that I did. There were the samples collected during the Monell pheromone studies over two autumns and a spring, described in Chapter 6. Now it was my turn to call on Dr. Friedmann. She agreed to share her data to permit further analysis.

When I looked at Dr. Schleidt's graph and saw the beautiful ever-changing curve, from its height at full moon to its lowest point a few days after new moon, I was really excited.

The idea that the moon was moving around the earth in a natural, harmonic way was something I had always vaguely perceived. In addition to that understanding I now had a new harmony to consider. Women's menstrual cycles were moving along with the moon. Each day there was a slight change compared with the day before. And as the days approached the next full moon, more and more fertile women were starting to menstruate.

In other words, as the moon was waxing to its fullness, the number of fertile women starting their periods was waxing to the maximum. And as the moon was waning, the number of fertile women whose menstrual periods were beginning was decreasing. We had another harmony in nature, another cosmic reality.

In order to plot the additional four years of data correctly, I needed to find someone with statistical skill and a facile ability with a computer. Dr. Robert Stine, an assistant professor of statistics in the Wharton School of the University of Pennsylvania, agreed to help. He had a computer facility and a statistical competence that enabled him to grasp the value of what Dr. Schleidt had proposed. And he now improved upon the method.

Dr. Stine was interested in the use of statistics as they could be applied to biological questions. He agreed to help me chart and test these data on his computer system.

He suggested a variation on Dr. Schleidt's method. He liked the idea of the moving average. But he wanted to "center" the moving average and to "weight" it in order to give the heaviest weight to the item in the center, with lesser weights to the peripheral data points. He created the graphs shown in the next four figures—and our eyes confirmed what his statistics proved. Menstruation of fertile-length cyclers was synchronized close to the full moon.

The pilot study from 1976 (Figure 8-6) showed the harmony. Most women got their periods around a day or two before full moon. The fewest women got their periods around a day or two before new moon.

FIGURE 8-6: The 1976 Study

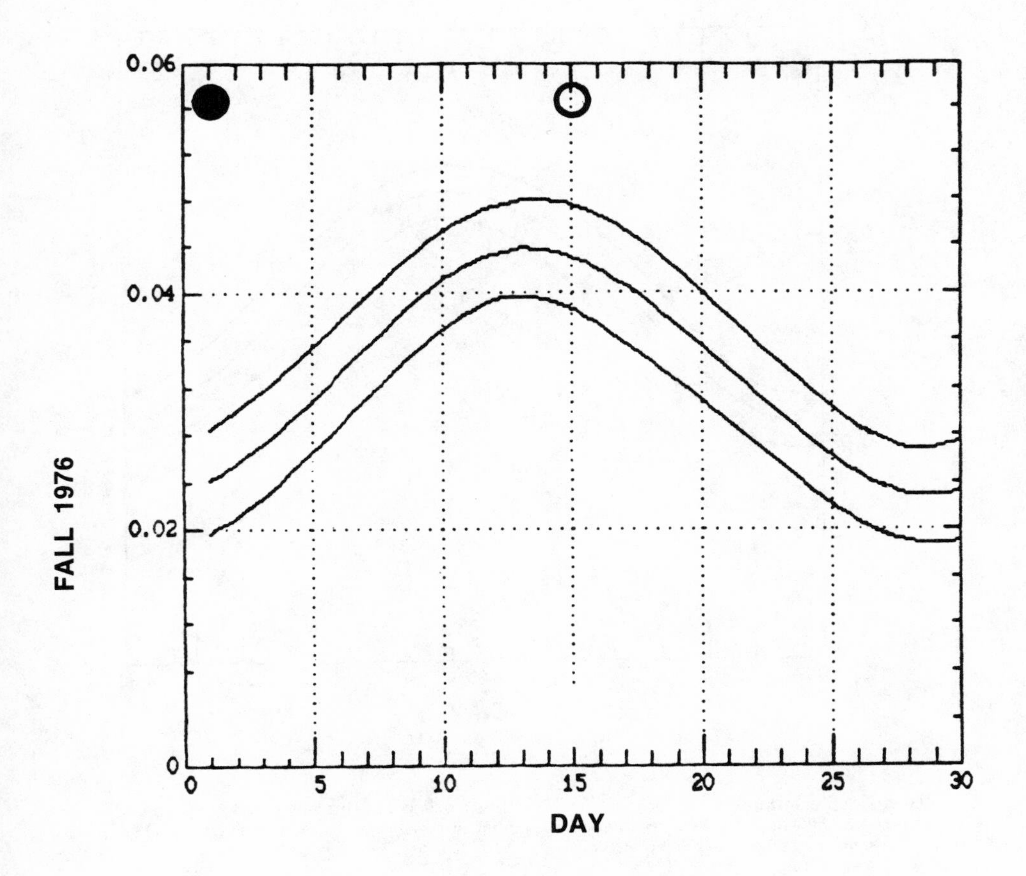

Lunar- and menstrual-phase locking, fall 1976. The Y-axis is scaled so that the total area under the curve will equal 1. The X-axis indicates number of days since new moon; therefore Day 15 represents the time of the full moon.
● = New Moon ○ = Full Moon

The larger, 1977 sample is plotted as Figure 8-7. This is equivalent to Dr. Schleidt's graphing of it in Figure 8-4 and to my earlier one in 8-3. But the center weighting of the moving average reveals it differently.

FIGURE 8-7: The 1977 Study

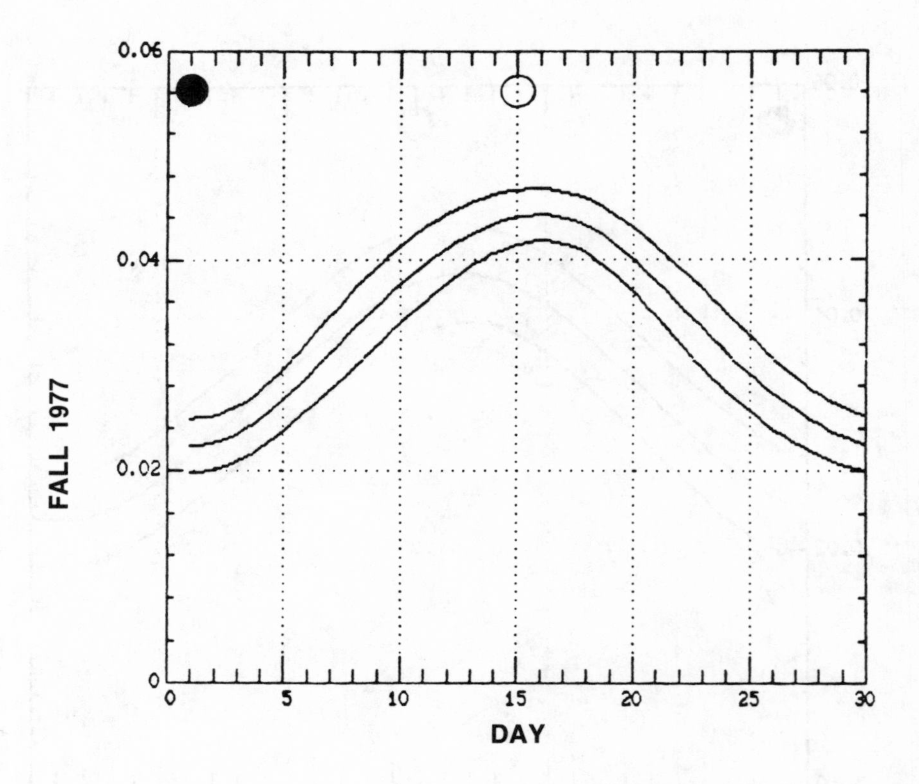

Lunar- and menstrual-phase locking, fall 1977. The Y-axis is scaled so that the total area under the curve will equal 1. The X-axis indicates number of days since new moon; therefore Day 15 represents the time of the full moon.
● = New Moon ○ = Full Moon

Figure 8-8 is the graphing of Dr. Friedmann's fall 1979 sample. Figure 8-9 shows the spring 1983 sample.

We had wondered whether things changed from fall to spring. Our data seemed to suggest that they don't. Our data seem to suggest that women menstruate in phase with the moon in pretty much the same pattern in any time of year. But we have only collected one spring, so we can't yet be sure.

FIGURE 8-8: Dr. Friedman's Sample

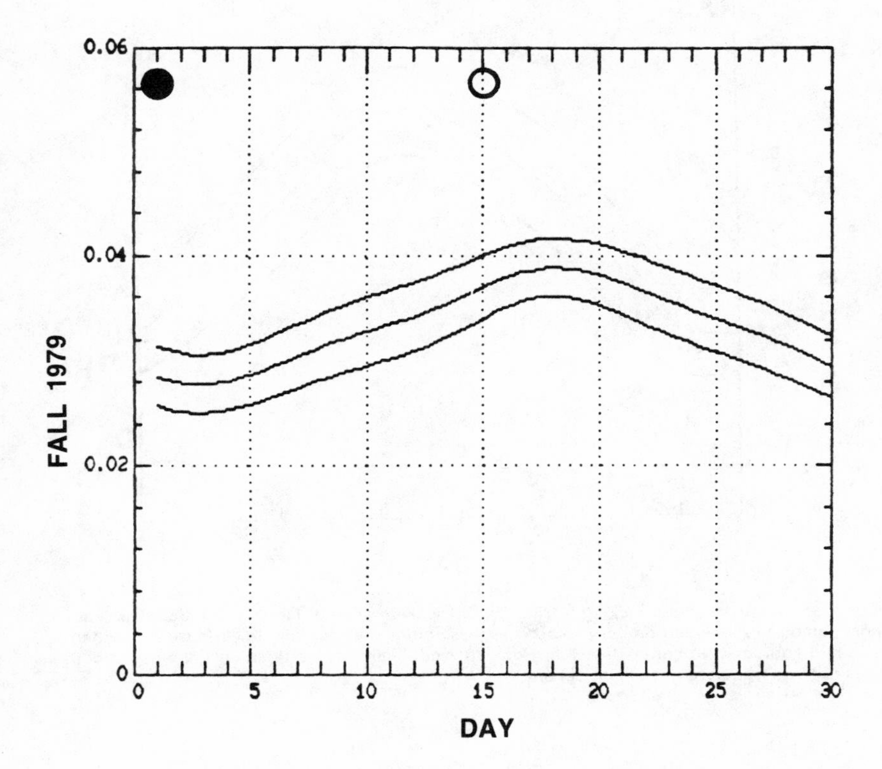

Lunar- and menstrual-phase locking, fall 1979. The Y-axis is scaled so that the total area under the curve will equal 1. The X-axis indicates number of days since new moon; therefore Day 15 represents the time of the full moon.
● = New Moon ○ = Full Moon

One more question intrigued me. What about the women in Phase 8 of the Monell pheromone study, the 29.5-day cyclers who received female essence collected during the full moon of the year before? Would they show a menstrual lunar-phase-locking effect? Would it relate to the full moon of the current year or the full moon of the essence year? Figure 8-10 shows what we saw.

We must await more data, to be sure, but it looks like there is a phase-locking effect. And it seems to be dominated by the year of the essence

FIGURE 8-9: Spring 1983

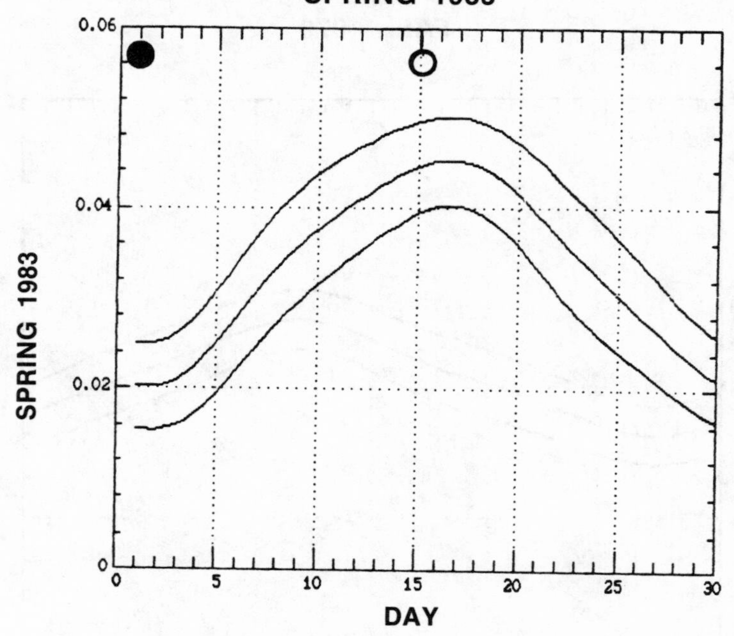

Lunar- and menstrual-phase locking, spring 1983. The evaluation of the spring 1983 menstrual/lunar relationship followed similar protocols to those of the autumn data. In this case the second cycle, mid-February through mid-March, was analyzed. Similarly, the days since the new moon were captured from reference to an almanac.
● = New Moon ○ = Full Moon

collection, because the peak number of menses onsets was twelve days out of phase with the current year's full moon. It coincided with the almanac timing of full moon of the previous year, when essence was collected from women whose menses began within one day of full moon.

Does this mean that the moon, in driving the neuroendocrine rhythm, drives the reproduction pheromone? And that pheromones from fertile women communicate and drive the cycles of the less-fertile women to bring them into sync with the full moon? We do not yet know, but it is the kind of question that scientific experiments might address someday.

As I began to reflect on the possible meaning of this discovery, I was struck by the fact that a woman's menstrual cycle, if she was ideally fertile, tended to synchronize to the same length as the moon cycle.

FIGURE 8-10: Fall 1983

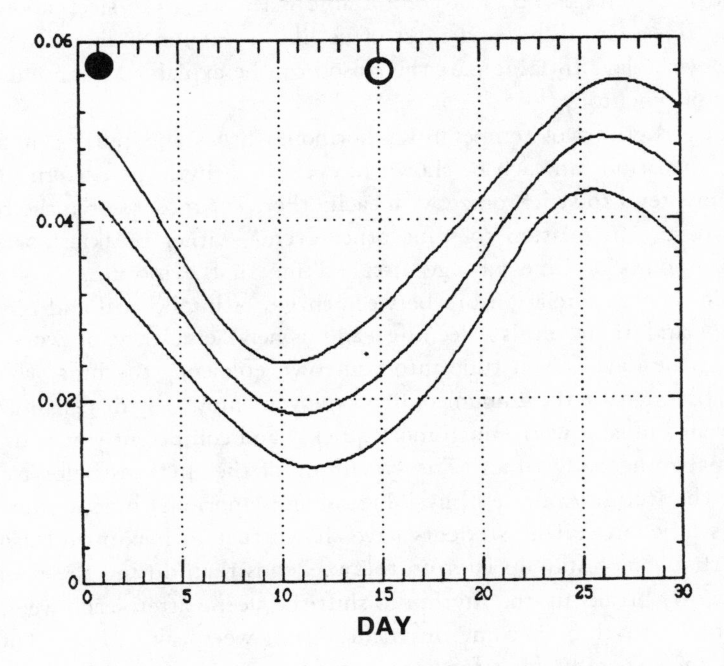

The Y-axis is scaled so that the total area under the curve will equal 1. The X-axis indicates number of days since new moon; therefore Day 15 represents the time of the full moon.
● = New Moon ○ = Full Moon

One thing is now clear. The reproductive biology of a fertile woman shows a harmonic relationship between her cycling body and the moon's cycling body. In the eleven years that have passed since this first 1980 publication, I became convinced that a profoundly important pattern exists.

The lunar studies described in this chapter focused on women in their twenties and thirties. These are the years of the middle stage of a woman's reproductive life, and these are the years for which we now have proof of moon/menses relationships. Within this middle stage the monthly hormonal cycles follow a very specific and orderly pattern when women are fertile. In other life stages and in men the rhythm of the moon is also revealed. To understand that research, it helps to look at *daily* biological rhythms.

SYNCHRONIZING THE RHYTHMS

Just as the ocean ebbs and flows in daily tides via its wave motion along the shore, body rhythms ebb and flow in a regular and repeating pattern. Most of us tend to be most comfortable when we go to sleep about the same time each night, and most of us like to eat lunch about the same time every day. In large part the reason can be explained as an enhancement of synchrony.

Body rhythms of temperature, hormone surges and purges, appetite and exhaustion can each be shown to cycle in a rhythmic pattern. These rhythms tend to synchronize with each other. For most people the bowel urge occurs in relation to some other event—either breakfast or some fixed amount of time passage after waking in the morning. For most people there is a relationship between surges of intellectual and physical energy and their meals, sleeping, and general efficiency of work flow within the day. If you tune into your own efficiency rhythms, you can probably increase the amount of free time you enjoy—by dispensing with different kinds of work much more quickly and competently at your own optimal mind/body times. The synchrony of the rhythms helps to promote the stability of one's physiological and emotional functioning. Recent studies in college students have shown that to maximize learning, they should not stay up later on the weekends than during the week. It apparently breaks up the rhythm to shift the sleep pattern each weekend. When the rhythms become unsynchronized, we speak of being "out of sync" and generally don't function at our best.

DISRUPTING THE SYNCHRONY

Jet lag, in lesser or greater degree, occurs when we make a rapid shift in time zones. Jet lag occurs because some of the body rhythms fall out of phase with others, and the resulting desynchronization disrupts our sense of equilibrium. Until time passes and the body readjusts to the new time zone, we usually feel awful. The cause of jet leg appears to relate to the two-sided aspect of body rhythms. One set of rhythms responds to the motion of the cosmos, the relative positions of the earth, sun, and moon; the other responds to our sleep/wake cycle. Under normal circumstances the two rhythms synchronize with each other because we tend to awaken at the same relative position on earth each day, at the same sun and moon time. When we change our cosmic position by jet travel, the coordination between the sun and moon rising and setting time has changed with respect to our sleep and wake time. The body and the mind become confused as the rhythms lose their synchrony.

THE DAILY SUN CYCLE INFLUENCES BIORHYTHMS

For those of us who are not traveling across time zones, the twenty-four-hour day beats out an important rhythm in the biology of both men and women. For those of us who go to sleep every night at about the same time, experts can pretty well predict the specific hormones of the body that will reach maximum levels on a regular basis at the same time of day at particular stages of life. The rising and falling of different hormonal secretions is in tight synchrony within the body. As one hormone is rising, another is falling with predictable regularity. Experiments have been set up in which people were removed from the sunlight and removed from clocks so that they became unable to tell when it was noon and when it was midnight. Major body changes resulted.

As we gain understanding of the natural rhythms that direct our biological cycles, we enhance our capacity for increasing the loving in our lives. To expand this intuitive level of understanding, let us again look toward the cosmos to see how lunar day was revealed to dominate natural secretions.

COSMIC CONTROL

What Does the Sun Have to Do with It?

Individuals who are regularly exposed to sunlight have tightly bound synchrony of their body rhythms. And timed treatment with full-spectrum "vita-lights" are being experimentally used to overcome certain kinds of seasonally occurring depression. People who are removed from the awareness of the sun's cycle experience significant changes. Men and women have demonstrated this experimentally by going underground to live for a month or so at a time. Such experimental subjects experience some major shifts in their biorhythms. First of all, the "day," or span of time between starting one new day and starting the next, changes. When removed from visual exposure to the sun and clocks, people no longer live on a 24-hour cycle, but they do tend to form another stable biological rhythm. They generally sleep and wake to a different time span. In most cases they develop a "lunar day"—a 24.87-hour rhythm instead of a 24-hour one. To understand why that span is important, and how it was discovered, it helps to consider West German experiments on electromagnetic radiation.

The Extraordinary Effect of Electromagnetic Radiation

In the late 1960s and early 1970s scientists working at the Max Planck Institute in West Germany had two underground apartments built to help them study synchrony of biological rhythms. These apartments appeared identical, but were different in terms of cosmic transmissions. Each apartment contained a kitchen, a bathroom, and a living room with a bed, armchair, television, and dining table and chairs. There was one difference. Electromagnetic shielding was built around the walls, floor, and ceiling of one. No such shielding was placed around the other. This shielding was actually composed of sheets of electromagnetically pulsable coils that were embedded into the earth before the apartment construction began. The coils were connected to electrical circuits that could be programmed to emit various wavelengths of electromagnetic radiation. Something like masking noise with the use of "white noise," its purpose was to block the natural radiation pulses that move through the earth.

Unlike ocean waves, which are big and slow enough to discern, the waves of electromagnetic radiation beaming at our bodies are imperceptible. We seem to need these electromagnetic beams, however. When a force field is built to shield a person from them, the person's body rhythms change. When people are shielded from the electromagnetic radiation that normally reaches the earth, as in the underground-apartments experiments, they experience a desynchronization of body rhythms.

The electromagnetic shielding was turned on in one apartment and was not available in the other. Research subjects went down to live in these apartments for about four weeks at a time. (Most of them reported that they enjoyed the experience and asked if they could be volunteers again.) Tests conducted on these subjects have provided some startling conclusions about the nature of biorhythms as they relate to electromagnetic radiation. Although their love cycles were not studied, these subjects' basic biological rhythms were. The people tested in the unshielded apartment showed a very different body-rhythm pattern than the people tested in the shielded apartment. The results are important for understanding the underlying principles upon which the love cycles are built.

When the natural pulses of electromagnetic radiation were blocked, the body temperature, the sleep/wake cycle, and hormonal and other secretions all became desynchronized. The result was akin to the experience of jet lag. In those shielded underground apartments there was no clear continuity from one person to the next in the length of their "day." Somehow they needed the radiation to keep their body rhythms "in sync."

In contrast, people who lived underground *without the shielding* did not suffer the jet-lag effects. For each person the various body rhythms synchronized into a predictable phase relationship, but instead of being synchronized to a 24-hour clock, they developed 24.87-hour cycles. The 24.87-hour "lunar day" became the dominant rhythm among individuals living in unshielded underground apartments. When electromagnetic radiation can pass through the earth, into the bodies of people living underground, their body rhythms synchronize to the rhythm of the lunar day. Although the details are elusive, the conclusion is not. There is an energy passing into our bodies that synchronizes our body to a 24.87-hour rhythmic cycle—the lunar day—when we cannot see the 24-hour sun rhythm.

HOW COULD THE FORCE BE TRANSMITTED?

For now, as an introduction to this complex radiation force of electromagnetic pulsation, it helps to think of it as the wave motion of radiating energy—energy that leaves a source such as the sun or the moon and moves out into space away from its own origin. Eventually this wave motion of energy gets to us. When it does, it may dominate our own biorhythms, as the strobe light does the dancer. It may help to understand what wave motion is by thinking about the motion of waves on the beach. Waves ebb and they flow as they beat against the sand and form the tides. Given time, the depth of the shoreline and the shape of the sand noticeably change as a result of the energy of the waves moving in and out upon them.

Other forms of wave-motion energy create electromagnetic radiation. Fields of energy within the electromagnetic spectrum form oceanlike waves that ebb and flow, traveling in space and impacting upon the living things which they bombard.

Although these cosmic rhythms are without substance, they affect the rhythms of our bodies. Let's look at the internal human machinery to identify something of the mechanical system that gets buffeted by those external forces.

THE INTERNAL HUMAN NETWORK

Hormones secreted in the gonads travel in the bloodstream to influence the workings of both the brain and the pituitary gland. Certain electrical and chemical impulses from the brain trigger the brain and the pituitary to release hormones that travel to the gonads and in turn stimulate hormonal secretion from them. Altogether the brain, pituitary, and gonads form an elegantly functioning interconnected sexual system. If the mind or the body triggers one of these three elements via a sexual thought, a pituitary or brain hormone secretion, or a gonadal hormonal secretion, the electrochemistry of the other two is inevitably affected.

The system is complex, and superimposed upon it are external forces. Cosmic rhythms affect all three of these critical love-cycle structures. These *external* rhythms (the electromagnetic radiation and the cosmic motion of earth, sun, and moon) significantly influence our *internal* biorhythms. Consider, for example, the hormone melatonin, whose role in controlling the internal synchronization of hormonal rhythms has been studied extensively in the last few years.

THE HORMONE INVOLVED IN INTERPRETING COSMIC TIME

Melatonin, a hormone produced in the pineal gland, has generated a great deal of interest for its potential involvement in the body's physical reaction to changes in the relative position of the sun and the length of the day.

Box 8-1: Melatonin Levels in Humans

Produced in the pineal gland in the brain, the hormone melatonin is a derivative of another hormone called serotonin (Brzezinski, 1988). But removal of the pineal gland does not eliminate the production and circulation of melatonin. Therefore, it must also be made elsewhere. Investigators have reported that in some animals melatonin is produced in the retinal cells of the eye; see Morgan (1989). Scientific investigation of the eye made sense because the amount of daylight seems to influence the secretion of the hormone.

Melatonin circulates in the blood, and its concentration can be measured by a laboratory test on a sample of blood. The hormone's metabolites can be measured in urine as well.

(continued)

A number of recent scientific papers note that the levels of melatonin change in synchrony with the times of year, the times of day, and the phases of the menstrual month. Abnormally low melatonin levels in sufferers of seasonal affective disorders and major depressive disorders suggests that melatonin is also involved in depression somehow.

The circadian (around the day) secretory rhythm of melatonin is the most pronounced of the rhythms. During the night melatonin levels rise four- to sixfold higher than daytime levels. The peak level of melatonin found in the blood is usually around 2:00 A.M. in both men and women. The normal nighttime elevation can be disrupted by jet lag or by abrupt changes in one's sleep/wake cycle. When a person makes a sudden shift in his or her day/night cycle, such as by shining bright light in the middle of the night, shifting work and sleep times (see page 95, "Sex Hormones and Sunshine," in Chapter 4), or through jet travel, it takes five to seven days to recover the normal nighttime elevation cycle of melatonin. The brighter the artificial light shined at night, the more suppressed the normal nighttime elevations of melatonin will be, until you turn out the light. Darkness makes the levels immediately bound back up. Bright light is being experimentally tested to see if it can help more rapid adaptation to the effects of jet travel.

In fertile-age women a seasonal influence on the monthly melatonin rhythm has been reported. This influence, restricted to geographical regions close to the Arctic Circle, appeared only on two particular days of the menstrual cycle (cycle day 2 and cycle day 10) and was represented by a slightly higher peak in winter than in the summer. The sex steroids of the ovary were lower in the winter in the luteal phase in this far-northern latitude of Finland. In other words, the ever-vulnerable luteal phase is the one stage of the monthly cycle where melatonin seems to inhibit ovarian sex-hormone secretion.

Recent research has shown that the follicular cells of the ovary tend to concentrate about three times the circulating blood levels of melatonin. Therefore, it seems that one influence of melatonin secreted in the brain may be mediated at the ovary. How it does so remains to be discovered.

There are seasonal effects on the circadian melatonin secretion as well. Although the peak levels were found in both winter and summer in Australia and in Finland, the timing of the peak is later in winter, according to studies published in Finland. One other seasonal effect has been clearly noted: The early-morning (8:00 A.M.) plasma levels of melatonin are higher in January and July than in April and October, according to a study published in Switzerland. Thus in these studies the spring and fall have *trough* levels in common while the winter and summer have in common *peak* levels of what seems to be the "timing" hormone.

The investigators believed that the amount of light was responsible for these seasonal changes in melatonin. To me, however, it seems more likely that *two* different earth-sun relationships combine their con-

(*continued*)

trol: (a) the position of the earth on its axis (time of day/amount of daylight); and (b) the earth's distance from the sun regardless of which hemisphere one lives in.

The elliptical orbit brings the earth closest to the sun in April and October and takes it farthest from the sun in January and July.

Since January and July do not have the same length of day but do have the same relatively large distance between earth and sun,* the discovery that these dates share peak melatonin levels of the year suggests that orbital position rather than day length is pivotal to melatonin changes. The April and October months share both day length *and* earth distance from the sun—and they both show trough levels of melatonin for the year.

*Actually, in January the sun is three million miles closer to Earth than in July, when it is farthest. However, these opposite stages of season reach their semiannual peak distances apart in their orbiting journey.

These studies in Box 8-1 suggest to me that melatonin is a significant hormone in its communication of day versus night, or motion of the earth with respect to the sun. Future research should elucidate how the mechanisms (e.g., pineal gland and gonads) mediate melatonin's effects.

Thus I conclude that cosmic motion has a powerful influence on the hormonal responses of life on earth and that the three influences—circadian, monthly, and annual—transmit themselves via melatonin on the human condition.

SUN OR MOON: WHICH IS DOMINANT?

The evidence seems to point to the moon as the dominant force in the pulse of human love cycles. It is not that the sun is irrelevant. In fact, annual cycles of sperm production, sperm motility, and testosterone level have been clearly demonstrated in men and male monkeys. In other words, different mammals—from rats, to monkeys, to humans—have been studied in different ways, but all reveal the influence of the sun and moon on sexual cycles. Sexual cycles of rats revolve to the rhythm of the sun, whereas the sexual cycles of monkeys and humans seem synchronized to the timing of our moon's revolution around the earth. Although the evidence of physicists from the experimental underground apartments does help to explain something about the force of the moon's (24.87-hour) daily rhythm on people, the data are incomplete. Science can penetrate a little of the mystery, but the magic remains.

Other evidence for lunar effects on humans supported my growing sense of a cosmic role of the moon in the ecology of life. The reproductive cycle of the female of the species is, without question, the central factor in the survival of each species. Reproductive cycles stimulate our love cycles, and our love cycles promote our survival as human beings. Our bodies and our minds are designed for sexual loving and for its natural result—pregnancy and gestation by the female (with support and protection by the male) and the continuation of the species. Just like in our jet travel, we can appear to overpower these rhythms. In our reasonable desire for fuller lives we can delay and postpone our loving, but the pulse beats on—and we always pay a price when we try to dominate our own biorhythms.

CONCLUSION: THE MOON AFFECTS HUMAN BEINGS

The compelling mechanism appears to be the electromagnetic radiation that moves through the atmosphere and into the earth. Whether the moon is the source of this radiation or merely a reflector of it has not yet been determined. What is clear is that the cycle length of the moon (every 29.5 days, every 24.87 hours) coincides with certain hormonal- and sexual-cycle patterns in human beings.

Why should this matter? Once we begin to perceive that the motion of the earth, the electromagnetic radiation coincident with the moon cycle, or the timing of the transition from waxing to waning moon is played out in the rhythm of our bodies, we can begin to understand the value of finding harmony with these rhythms. Imagine falling off a sailboat in rough ocean waters and floating on top of the waves until they bring you onto the shore. Compare this action of the coming into harmony with the natural forces of the ocean waves with the person who starts fighting the waves, flailing his arms, trying to stay on top in the air, and panicking. Such behavior will rapidly lead to exhaustion, drowning, and death. So it is in love cycles: the science of intimacy. When we understand the tremendous power of the external forces on the rhythms of the sex-hormone secretions of our bodies, we begin to be able to "ride the waves" instead of fighting them. And when we learn how to ride the waves, we can more easily get to where we want to go.

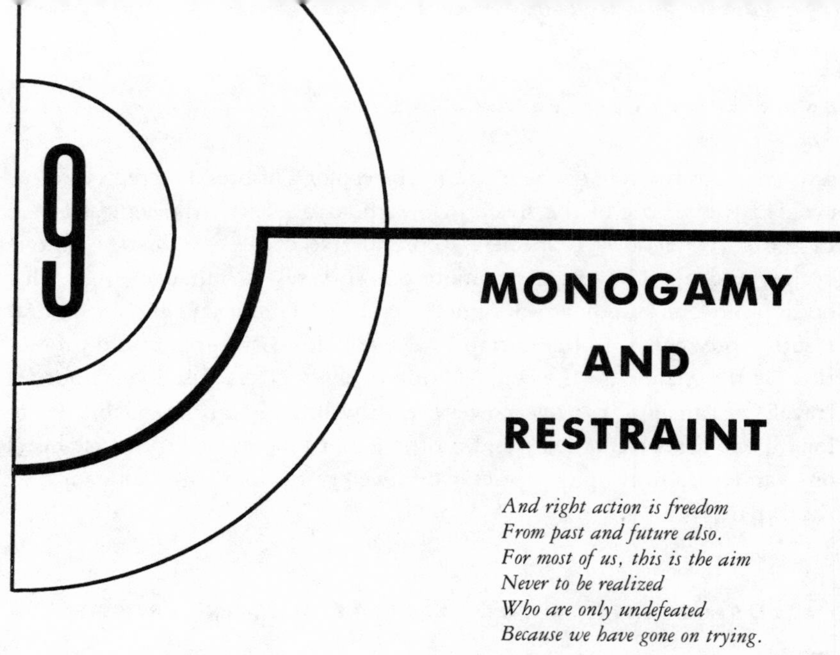

9 MONOGAMY AND RESTRAINT

And right action is freedom
From past and future also.
For most of us, this is the aim
Never to be realized
Who are only undefeated
Because we have gone on trying.

—T. S. ELIOT, "The Dry Salvages," *The Four Quartets*

Human cycles of love, sex, intimacy, and sensuality are not confined to the privacy of the bedroom. They reverberate in the patterns of the culture. They define the social life and the work order. They determine how clerics, therapists, and physicians counsel their penitents and their patients. They profoundly influence an individual's spiritual, psychological, and physical well-being.

A full range of sexual life-styles in different civilizations is well documented in both the historical record and present-day cultures. From monogamy at one extreme to complete promiscuity at the other, wherever reproduction derives from sexual contact, some variation in the sexual roles of male and female can be expected.

Because scientists cannot experimentally manipulate human lives to test for outcome, I need to combine the scientific analysis of data with the perspective of educated opinion in order to form global conclusions. Even so, a thoughtful consideration provides a clear message. My research on the health and well-being of women has made it ever more clear: Sex is a precious resource.

SEX IS A PRECIOUS RESOURCE

Any behavior powerful enough to alter the hormones and promote health is a resource to cherish. And, as with most resources, sexual ones are often squandered. Still, the individual who treats resources as precious, who preserves them, usually enjoys a sense of security, self-confidence, and spiritual wealth.

The message extends both to couples and to single people, to men and to women. For those who choose it, the practice of monogamy and restraint can produce a rich, healthy, and loving partnership—the ultimate love cycle.

THE NEED FOR AN INTIMATE PARTNER

Whatever the duration, the need for an intimate partner is real, and it is pervasive. In 1990, publishing a scholarly chapter in a new book, *Emotions in the Family,* psychologists Elaine Hatfield and Richard Rapson reviewed the results of their interviews with thousands of individuals. From dating couples to newlyweds to elderly people, almost all listed a similar range of benefits and rewards they gained through a committed intimate bond. Let's take a close look at what monogamy and restraint can provide.

Box 9-1: The Rewards of Intimacy

Personal rewards: A partner who	Is an asset in social settings Provides intellectual stimulation Provides an attractive image to enjoy day after day.
Emotional rewards:	Liking and being liked Loving and being loved Understanding and concern Acceptance by a partner (which leads to the freedom to grow) Being appreciated Physical affection (touching, hugging, and kissing) Sex Security The pleasure of sharing plans and goals
Day-to-day rewards:	Financial stability Sociability A partner to share decision making and accept a fair share of the daily responsibilities Being remembered on special occasions

Monogamy and commitment can provide many rewards that are unavailable to those who live a "freer" life-style. Both partners benefit: The woman gains because she is more likely to have regular sexual contact in a committed relationship. This regulates and perpetuates the regular cycling of her hormones, preserving and promoting both sexual and reproductive-system health. The man also wins—the regular cycling of her hormones affects his own hormonal dynamic. And he gets the opportunity to experience for himself the symphonic complexity of a woman. Together both man and woman gain the complex personal, emotional, and day-to-day rewards that are shown in Box 9-1.

There is a catch. The greatest rewards come when monogamy and restraint are freely chosen. But free choice is possible only in those cultures that provide equality of opportunity for its men and its women. Most do not. History shows that in a feudalistic environment, where the physical survival of the group is at risk, the strongest men often garner for themselves a bevy of wives. Wherever brute force controls culture, the freedom of its women is impaired. A more subtle force is seen in the control of money. It, too, can limit freedom.

The comments that follow are not relevant to cultures in which slavery, forced servitude, and inequality of the sexes cannot be overcome. But where there is freedom, there is choice. Where there is choice, monogamy and restraint allow access to an enriched life-style for both women and men.

ECONOMIC SOLVENCY OR SLAVERY?

Women in the 1990s have an opportunity that women in the 1950s and 1960s usually did not have. Economic independence is now a real option for women. Solvency of the individual as well as the couple is a worthy goal.

With economic solvency comes a new level of freedom, a different negotiating process in either the search for a partner or the continuing relationship with a current one. Economic solvency permits a woman to spend her money on actions that promote the loving quality of her life. But solvency is tough to attain. The possibility of widespread female solvency is a product of the women's liberation movement but has been precariously achieved and by only a small proportion. Still the possibility now does exist. Consider conditions just thirty years ago.

During the 1960s and 1950s a single woman could search to maximize her condition, but this usually required marrying a good provider. Once tied down with children and lacking a way to earn the minimum married standard of living, she was dependent on her husband's continued goodwill toward her. This economic dependence permitted a kind of slavery. The relationship lacked a fair balance of power. Men expected to have a full family *and* an independent economic life. Women usually were not allowed nor did they expect to "have it all." Under such conditions monogamy and restraint provided a woman's major tool for leverage. By withholding sex until she secured commitment, she could negotiate for marriage. And marriage brought a legal commitment that provided the highest level of economic security that the culture permitted for women who wanted to nurture children. As long as the relationship held, she could expect support. If a divorce became necessary, she could sue for lifetime alimony.

BIOLOGY SETS THE LIMITS

The society of the 1990s appears to offer more options for women. As women soon discover, it is their reproductive biology, not their partner's domination, that usually interferes with their freedom of choice. In the 1990s, as in the 1950s, monogamy and restraint still maximize the opportunity to develop and maintain love cycles. Monogamy and restraint permit a loftier kind of love relationship than the self-seeking, "me-first" pleasures so widely touted in much of the media, but there is a price.

Getting there takes time, self-control, and sustained effort. In the journey from first meeting to fulfilling the potential of an elevated long-term relationship, monogamous partners can learn to synchronize their individual sensuality cycles through mutual effort, communication, and commitment. Monogamy is an exclusive contract between partners, one with tremendous possibilities for a life of high art. It is a journey played out within the movement of the individual throughout the days of her or his adult life, with its origin as counterpoint in the journey all life makes in moving through the cosmos. Movement and change occur even when we perceive that we are standing still. Becoming aware that one is in motion can elevate the cognition.[1] With an elevated perceptual framework one can use the sense of stillness, which some call boredom, as an opportunity for rest and renewal, a chance to shift from major to minor key.

The ideals of monogamy and restraint may seem out of synchrony with

[1] What I'm describing is the perceptual difference between hedonic pleasure of the moment versus seeing one's life as a journey—comparable to the cosmic journey.

a culture in which sexual excitement and novelty are held up as icons. But now, more than ever, there are sound biological arguments to be made in favor of sexual, emotional, and physical exclusivity. With the AIDS pandemic and the extraordinary increase in sexually transmitted diseases, prudence protects health.

PHILOSOPHY MEETS BIOLOGY

Philosophy meets biology in the message of monogamy. *Monogamy* is not synonymous with *monotony; restraint* is not another way of saying "stuck in a rut." In fact, monogamy and restraint bring the rewards of good health, emotional stability, and nurturing support. Monogamy and restraint enhance health and well-being.

Even so, periods of unrestrained and nonmonogamous behavior are common and maybe even optimal at different life stages. Consider that more than half of teenagers are sexually active while they are yet in high school (see Figure 7-2) and that most do not marry their first lover. Adolescent activity provides important experience. It forms the basis by which young people learn to develop sound judgment. It leads to a familiarity with the laws of cause and effect. The problem for adolescents and for those who guide them is in trying to decide where to draw the line between self-defeating promiscuity and necessary life experience. In spite of family and other social pressures, we each decide for ourselves. At various stages "serial monogamy" is a common solution. Defined pretty simply, serial monogamy refers to intimacy with one partner at a time.

For some the duration of a commitment is measured in years, for others in an evening. The facts are that nature discourages human promiscuity and rewards stable sexual behavior. For women regular weekly sex (which is more likely over the long term) promotes higher levels of estrogen and longer-length luteal phases during their twenty to thirty premenopausal years. Consider one recent example of the way this can affect a woman's health.

In 1991, in the *New England Journal of Medicine,* Dr. Jerilyn Prior published the first demonstration that in *young* athletic women the proportional length of the luteal phase of the menstrual cycle was directly proportional to the changing density of the bones in their spine. The longer the luteal phase, the better the bones. The shorter the luteal phase, the more bone was disintegrating—even in their twenties.

Since the pattern of woman's sexual behavior can affect the length of her luteal phase (see pages 16, 135–136), I conclude that the density of her bones is vulnerable to the choices she makes about sex. If she has sporadic sexual behavior, shorter luteal-phase lengths are likely. Weaker bones can be expected. According to 1990 worldwide data, 40 to 50 percent of fair-skinned women can expect to fracture a bone by the time they reach their sixties. Half of these will be crippled for life.

For both men and women the studies suggest that there are multiple hormonal benefits to a regular weekly sex life.

For women:

- Higher levels of estrogen are associated with lower levels of cholesterol and better cardiovascular health.
- Longer luteal phases appear to reduce the risks of fibrocystic breast disease and uterine fibroid tumors, conditions that affect close to 35 percent of all women by the age of forty-five.
- Higher levels of estrogen also improve the health of the urogenital tissue: less cystitis, stress incontinence, and vaginal atrophy.

For both men and women:

- Monogamy provides the space and the time to develop one's expertise. One can really learn the unique sensual "wiring" of one's partner.
- Regular sexual contact is available when partners make mutually exclusive commitments.
- Their sleep/wake cycle will be maintained, assuring the capacity for optimal concentration at work.
- The potential for sexually transmitted diseases will be minimized.
- The energy of the individual can be directed away from the time-consuming search for a sexual partner and toward enhancing the lives of others.

Monogamy and restraint are as valuable for single people as they are for those already in a committed relationship.

SEX IS NEUTRAL—EITHER GOOD OR BAD DEPENDING ON THE CONTEXT

Depending on how it is used, sex can offer a magnificent enhancement or a devastation. It has major power either way. Sex can serve for health, eros, trust in intimacy, and bonding. Or it can titillate, disrupt, and destroy. Biology is neutral, unemotionally playing out its rules of cause and effect. When you understand the rules before the game begins, you have an opportunity to choose to play like a winner. If you have to bumble along and discover the sexual rules of cause and effect through trial and error, chances are you will suffer many losses before you learn how to win. Even when you try to act like a good sport, losing is painful.

The principles of Western religion and of biology converge on the subject of monogamy and restraint, but there is a difference. Although religions and other moral ethics may set down the rules, teaching that certain behaviors should be avoided because they are sinful and others engaged in because they are spiritually healthy, biology teaches its lessons without the labels. Cause and effect, action and reaction—some sequences are just that inevitable once they have been set in motion. Consider first the bad side of sex.

SEXUAL INTERCOURSE DURING MENSES MAY ABUSE THE BODY

Both among young women and among those who are approaching their menopause, abstaining from sex when the uterus is bleeding appears likely to promote the health of the woman. For young women it may reduce the risks of endometriosis (see page 69). For those in their mid to late thirties and beyond, it may reduce the risk of being prescribed a hysterectomy.

During the seven premenopausal transition years blood flows differently than before. As women approach menopause, some bleed more lightly than their usual pattern. Others bleed more heavily. Some have longer cycles. Others have shorter cycles. All of these patterns have been categorized as being within the "normal" range of bleeding change. But heavy bleeding can lead to the prescription for a hysterectomy.

If a woman is bleeding more often or more heavily, she suffers a loss of energy. If her physician is not aware of the recent studies that demonstrate the value of the uterus—in sexual function, cardiovascular health, and hormonal stimulation—he or she may suggest solving the problem of bleeding by removing the organ that bleeds. The United States has five times the rate of hysterectomy of the six nations in Europe for which data have been gathered. Heavy bleeding accounts for a significant fraction of these hysterectomies.

THE STANFORD STUDY

During my postdoctoral studies at Stanford University I collected data from fifty-six perimenopausal women about whether they did or did not abstain from sex when they were menstruating. The group broke down just about fifty/fifty. Twenty-nine said they did have intercourse while they were menstruating. Twenty-seven said they abstained when they were bleeding. I compared these data with the bleeding changes the women were experiencing as they were approaching menopause.

Results: Sex during menses appears to increase the risk of heavy bleeding. I found a *highly significant difference* in the incidence of increased bleeding that went with a behavior pattern of sex during menses.

- 66 percent of the women who had menstrual sex habits were excessively heavy bleeders at perimenopause.
- 15 percent of the women who abstained experienced heavy bleeding as menopause was approaching.

In other words, women who had sex when they menstruated had four and one half times the rate of a heavy bleeding pattern as menopause approached.

Those women who abstained when they were menstruating had nine times the incidence of a lightened blood flow during menstruation as menopause was approaching.

My theories and these preliminary studies cannot prove that sex during bleeding causes problems. But these are the only data that I have seen which address the possibility. Until more studies are reported, restraint during menses may be a wise course of conduct. This would mean abstaining from orgasm, which might trigger uterine contractions. Other forms of problems with restraint are better documented, such as the problems of sexual abuse.

SEX CAN BE DANGEROUS TO YOUR WELL-BEING

Sexual abuse can take many forms—physical, psychological, and social. Anal sex is physically abusive sex. It is simply a fact of biology. Anal sex abuses the intestinal lining of the receiver, triggering tears and cuts, resulting in a variety of intestinal problems that at the most extreme increase the propensity for fatal diseases. Considering that one in three women polled (see page 185) acknowledged engaging in anal sex, these facts should become more widely taught.

Sex can also be psychologically abusive. A prime example: pornographic films that portray women in roles that damage them. Coercive sex is a more subtle form of sexual abuse. "Date rape" with an unwilling partner is physically and psychologically abusive. Among educated young women (in colleges that collect the data), more than 75 percent have reported experiencing attempted coercive sex by their dates, even when they clearly express an unwillingness to be more intimate. Typically in such cases a young woman who consents to a kiss finds that her date assumes she has consented to intercourse, even when she clearly states she is unwilling. According to a 1990 article about a Rutgers University study, two thirds of these unwilling women managed successfully to fend off the advances. Still, 22 percent of the college women polled acknowledged that they had been raped by a date. Similar data were reported at Swarthmore College. Coercive sex is abusive to women. Even if the arousal mechanism has been activated, when willingness does not follow, forcing what she is unwilling to consent to is low-level behavior: rape.

As described in Chapter 7, it is the women who learn that they must exercise the restraint. They must acknowledge the reality of their culture and recognize the danger. Where there is danger, wise women will avoid the appearance of availability for sexual intercourse by exercising caution until they have decided to be available. And when they decide to be available, they will define what is and what is not acceptable sexual activity.

This isn't just a matter of morality or common sense, it is a biological reality. Men have about 25 percent more muscle mass than women. Their bones are heavier, their limbs longer. They are usually stronger and faster. Most men can overpower their dates with the use of force. A wise woman, in touch with these facts, will anticipate and protect herself from danger. Sexual arousal can be dangerous. Smart women explore new relationships slowly and cautiously.

In the proper context a stronger partner can serve as a protector, and sensual pleasure is safe. In the proper context both sensual pleasure and a stronger partner can conserve, and enhance, a woman's health and her integrity.

SEX AS AFFIRMATION OF LIFE

As discussed in the earlier chapters, a woman who engages in regular weekly sex receives a variety of benefits to her endocrine system. And these hormonal effects serve her health and her appearance. A healthy reproductive system reflects a healthy physical system. It shows that her hormones are cycling in ways that promote the strength of her bones, the health of her cardiovascular system, an increased joy in life, and abundant energy. But sometimes a partner is not available; sometimes celibacy is a reasonable choice.

THE REWARDS OF CELIBACY

Celibacy is *not* a penance or a punishment. It represents an active choice, not a passive position. Restraint and timing count. In the search for a partner, restraining one's desire for attachment is a strategy that yields the highest rewards. It releases time—time to allow the other's true character to emerge, time to test one's own feelings, time to establish the friendship. It can speed the search for a partner who is willing, ready, and able to commit.

Restraint can preserve a woman's hormonal health. In women (and possibly in men) it quiets the chaos of an overstimulated neuroendocrine system. It provides a regenerative, healthy respite. And it provides time for developing one's attractive qualities.

Taking action—to develop a friendship, to end an unwholesome affair or a string of brief connections—often means choosing celibacy for a while. Contrary to some of the media hype, celibacy is a positive action, a free choice. It is usually found to be peaceful after one, two, or three menstrual cycles have passed. Once the endocrine cycle restabilizes, women usually discover that they feel well, often better than they have felt in a long time. I suspect that the more chaos that preceded the onset of the celibate time, the longer it takes to calm down the neuroendocrine system. Celibacy does calm the neuroendocrine system down. The hormones cycle in a different rhythm, and there is a shift to a more gentle state. According to my research, the hormone pattern of the celibate woman was better than that of the sporadically active woman. Her estrogen was lower than among weekly active women, but not perilously low. Among sporadically active women, about half had such low estrogen levels in their twenties that their hormonal pattern looked like that of a menopausal woman.

Somehow there is magic in the mechanism. For women a period of celibacy seems to serve and indeed preserve the body. The nervous system and hormonal status of the celibate person should promote a state of being that permits *judgment* to override passion.

How about celibacy for men? The mechanisms may be similar. Celibacy seems to lower testosterone in men, whereas regular sexual activity or its anticipation seems to raise it (see pages 107–8). Whether this lowering affects the emotional state has not yet been evaluated in scientific reports.

Consider this: Calm people enhance their capacity to intelligently observe the romantic partners they encounter.

Finally celibacy "gentles" you. Being gentle is a good way for single people to encounter new people. By exploring for friendship first, you take the time needed to know another person before deciding on more intimate connections. Be aware. What you see is not what you get, because love, whenever it comes, changes people. Move slowly and you give yourself time to evaluate before you risk intimacy.

Since sex is so important, a potential new partner needs to be carefully evaluated. One's very life is as much at stake in the nineties as it was in the fifties by the choice of bed partners.

In maintaining high standards and being selective, a woman demonstrates the value she places on herself. And this message is communicated to the men who approach her. She commands the respect of someone of high value.

WHO SHALL LEAD? THE LEADER IS LED BY THE FOLLOWER

It appears to be built into the design of human biology that the female sets the limits as cultures evolve to a civilized state. Society and culture may be male-centered, but in the power dynamics of sexual connection it is the woman who has the power to set the pace, define the limits, and regulate access to her person. A woman's true power lies in her capacity to be selective, to decide with whom she will spend her time, with whom she will consort.

The power to lead and to follow in the intimate dance has its counterpart in biology. Given the design of reproduction, the animal with the thruster generally makes the advances for a successful *coital* connection. However, the animal who is the receiver has power too. She decides whom she will receive. The power is reciprocal. A female has the power to lure and she has the right of veto.

A male will not know that a female is available unless she gives off clear signals. It is part of the dance of intimacy. Hers is the power to lure. His is the power to approach. Once she has been approached, it is the woman who sets the pace. It is a very powerful control that females of all species exert.[2]

Two scientific experiments convinced me of the fundamental nature of this biological pattern: the Rape Cage (a feudal society) and the Male-Tethering Experiment (a balancing of power).

THE RAPE CAGE

In the mid-1970s, while attending the annual Conference on Reproductive Behavior, I was unsettled by the approach of the mostly male scientists to the study of sexuality in small mammals. From their perspective *only* the males were in control of the sexual dynamic. At the University of Pennsylvania, when I asked the experimenters about the female rats—whether they had preferences or experienced orgasm—their answer was that sex is a male prerogative: Only males choose; only males experience orgasm.

What I could see when I observed the sexual behavior in the cages or reviewed the videotapes was an experimental method I privately named the rape cage.

There was a cage with a large male rat in it. The experimenter dropped a female into the cage. The females were half the size of the males. They looked terrified to me—all alone in a small cage with a giant male.

Within seconds the male always did the same thing. He pounced. He mounted the female from the rear and began his copulatory sequence. Meanwhile she froze.

Immobile, her tail was deflected to the side, an action that permitted the penis to enter her vagina. The sequence would continue in the usual rat style: ten or twelve thrusts, dismounting, female frozen, the male circles her five or six times, then pounces and begins the sequence again. After three to eight "copulatory bouts" he would ejaculate. And the experimental rape was over.

I was dumbfounded. I listened to the explanations the scientists gave me for what they thought they were observing. They believed they were studying male-rat sexual behavior. They were measuring things such as how many thrusts it took before ejaculation occurred, how many copulatory bouts it took to produce pregnancy. To me they had set up a rape cage. I asked them whether all males behave this way, and the answer

---

[2] Not to say this can't be turned around. But I believe that when it is, the sexual dynamic is different. And that, with experience, most women (and men) find the dynamic not to be what they want.

was no. Only the "properly behaving" males, called stud males, were selected for these experiments.

I asked about females. The answer was, "They *always* behave the same way." This was evidence to the male psychologists of the general principle that the females got no pleasure from sex. I thought the experiments in rats were important for understanding anatomical and physiological fundamentals, but I did not agree with the perspective the male researchers were giving to their results. Neither did another woman who was in a position to demonstrate why, with her own experiments.

THE MALE-TETHERING EXPERIMENT

A few years later a remarkable scientist added a new perspective. Dr. Barbara Fadem, then a graduate student at the Institute for Animal Behavior and now a professor of psychology at Rutgers University Medical Center, showed what was really happening with the female. Her presentation of an experiment at the Conference on Reproductive Behavior was remarkable because it changed the perspective of the scientific community.

Dr. Fadem reasoned that in order to study the sexual behavior of rats, it was important to create conditions that mimicked their natural environment. Instead of a small, bare wire cage, she set up a larger cage with a "seminatural" environment that had nooks and crannies. She put a little leash around the neck of the male stud rat, tethering him to a clothesline suspended at the top of the cage. The male could run around most of the cage. But the female had places to go where she could get away from the male—places where the leash did not permit the male to reach.

This environment was closer to what females experience in nature. Free-living females usually do not have to be alone with a male. They are free to hide in small spaces where the male cannot enter because of his size. Under these experimental conditions the females demonstrated power, a reciprocal control of the sexual dance. They wiggled their ears. They darted and hopped. Then they ran away. They teased the males. They jumped into the male space, then jumped away.

What Dr. Fadem showed was that, given the opportunity, the females exercised choice. They had preferences. They gravitated toward some males and away from others. Each female chose which male to lure and she chose when and with which male she would mate. When the males were tethered, the females had some control. They engaged in sex with the male, but only when they both wanted to.

This experiment reflects a subtlety in nature, in the power dynamics of male/female sexual connection. The female sets the limits among the males who want her—as long as she's not held within a "rape cage." This power can lead to negotiating for monogamy, a choice that can enhance the quality of life for those who choose it—a choice that is sustainable as long as each individual continues to provoke that wanting in the other.

MONOGAMY AND RESTRAINT IN THE LONG-TERM COMMITTED RELATIONSHIP

When couples commit to monogamy, the man may actually have the best of the deal. He gets access to an instrument capable of a hormonal symphony, one that he can learn to manipulate to his own never-ending fascination. Rather than having to endure a boring sameness because he himself fails to penetrate the complexity, he can study the instrument and learn to make beautiful music—which he can enjoy. A man who wants to please—and to continue to please—a hormonally cycling woman is engaged in a complex process because the woman's needs and sensitivities keep changing. Slow strokes and gentle caresses may please her during one phase, more urgent strokes and bold caresses at another. She is many women in one. Even when hormone cycles give way to steady-state hormones after their fifties, women can continue to set the pace.

THE SEXUAL CONNOISSEUR

Depending upon the level of arousal, the subtleties of neuronal firing, sensations, and perception will vary. In a new romance, one filled with the excitement of discovery and the optimism of new lovers, almost any kind of clumsy bumbling will probably serve arousal. What someday might irritate, may now titillate. The early perceptions may be mostly projections, assumptions that turn out to be false. The early passion can indeed be temporarily blinding. Individuals can be so filled with the joy of a new connection that the fine points of sexual interaction cannot yet be addressed. Our nervous systems have limits. They can process only a limited amount of information at a time.

Once the relationship settles down to a stable pattern, the timing subtleties will be ready for development. The reward: a highly refined sensual interaction for couples who want to experience it. It is rather like becoming a connoisseur. Before one first learns about fine dining, one tends to miss the complex and subtle flavors. As the palate becomes more sensitive, one becomes more demanding, harder to please. So it often is with sensual pleasures. Sensuality means different things at different times in a person's life as well as in the life of relationships. This marvelous reality can enhance the loving experiences of couples because it is ever changing. At times sexual connection can be a secondary aspect of relationships, quickly completed yet ever present. At other times its importance can bloom. Sophisticated lovers continue to learn. They do not need to settle into routines they find boring. They can reach for the *flavor*.

So, a lack of refined polish is okay for beginners. It can provide a source of humor and fun. And it is an almost inevitable process until the couple get to know each other. For those who commit to monogamy and restraint, "knowing" can be a lifetime process, improving with the growth and the creativity of the individuals.

Couples who monogamously commit themselves can discover how the symphony works. People who keep flitting from one to another never really get to penetrate the complexity of the cycle, and in consequence they may fail to experience the astonishing range of sensuality of which a woman is capable.

THE SENSUALITY CYCLE

Sensuality does cycle: throughout the stages of life, throughout the days of the month in menstruating women, and throughout the course of changing relationships. Each couple must find their own rhythm, and in order to enhance that cyclic rhythm, they need to learn the facts of male and female anatomy and physiology, as discussed in Chapters 3, 4, and 7. Consider touching. To expect touching to be right—without communicating what you want—is to invite failure. People vary, and each requires an individual evaluation and approach. In the safe harbor of a long-term commitment the needs and secret desires can find the time for expression—and possibly for fulfillment.

SLOW, MUTUAL EXPLORATION—AN EXPLORATION THAT TAKES PLACE ACROSS THE YEARS—KEEPS SEXUALITY FRESH

Boredom is not inevitable, but it is also not always avoidable. What can keep boredom from taking over is combining a mutual desire to please each other with an awareness of the inevitability of cycles in the sensuality of each partner. This awareness may sometimes be in the forefront, other times relegated to the background. When couples address the complexity of growth, helping each other to develop meaningful work, recreation, and/or parenting children, then life can be richly complex. One of the elements that make it rewarding is the knowledge that your efforts serve your partnership. When love partners learn to *appreciate* each other, they continue to cultivate the relationship. To maintain the strength, health, and vigor of the partnership, they need to commit to monogamy. Slow, mutual exploration—an exploration that takes place across the years—keeps sexuality fresh. We may continue to have the same partner, but our bodies and mental capacities are ever changing. The imagination, the word play, the images of shared repertoire can serve your intimacy.

Courtesy also counts. When two people share a bed, they enhance their sensual experience by considering what they bring to it.

Sexual dalliance for variety is not a solution. It leads to suffering; it skims the cream away from the partnership and dissipates the energy. When one sets out to overcome boredom with dalliance, one is moving toward a dead end. The lies that become necessary destroy the trust. The primary relationship is compromised; and the cost is paid out over the years that follow. It doesn't work in the long run, and in the short run it does tremendous damage to all members of such triangles.[3]

Monogamy offers true freedom, releasing the spirit for growth because the fundamental partnership provides a solid footing. It is liberating, enriching, and enlightening. Consider this: Self-confidence enhances sexual expression, particularly when one is coupled to a loving partner who understands one's unique sensual wiring. Self-esteem grows through a gradual forging of mutual trust. How can you believe you are worthy, how can you take risks, if your lover is in someone else's bed or treating you disrespectfully the next night? That idea being accepted, a question follows.

WHAT ABOUT TEMPTATION?

Sophisticated people recognize the relationship between arousal and the ensuing inclination for willingness. Chapter 7 describes it in biological detail. With that sophistication a wise person who wants to maintain

[3] For those caught within this difficulty, a helpful guide out is available. See Frank Pittman, M.D., *Private Lies* (New York: Norton, 1990).

monogamy and restraint uses his or her energy to prevent temptation. Consider the sensuous state of sexual arousal—blood flowing, nerves tingling, panting, and so forth. It takes less energy to avoid temptation than to fight arousal.

Although former president Jimmy Carter admitted that he had experienced temptation—"I lust in my heart," he once said in a self-castigating reference—I have a different view. The faithful wife or husband on a business trip who lusts for the steward(ess) doesn't damage the marriage; they do if they go out for dinner and share intimacy. Acknowledging temptation and giving in to it are *not* the same. Attraction and action are very different. One behavior preserves and the other behavior violates monogamy.

When one recognizes the relationship among libido, arousal, willingness, and sexual fulfillment, one chooses the terrain in which these elements can be healthfully experienced. The reward is wonderful. The benefits that accrue to those who achieve restraint include the possibility of an enriching and deeply loving partnership. The capacity to look your partner in the eye and truthfully say, "You can trust me, I honor my commitments": What a beautiful gift true lovers can give.

Monogamy and restraint keep eros alive. How can this be so, in a culture that is so relentlessly in search of the new? Enlightened partners do not search for the new sexual connection outside the partnership. They renew themselves regularly. Enlightened partners who have chosen monogamy thoughtfully consider how gracefully to balance the energy in their personal lives with that of their professional lives. The same principles are the marks of both a thriving dynamic career and an intimate partnership. They include:

- Staying challenged
- Keeping control
- Maintaining a sense of competence
- Staying connected and in communication
- Preserving time for rest and renewal

A TIME APART

By establishing regular times for sexual abstention—"a time to refrain" —couples can enhance the quality of their "time to embrace." Why? A regular time apart allows each partner time for inner reflection, a chance to appreciate the partner and to look forward to the other's return. If, when alone, we take action to avoid temptation, we preserve our energy. We allow ourselves the time and the space to rest, restore, and enhance our appreciation of our partner.

TO APPRECIATE MEANS "TO ENHANCE THE VALUE OF"

Appreciation of those near us means finding what is valuable and—as in the cultivation of a garden—providing the nutrients that make it grow. When we appreciate our partners, we enhance the relationship. I see it as a two-pronged cognitive action:

- Recognizing the value of one's partner enhances one's own sense of the partnership.
- Expressing that appreciation enlarges the joy of the recipient and echoes back to the donor.

It takes practice. It leads to joy. Imagine the garden you can grow if you use the time apart first to restore your own energy and then to focus that energy on plans for cultivating the partnership.

To be most effective, the retreat should be regular. The monthly cycle and its menstrual-flow days can be used as the clock to time the retreat during the years that a woman is menstruating. Women who do not menstruate can use the moon. Psychological and physical benefits can result. For some women abstaining from sex during menstruation may prevent endometriosis. For others it may prevent the excessive bleeding that is characteristic of close to half of women in their forties as their menopause approaches. In both cases the restraint of regular times apart can serve to enhance the luster of monogamy.

CONCLUSION

Philosophy does meet biology in the message of monogamy. Both for physical and for psychological health a time to refrain from sexual activity may serve both partners. Within the context of a loving relationship a time apart is usually followed by a renewed hunger for one's lover. Something like the spacing of meals to develop an appetite or the vacation that serves to renew the spirit, a sexual refrain can heighten romance.

Biology affects our destiny. Biology is neutral. Without "moral" imperatives, it plays out its rules of cause and effect. The lessons of history and science teach that all life elements are in flux. The earth rotates on its axis every day as it journeys on its annual orbit around the sun. The moon moves through its monthly orbit around the earth, and all heavenly bodies return to their origin before setting out again. The great paradox is this: that nothing—and everything—stays the same. Human beings evolve, societies rise and fall, intellect is amassed, manners and customs are acquired and discarded. Biology remains the constant. Its lessons have a true relevance, then as now, because they are geared to the perpetuation of life.

Truly committed partners create a universe into themselves. Their capacity for continued growth and satisfaction is limited only by their capacity for imagination.

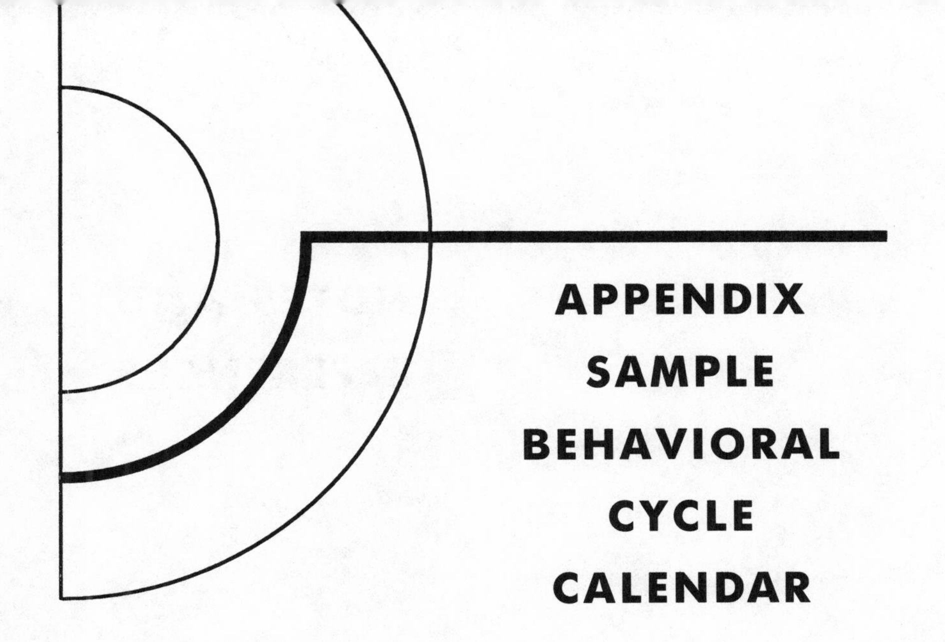

APPENDIX SAMPLE BEHAVIORAL CYCLE CALENDAR

NAME:____________________________ AGE:______ CURRENT YEAR:____________

Month	1	2	3	4	5	6	7	8	9	10	11	12	13	14	15	16	17	18	19	20	21	22	23	24	25	26	27	28	29	30	31
JAN																															
FEB																															
MARCH																															
APRIL																															
MAY																															
JUNE																															
JULY																															
AUG																															
SEPT																															
OCT																															
NOV																															
DEC																															

NOTES AND REFERENCES

A NOTE TO THE READER

The system of referencing used in this book has two parts. Each chapter lists in normal, scientific fashion the scientific papers by author, year, title of article, name of journal, and volume and pages.

In front of this alphabetical list, however, is a system of notes to indicate which references are relevant to chapter topics in those cases where the chapter topics have not already listed the name of the first author in an unambiguous fashion.

PREFACE

My four prior books are the following:

Cutler, W. B.; Garcia, C. R.; Edwards, D. A. (1987) *Menopause: A Guide for Women and the Men Who Love Them,* 1st ed. (New York: Norton).

Cutler, W. B., and Garcia, C. R. (1984) *Medical Management of Menopause and Premenopause: Their Endocrinologic Basis* (Philadelphia: Lippincott).

Cutler, W. B. (1990) *Hysterectomy: Before and After* (New York: Harper & Row).

Cutler, W. B., and Garcia, C. R. (1991) *Menopause: A Guide for Women and the Men Who Love Them,* rev. 2nd ed. (New York: Norton).

CHAPTER 1: A TIME TO EMBRACE

Cycle Length and Fertility

Although I had read Dr. Vollman's impressive volume, cited below, it was at his seminar—presented to the medical faculty of the Department of Obstetrics and Gynecology, the School of Medicine, University of Pennsylvania—that I met Mrs. Vollman. I sat with her as her husband gave the seminar and had the chance to talk with her at length about the work she had done. Later, when I met her husband, he confirmed what she had told me about her pivotal role in the data collection, tabulation, and analysis.

Fertility Is a Reflection of Good Health

For example, see Cutler (1980), Cutler (1981), Cutler (1988), Demyttenaere (1990), Drinkwater (1990), Emans (1990), Hamilton (1987), Mauvais-Jarvis (1985), Prior (1990), Sitruk-Ware (1977).

Setting the Cycle

The studies published in relation to sexual behavior and menstrual-cycle length include the following: Cutler (1979), Cutler (1979), Cutler (1983), Cutler (1985), Cutler (1986), Cutler (1980).

The Positive Side of Celibacy, the Negative Side of Promiscuity

Irregular periods do indicate deficiencies in hormones, according to a number of studies. For example, see Cutler (1986), Prior (1990).

The results showing that sporadic activity was highly correlated with subnormal estrogen levels was reported in Cutler (1986).

The finding that young women with abnormal menstrual cycles showed problems with bone mass has been reported by Drinkwater (1990), Emans (1990), and Prior (1990).

Timing Is Important for Men Too

Dr. Jeanette Chen, who shared an office with me when we were both postdoctoral fellows at Stanford, conducted these preliminary studies on sexual potency and sexual denial. For other studies showing the importance of timing for men's hormones and fertility, see Chapter 4, "The Hormonal Symphony of Men."

The More Subtle Implications of the Benefits of Restraint

For details showing that sporadic patterns do not favor excess sexual activity, see Cutler (1980).

Why the Timing of the Luteal Phase Is Critical

As an example of recent research showing that luteal-phase deficiencies are associated with reduced progesterone, see Miller (1990).

The Estrogen Connection

The studies in premenopausal women include the following: Cutler (1983), McCoy (1985).

The studies in younger women are cited in Cutler (1986).

THE REFERENCES

Anderson, C., and Mason, W. (1977) Hormones and social behavior of squirrel monkeys. *Hormones and Behavior* 8:100–106.

Cutler, W. B.; Garcia, C. R.; Krieger, A. M. (1979) Sexual behavior frequency and menstrual cycle length in mature premenopausal women. *Psychoneuroendocrinology* 4:297–309.

Cutler, W. B.; Garcia, C. R.; Krieger, A. M. (1979) Luteal phase defects: A possible relationship between short hyperthermic phase and sporadic sexual behavior in women. *Hormones and Behavior* 13:214–18.

Cutler, W. B.; Garcia, C. R.; Krieger, A. M. (1980) Sporadic sexual behavior and menstrual cycle length in women. *Hormones and Behavior* 14:163–72.

Cutler, W. B. (1981) Premenopausal hysterectomy and cardiovascular disease. *Am. J. Obstet. Gynecol.* 141:849.

Cutler, W. B.; McCoy, N. Davidson, J. M. (1983) Sexual behavior. *Neuroendo. L.* 5:3:185.

Cutler, W. B.; Preti, G.; Huggins, G. R.; Erickson, B.; Garcia, C. R. (1985) Sexual behavior frequency and biphasic ovulatory type menstrual cycles. *Physiol. Behav.* 34:805–10.

Cutler, W. B.; Garcia, C. R.; Huggins, G. R.; Preti, G. (1986) Sexual behavior and steroid levels among gynecologically mature premenopausal women. *Fertil. & Steril.* 45:4:496–502.

Cutler, W. B.; Karp, J.; Stine, R. (1988) Single photon absorptiometry imaging as a screening method for diminshed dual photon density measures. *Maturitas* 10:143–55.

Demyttenaere, K. (1990) *Psychoendocrinological Aspects of Reproduction in Women* (Louvain, Belgium: Peeters Press).

Drinkwater, B. L.; Bruemner, B.; Chestnut, C. H. III, (1990) Menstrual history as a determinant of current bone density in young athletes. *JAMA* 263:4:545–48.

Emans, S. J.; Grace, E.; Hoffer, F.; Gundberg, C.; Ravnikar, V.; Woods, E. (1990) Estrogen deficiency in adolescents and young adults: Impact on bone mineral content and effects of estrogen replacement therapy. *Obstet. Gynecol.* 76:585–92.

Hamilton, C. J. C. M.; Evers, J. L. H.; DeHaan, J. (1987) Ovulatory disturbances in patients with luteal insufficiency. *Clinical Endocrinology* 26:129–36.

Mauvais-Jarvis, P.; Sitruk-Ware, R.; Kuttenn, F. (1985) Luteal phase defect and benign breast disease: Relationship to breast cancer genesis. *Breast Diseases —Senologia* 1:58–66.

McCoy, N.; Cutler, W. C.; Davidson, J. M. (1985) Relationships among sexual behavior, hot flashes and hormone levels in perimenopausal women. *Archives of Sexual Behavior* 14:385–94.

Miller, M. M.; Hoffman, D. I.; Creinin, M.; Levin, J. H.; Chatterton, R. T.; Murad, T.; Rebar, R. W. (1990) Comparison of endometrial biopsy and urinary pregnanediol glucuronide concentration in the diagnosis of luteal phase defect. *Fertil. & Steril.* 54:1008–1011.

Prior, J.; Vigna, Y. M.; Schechter, M.; Burgess, A. (1990) Spinal bone loss and ovulatory disturbances. *NEJM* 323:1211–72.

Prior, J.; Vigna, Y. M.; Schulzer, M.; Hall, J. E.; Bonen, A. (1990) Determination of luteal phase length by quantitative basal temperature methods: Validation against the midcycle LH peak. *Clinical Investigative Medicine* 13:123–31.

Sitruk-Ware, L. R.; Sterkers, N.; Mowszowicz, I.; Mauvais-Jarvis, P. (1977) Inadequate corpus luteal function in women with benign breast diseases. *J. Clin. Endocrinol. Metab.* 44:771–74.

Sitruk-Ware, L. R.; deLignieres, B.; Mauvais-Jarvis, P. (1986) Progestogen treatment in postmenopausal women. *Maturitas* 8:95–100.

Stephenson, N. (1990) In Finland's North. *Phila. Inquirer* 12/29 (via Reuters).

Vollman, R. F. (1977) *The Menstrual Cycle,* vol. 7. In *Major Problems in Obstetrics and Gynecology* (Philadelphia: W. B. Saunders).

CHAPTER 2: THE LIFE CYCLE IN THE LOVE CYCLE

Shedding Light on Reproduction

For examples of the influence of the timing of the light/dark cycle on laboratory mammals, such as hamsters and rats, see the references below: Berndtson (1974), Brown-Grant (1973), Deguchi (1979), Elliot (1972), Gray (1978).

Subsequent analysis of the role of the pineal gland as stimulated by the change transmitted through the retina of the eye in response to timing of light and dark has been the subject of an enormous scientific investigation worldwide. Two examples of the classical nature of this research are: Elliot (1972), Moore (1974).

Studies showing the September and October human testosterone peak are reviewed in Chapter 4, "The Hormonal Symphony of Men."

For more detail on the role of the pineal (and/or retinal) production of melatonin, a hormone involved in this mediation of the light/dark cycle, see later chapters, especially Chapter 8, "Cosmic Cycles."

Significant research in comparative biology has revealed that the primates respond differently than what has been shown to be a tight coupling between light/dark cycles and hormonal rhythms of the lower mammals, such as rats and hamsters. The primate exceptions to the rules are cited in: Knobil (1980), (1974), (1978), (1978), Krey (1975), Plant (1978).

This primate escape from a strict light/dark control does not mean that the role of light and dark is irrelevant. Rather, as the data are developed and explained throughout the text, research studies suggest that mammals with more complex structures, such as the monkeys and humans, have a more complex mediation of the control of the cycle—and light/dark timing plays a less severe role in influencing primate fertility. Yet there do remain some subtle effects of light/dark and timing changes, as described in Chapters 4 and 8.

Connecting Vision to Sex

The "extraordinary committee of University of Pennsylvania professors" comprised five scholars who directed my studies in the diverse areas described throughout this book. Dr. Elliott Stellar, now a professor of anatomy, then provost of the university, was one of the early discoverers of the hedonic value of the hypothalamic region of the brain. He continually exposed me to areas of research that affected my own thought process and subsequent studies. Dr. Ingrid Waldron, professor of biology, had initially focused her energies on neurophysiology in studying the way that nerve cells respond to external stimulation. At the time that I began working with her, she had recently shifted her field from these cellular studies to focus more broadly on women's health and how behavior impacted on health in the cardiovascular system and with respect to stress. Dr. Alan Epstein, a physician and professor of biology, was heavily involved in studies of brain and behavior and had made significant contributions to knowledge in the areas of kidney function and the biology of physiologic regulation. Both as his student in several advanced courses and as his teaching assistant in an undergraduate course, I began to develop an understanding of the breadth of the field of brain and behavior. Dr. Abba Krieger, now a professor of statistics in the Wharton School of the University of Pennsylvania, helped guide and design the statistical analysis of the data. Dr. Philip George, professor of biology, chaired my committee. Dr. Celso R. Garcia, professor of gynecology, is discussed throughout the text.

Length of gestation actually had been documented. See Gibson (1950), (1952).

Sexual Initiation and Infertility

See Cutler (1979).

Maturity: The Potential for Sexual Liberation

For data showing the increased rates of congenital abnormalities with increasing age of the parents, see Hook (1981), Goldman (1989).

For a review of the symptoms of the menopause and perimenopause as well as the relationship of hormone-replacement therapy to it, see Cutler, McCoy, Garcia (1987) and Cutler and Garcia (1991).

The study of perimenopausal reductions in estrogen in association with less than weekly sexual activity is described in Cutler (1983).

New Sensual Riches

See, for example, McCann (1989) showing how attitude affects experience.

Sex Grows Up

Because the details of these concepts are described later in the book, they are not referenced.

THE REFERENCES

Berndtson, D. (1974) Circulating LH and FSH levels and testicular function in hamsters during light deprivation and subsequent photoperiodic stimulation. *Endocrinology* 95:195.

Bretschneider, J. G., and McCoy, N. L. (1988) Sexual interest and behavior in healthy 80–102-year-olds. *Archives of Sexual Behavior* 17:2:109–129.

Brown-Grant, K.; Davidson, J. M.; Greig, F. (1973) Induced ovulation in albino rats exposed to constant light. *J. Endocr.* 57:7–22.

Cutler, W. B.; Garcia, C. R.; McCoy, N. (1987) Perimenopausal sexuality. *Archives of Sexual Behavior* 16:3:225–34.

Cutler, W. B., and Garcia, C. R. (1991) *Menopause: A Guide for Women and the Men Who Love Them,* rev. 2nd ed. (New York: Norton).

Cutler, W. B.; Garcia, C. R.; Kreiger, A. M. (1979) Infertility and age at first coitus: A possible association. *J. Biosoc. Sci.* 11:425–32.

Cutler, W. B., McCoy, N.; Davidson, J. M. (1983) Sexual behavior. *Neuroendo. L.* 5(3); 185.

Deguchi, T. (1979) Circadian rhythm of serotonin N-acetyltransferase activity in organ culture of chicken pineal gland. *Science* 203:1245–47.

Elliot, J. A., Stetson, M. H., Menaker, M. (1972) Regulation of testes function in golden hamsters: A circadian clock measures photoperiodic time. *Science* 178:771.

Friedmann, E.; Katcher, A. H.; Lynch, J. J.; Thomas, S. A. (1980) Animal companions and one-year survival after discharge from a coronary care unit. *Public Health Reports* 95:307–12.

Friedmann, E.; Katcher, A. H.; Thomas, S. A.; Lynch, J. J.; Messant, P. R. (1983) Social interaction and blood pressure—influence of animal companions. *Journal of Nervous and Mental Disease* 171:8:461–65.

Friedmann, E. (1990) "Pets, benefits and practice" from the proceedings of a symposium held in Harrogate, England, on April 19, 1990, in conjunction with the First European Congress of the British Small Animal Veterinary Association, 8–17.

Gibson, J. R., and McKeown, T. (1950) Observations on all births (23,790) in Birmingham (1947). I. Duration of gestation. *Brit. J. Social Med.* 4:221–33.

Gibson, J. R., and McKeown, T. (1952) Observation on all births (23,970) in Birmingham (1947). IV. Birth weight, duration of gestation, and survival related to sex. *Brit. J. Social Med.* 6:152–58.

Goldman, N., and Montgomery, M. (1989) Fecundability and husband's age. *Social Biology* 36:146–66.

Gray, G. D., Sodersten, P.; Tallentine, D.; Davidson, J. M. (1978) Effects of lesions in various structures of the suprachiasmatic-preoptic region on LH regulation and sexual behavior in female rats. *Neuroendocrinology* 25:174–91.

Hook, E. B. (1981) Rates of chromosome abnormalities at different maternal ages. *Ob/Gyn* 3:282–85.

Knobil, E. (1974) On the control of gonadotropin secretion in the rhesus monkey. *Recent Progress in Hormone Research* 30:1–37.

Knobil, E. (1978) Personal communication.

Knobil E., and Plant, T (1978) Neuroendocrine control of gonadotropin secretion in the female rhesus monkey. *Frontiers in Neuroendocrinology,* vol. 5, 249–61. Ed. by L. Martini and W. F. Ganong (New York: Raven Press).

Knobil, E.; Plant, P. T.; Wildt, L.; Belchetz, P. E.; Marshall, G. (1980) Control of the rhesus monkey menstrual cycle: permissive role of hypothalamic gonadotropin-releasing hormone. *Science* 207:1371–3.

Krey, L. C.; Butler, W. R.; Knobil, E. (1975) Surgical disconnection of the medial vasal hypothalamus and pituitary function in the rhesus monkey. I. Gonadropin secretion. *Endocrinology* 96:1073–87.

Lao Tsu. *Tao Te Ching* (1972 translation by Gia-fu Feng and Jane English) (New York: Vintage Books).

McCann, J. T., and Biaggio, M. K. (1989) Sexual satisfaction in marriage as a function of life meaning. *Archives of Sexual Behavior* 18:1:59–72.

McFalls, J. A., Jr. (1979) *Psychopathology and Subfecundity* (New York: Academic Press).

Moore, R. Y. (1974) Visual pathways and the central neural control of diurnal rhythms. In *The Neurosciences Third Study Program,* pp. 537–542. Ed. by F. O. Schmitt and F. G. Worden (Cambridge: MIT Press).

Plant, T. M.; Krey, L. C.; Moossy, J.; McCormack, J. T., Hess, D. L.; Knobil, E. (1978) The arcuate nucleus and the control of gonadotropin and prolactin secretion in the female rhesus monkey. *Endocrinology* 102:52–62.

Swan, S., and Brown, W. (1981) Oral contraceptive use, sexual activity, and cervical carcinoma. *Am. J. Obstet. Gynecol.* 139:52–57.

Treloar, A. E.; Boynton, R. E.; Behn, D. G.; Brown, B. W. (1967) Variations of the human menstrual cycle through reproductive life. *I. J. Fertil.* 12:77–126.

CHAPTER 3: THE HORMONAL SYMPHONY OF WOMEN

The New Facts of Life

The bulleted list of hormonal relationships is derived from an enormous literature:

- Changes in physiology, spatial and motor skills, mood, etc.: see, for example, Arafat (1988), Carlstrom (1988), Hampson (1990), Hanson (1975), Khan-Dawood (1988), Laessle (1990), Lenton (1988), Reinisch (1977), Reinisch and Karow (1977), Reinisch (1987), Sanders (1985), Vogel (1971).

  However, there is not universal agreement on the conclusion that hormones taken by pregnant women affect their pubescent offspring. See, for example, Meyer-Bahlburg (1977).
- Hormones and sexual pleasure or capacity for fertility: see, for example, Bancroft (1983), Bauman (1982), Bellerose (1989), Garde (1980), Minneman (1979), Myers (1990), and Sherwin (1985).
- Relation to diseases: see, for example, Arnbjornsson (1983), Mauvais-Jarvis (1985), Sitruk-Ware (1977), Steinberg (1985).
- Hormones and skin: see, for example, Brincat (1987), Wajchenberg (1989).
- Hair: see, for example, Brodie (1987), Rittmaster (1987), Ruutiainen (1988), (1988).
- Occupation and hormones: see Purifoy (1980).

The Dynamic Interaction—An Overview

For a detailed review demonstrating this extraordinary interaction, see Cutler (1980).

Hormonal Passages

References to the age of the menarche include Garde (1980), Kark (1943), Treloar (1967), (1974).

For a review of the studies on menopausal age, the most recent review would be the book I wrote with Dr. Garcia, *Menopause: A Guide for Women and the Men Who Love Them* (1991).

The Dual-Purpose Ovaries

For a thorough discussion of the value of the woman's ovaries and their equivalent value to a man's testes, see Cutler, W. B., *Hysterectomy: Before and After* (1988).

Ovarian Evolution

For studies relating to the changes in the ovaries throughout life, see, for example, Apter (1987), Finn (1988), Lenton (1988), Longcope (1986), Pache (1990), Rannevik (1986), Rozenberg (1988), Trevoux (1986).

Why Men Can Leave but Women Must Stay

For a discussion of the rapid acceleration of follicles as women enter into their early forties, see Richardson (1987).

Hormonal Patterns During Different Life Stages

For a very interesting study of the narcotic effect of natural progesterone, see Arafat (1988).

Window on the Womb

For studies showing the extraordinarily rich innervation of nerves into the muscle and at the cervix, see, for example, Krantz (1959), Morizaki (1989).

For a more detailed discussion of the matter in lay language, see *Hysterectomy: Before and After,* cited above, which cites numerous references that were published before Morizaki's study.

A New Meaning to Menstruation

See the classic research paper in 1937 by Bartelmez.

Discord in the Symphony: Endometriosis

For examples of autoimmune specificity in this disease, see Confino (1990).

For information about this disease, write to Endometriosis Association, International Headquarters, 8585 N. 76th Place, Milwaukee, WI 53223.

The Reproductive Gatekeeper—The Cervix

Two anatomical-physiological studies that demonstrated the rich innervation of the cervical region are Krantz (1973), Rodin (1973).

Box 3-1: How Tubal Ligation Disrupts the Hormonal Symphony

See Bhiwadiwala (1982), Cattanach (1988), Corson (1981), DeStefano (1983), Donne (1981), Hargrove (1981), Lu (1967), Radwanska (1981), Riedel (1981), Rocle (1981).

How the Sex Hormones Change Throughout Life

See, for example, Richardson (1987).

Estrogens and Emotional Energy: The Beta-Endorphin Link

For a review in much greater detail, the reader might want to read either of my two most recent books, *Hysterectomy: Before and After* or *Menopause: A Guide for Women and the Men Who Love Them.*

Menopausal Changes in Sex Steroids

Relevant studies providing the data include Buckler (1991), Rozenberg (1988).

Figure 3-8 has been redrawn from the data supplied in the research paper by Rozenberg et al. (1988).

Box 3-2: Hormone-Replacement Therapy (HRT)

This box reviews a composite conclusion I have drawn. For the arguments leading to those conclusions, the reader is referred to my two most recent books, cited above.

Aging and Androgens

Relevant studies include the following: Carlstrom (1988), Ismail (1968), Longcope (1986), Rannevik (1986).

The data from which Figure 3-9 was redrawn were taken from the research publication by Rozenberg (1988).

For Figure 3-10, the data were studied and a graph was redrawn from the paper by Lenton (1988).

Conclusions

The 5 percent figure for true PMS is provided by Walker (1989).

For a lucid report on the true symptoms of PMS, see York (1989).

THE REFERENCES

Apter, D.; Raisanen, I.; Ylostalo, P.; Vihko, R. (1987) Follicular growth in relation to serum hormonal patterns in adolescent compared with adult menstrual cycles. *Fertil. & Steril.* 47:1:82–88.

Arafat, E. S.; Hargrove, J. T.; Maxson, W. S.; Desiderio, D. M., Colston Wentz, A.; Andersen, R. N. (1988) Sedative and hypnotic effects of oral administration of micronized progesterone may be mediated through its metabolites. *Am. J. Obstet. Gynecol.* 15:5:1203–1209.

Arnbjornsson, E. (1983) Relationship between the removal of the nonacute appendix and the menstrual cycle. *Ann. Chir. Gynaecol.* 72:329:658–660.

Bancroft, J., and Sanders, D. (1983) Mood, sexuality, hormones and the menstrual cycle. III. Sexuality and the role of androgens. *Psychosomatic Medicine* 45:6:509–516.

Bartelmez, G. W. (1937) Menstruation. *Physiol. Reviews* 17:28–72.

Bauman, J.; Kolodny, R. C.; Webster, S. K. (1982) Vaginal organic acids and hormonal changes in the menstrual cycle. *Fertil. & Steril.* 38:572–79.

Bellerose, S. B. (1989) Body image and sexuality in surgically menopausal women. Thesis submitted to the Faculty of Graduate Studies and Research, Department of Psychology, McGill University, Montreal, June 1989.

Bhiwandiwala, P. P.; Mumford, S. D.; Felblum, P. J. (1982) Menstrual pattern changes following laparoscopic sterilization: A comparative study of electrocoagulation and the tubal ring in 1,025 cases. *J. Reprod. Med.* 27:5:249–54.

Brincat, M.; Versi, E.; O'Dowd, T.; Moniz, C. F.; Magos, A.; Kabalan, S.; Studd, J. W. (1987) Skin collagen changes in postmenopausal women receiving oestradiol gel. *Maturitas* 9:1:1–6.

Brodie, B. L., and Wentz, A. C. (1987) Late onset congenital adrenal hyperplasia: A gynecologist's perspective. *Fertil. & Steril.* 48:2:175–88.

Buckler, H. M.; Evans, C. A.; Mamtora, H.; Burger, H. G.; Anderson, D. C. (1991) Gonadotropin, steroid and inhibitory levels in women with incipient ovarian failure during anovulatory and ovulatory rebound cycles. *J. Clin. Endocrinol. Metab.* 72:116–24.

Carlstrom, K.; Brody, S.; Lunell, N. O.; Lagrelius, A.; Mollerstron, G.; Pousette, A.; Rannevik, G.; Stege, R.; von Schoultz, B. (1988) Dehydroepiandrosterone in serum: Differences related to age and sex. *Maturitas* 10:297–306.

Cattanach, J. F., and Milne, B. J. (1988) Post-tubal sterilization problems correlated with ovarian steroidogenesis. *Contraception* 38:541–50.

Confino, E.; Harlow, L.; Gleicher, N. (1990) Peritoneal fluid and serum antibody levels in patients with endometriosis. *Fertil. & Steril.* 53:242–45.

Corson, S. L., et al. (1981) Hormonal levels following sterilization and hysterectomy. *J. Reprod. Med.* 26:363.

Cutler, W. B., and Garcia, C. R. (1980) The psychoneuroendocrinology of the ovulatory cycle of women. *Psychoneuroendocrinology* 5:2:89–111.

DeStefano, F.; Huezo, C. M.; Peterson, H. B.; Ruvin, G. L.; Layde, P. M.; Ory, H. W. (1983) Menstrual changes after tubal sterilization. *Obstetrics and Gynecology* 62:6:673–81.

Donnez, J. et al. (1981) Luteal function after tubal sterilization. *Am. J. Obstet. Gynecol.* 57:65.

Finn, M. M., Gosling, J. P.; Tailon, D. F.; Madden, A. T. S.; Meehan, F. P.; Fottrell, P. F. (1988) Normal salivary progesterone levels throughout the ovarian cycle as determined by a direct enzyme immunoassay. *Fertil. & Steril.* 50:6:882–87.

Garde, K., and Lunde, I. (1980) Female sexual behavior: A study in a random sample of 40-year-old women. *Maturitas* 2:225–40.

Hampson, E. (1990) Estrogen-related variations in human spatial and articulatory-motor skills. *Psychoneuroendocrinology* 15:97–111.

Hanson, J. W.; Hoffman, H. J.; Ross, G. T. (1975) Monthly gonadotropin cycles in premenarchael girls. *Science* 190:161–63.

Hargrove, J. T., and Abraham, G. E. (1981) Endocrine profile of patients with post-tubal ligation syndrome. *J. Reprod. Med.* 26:339–62.

Ismail, A. A. A.; Harkness, R. A.; Loraine, J. A. (1968) Some observations on the urinary excretion of testosterone during the normal menstrual cycle. *Acta Endocrinol.* 58:685–95.

Kark, E. (1943) Menarche in South African Bantu girls. *S. Afr. J. Med. Sci.* 8:35–40.

Khan-Dawood, F.; Cai, H.; Danwood, M. (1988) Luteal phase salivary progresterone concentrations in ovulation-induced cycles. *Fertil. & Steril.* 49:4:611–15.

Krantz, K. E. (1973) The anatomy of the human cervix, gross and microscopic. in *Biology of the Cervix.* pp. 57–69. Ed. by R. J. Blandau and K. Moghissi (Chicago: University of Chicago Press).

Laessle, R. G.; Tuschi, R. J.; Schweiger, U.; Pirke, K. W. (1990) Mood changes and physical complaints during the normal menstrual cycle in healthy young women. *Psychoneuroendocrinology* 15:2:131–38.

Lenton, E. A.; Sexton, L.; Lee, S.; Cooke, I. D. (1988) Progressive changes in LH and FSH and LH: FSH ratio in women throughout reproductive life. *Maturitas* 10:35–43.

Longcope, C.; Franz, C.; Morello, C.; Baker, R.; Johnston, C. C. (1986) Steroid and gonadrotropin levels in women during the perimenopausal years. *Maturitas* 8:189–96.

Lu, T., and Chun, D. (1967) A long-term follow-up study of 1,055 cases of postpartum tubal ligation. *J. Obstet. Gynecol.* 74:875–80.

Mauvais-Jarvis, P.; Sitruk-Ware, R.; Kuttenn, F. (1985) Luteal phase defect and benign breast disease: Relationship to breast cancer genesis. *Breast Diseases —Senologia* 1:58–66.

Meyer-Bahlburg, H. F. L. (1977) Behavioral effects of estrogen treatment in males. Supplement, pp. 1171–77. Conference on Estrogen Treatment of the Young, Santa Ynez, Calif., Dec. 6–9.

Minneman, K. P.; Dibner, M. D.; Wolfe, B. B.; Molinoff, P. B. (1979) B1 and B2 adrenergic receptors in rat cerebral cortex are independently regulated. *Science* 204:866–68.

Morizaki, N.; Morizaki, J.; Hayashi, R. H.; Garfield, R. E. (1989) A functional and structural study of the innervation of the human uterus. *Am. J. Obstet. Gynecol.* 160:218–28.

Myers, L. S.; Dixen, J.; Morissette, D.; Carmichael, M.; Davidson, J. M. (1990) Effects of estrogen, androgen, and progestin on sexual psychophysiology and behavior in postmenopausal women. *J. Clin. Endocrinol. Metab.* 70:1124–31.

Niswander, G. D., Akbar, A. M.; Nett, T. M. (1976) The role of blood flow in regulating ovarian function. *The Endocrine Function of the Human Ovary,* vol. 7, pp. 71–79. Ed. by V. H. G. James, M. Serio, G. Giusti (London: Academic Press).

Pache, T. D.; Wladimiroff, J. W.; deJong, F. H.; Hop, W. C.; Fauser, B. C. J. M. (1990) Growth patterns of nondominant ovarian follicles during the normal menstrual cycle. *Fertil & Steril.* 54:638–42.

Purifoy, F. E., and Koopmans, L. H. (1980) Androstenedione T and free T concentrations in women of various occupations. *Social Biology* 26:179–88.

Radwanska, E., et al. (1982) Evaluation of ovarian function after tubal sterilization. *J. Reprod. Med.* 27:376.

Rannevik, G.; Carlstrom, K.; Jeppsson, S.; Bjerre, B.; Svanberg, L. (1986) A prospective long-term study in women from premenopause to postmenopause: Changing profiles of gonadotrophins, oestrogens and androgens. *Maturitas* 8:297–307.

Reidel, H. H., et al. (1981) Late complication of sterilization according to method. *J. Reprod. Med.* 26:353.

Reinisch, J. M. (1977) Prenatal exposure of human fetuses to synthetic progestin and oestrogen affects personality. *Nature* 266:5602:561–62.

Reinisch, J. M., and Karow, W. G. (1977) Prenatal exposure to synthetic progestins and estrogens: Effects on human development. *Archives of Sexual Behavior* 6:4:257–87.

Reinisch, J. M., and Sanders, S. A. (1987) Behavioral influences of prenatal hormones. In *Handbook of Clinical Psychoneuroendocrinology,* pp. 431–48. Ed. by C. B. Nemeroff and P. T. Loosen (New York: Guilford Press).

Richardson, S. J.; Senikas, V.; Nelson, J. F. (1987) Follicular depletion during the menopausal transition: Evidence for accelerated loss and ultimate exhaustion. *J. Clin. Endocrinol. Metab.* 65:6:1231–37.

Rittmaster, R. S., and Loriaux, D. L. (1987) Hirsutism. *Ann. Intern. Med.* 106:95–107.

Rodin, M., and Moghissi, K. S. (1973) Intrinsic innervation of the human cervix: A preliminary study. In *The Biology of the Human Cervix,* pp. 71–78. Ed. by R. J. Blandau and K. Moghissi (Chicago: University of Chicago Press).

Rozenberg, S.; Bosson, D.; Peretz, A.; Caufriez, A.; Robyn, C. (1988) Serum levels of gonadotrophins and steroid hormones in the postmenopause and later life. *Maturitas* 10:215–24.

Ruutiainen, K.; Erkkola, R.; Gronroos, M. A.; Kaihola, H. L. (1988) Androgen parameters in hirsute women: Correlations with bone mass index and age. *Fertil. & Steril.* 50:2:255–59.

Ruutiainen, K.; Erkkola, R.; Gronroos, M. A.; Irjala, K. (1988) Influence of body mass index and age on the grade of hair growth in hirsute women of reproductive ages. *Fertil. & Steril.* 50:2:260–66.

Sanders, S. A., and Reinisch, J. M. (1985) Behavioral effects on humans of progesterone-related compounds during development and in the adult. *Neuroendocrinology* 5:175–205.

Sherwin, B.; Gelfand, M.; Brender, W. (1985) Androgen enhances sexual motivation in females: A prospective, crossover study of sex steroid administration in the surgical menopause. *Psychosomatic Medicine* 474:339–51.

Sitruk-Ware, L. R.; Sterkers, N.; Mowszowicz, I.; Mauvais-Jarvis, P. (1977) Inadequate corpus luteal function in women with benign breast diseases. *J. Clin. Endocrinol. Metab.* 44:771–74.

Steinberg, A. D., and Steinberg, B. J. (1985) Lupus disease activity associated with menstrual cycle. *J. Rheumatology* 12:4:816–17.

Stergachis, A.; Sky, K. K.; Grothaus, L. C.; Wagner, E. H.; Hecht, J. A.; Anderson, G.; Normand, E. H.; Raboud, J. (1990) Tubal sterilization and the long-term risk of hysterectomy. *JAMA* 264:22:289.

Treloar, A. E. (1974) Menarche, menopause and intervening fecundability. *Human Biology* 16:89–107.

Treloar, A. E.; Boynton, R. E.; Behn, D. G.; Brown, B. W. (1967) Variations of the human menstrual cycle through reproductive life. *I. J. Fertil.* 12:77–126.

Trevoux, R.; DeBrux, J.; Castanier, M.; Nahoul, K.; Soule, J. P.; Scholler, R. (1986) Endometrium and plasma hormone profile in the perimenopause and postmenopause. *Maturitas* 8:309–26.

Vogel, W.; Broverman, D.; Klaiber, E. L. (1971) EEG responses in regularly menstruating women and in amenorrheic women treated with ovarian hormones. *Science* 172:388–91.

Wajchenberg, B. L.; Marcondes, J. A. M.; Mathor, M. B.; Achando, S. S.; Germak, O. A.; Kirschner, M. A. (1989) Free testosterone levels during the menstrual cycle in obese versus normal women. *Fertil. & Steril.* 51:3:535–37.

Walker, A., and Bancroft, J. (1990) The relationship between premenstrual symptoms and oral contraceptive use: a controlled study. *Psychosomatic Medicine* 52:1:86–96.

York, R.; Freeman, E.; Lowery, B.; Strauss, J. F. III, (1989) Characteristics of premenstrual syndrome. *Obstet. Gynecol.* 73:4:601–05.

CHAPTER 4: THE HORMONAL SYMPHONY OF MEN

The Male Sex Hormones

A decline in testosterone appears to be inevitable. For typical studies see Albeaux (1978), Baker (1976), Davidson (1978), Harman (1980), Hegler (1978), Nankin (1986), Vermeulen (1985).

Sex Hormones and the Time of Day

Deslypere (1984); see also Rubin (1975), Schiavi (1988).

Melatonin levels also show a daily, or circadian, cycle in men. Levels are highest at night and typically the very highest at about 2:00 A.M. plus or minus two hours. For references see Beck-Friis (1984), Brzezinski (1988), Kennaway (1986), Lewy (1980).

Sex Hormones and the Time of Year

For Australia see Smals (1976), who also cites Berger. For North America, see Reinberg (1988). For Paris, see Reinberg (1978). For Germany, see Doering (1975).

Regarding the 1985 study conducted in Germany on thirty-three healthy young men, see Christiansen (1985).

Sex Hormones Cycle in Men Individually

See Doering (1975).

It has been established that a single serum drawing is as good as a sample of three in determining male levels of hormones. See Bain (1983).

For a typical showing of the enormous variability from one man to the next in testosterone cycles, see, for example, Couwenbergs (1986).

Seasonal Variation in the Reproductive Capacity of Men

Studies include Kandeel 1988), Politoff (1989), Reinberg (1988), Touitou (1983).

Figure 4-2 on birth rates was redrawn from Levine (1988).

Cowgill in 1966 published her study showing the season of birth in man with reference to historical data throughout the world. For a particularly comprehensive review of seasonal variations in different parts of the world, see this study.

Seasonal variations in births have also been attributed to environmental changes that impact on the economics of the group. See, for example, Huss-Ashmore (1988).

Sex Hormones and Sunshine

See the following studies: Christiansen (1985), Knussmann (1985), Stephenson (1990).

The studies of melatonin include Beck-Friis (1984), Lewy (1980), Lynch (1975), Lynch (1978).

The Effect of Vasectomy on Seasonal Hormone Changes

See Reinberg (1988).

The Effect of Smoking and Stresses on Male Sex Hormones

See Aono (1972), Carani (1990), Davidson (1978), Deslypere (1984), Gray (1979), Parker (1985), Vermeuelen (1985).

Endurance athletics do not affect the sex hormones of men as they appear to affect those of women.

See Bagatell (1990).

Daytime noise and stress do affect nighttime prolactin levels and human growth-hormone levels; see Fruhstorfer (1985).

They also affect semen quality; see Giblin (1988).

The Effect of Personality and Eating Habits on Male Sex Hormones

See Deslypere (1984), Monti (1977), Van Kemenade (1989), Vermeuelen (1985).

Although in women obesity and estrogen are related, in men these two factors were not positively correlated. See Kley (1979).

Male Attributes and Male Sex Hormones

A series of early studies began to show relationships between male sex hormones and sexuality in men. Fox (1972), in a prolonged study on one subject, showed that the testosterone during and after coitus was significantly higher than in control conditions not related to sexual activity. Davidson and Trupin (1983), studying five women and seven men, found no postcoital change in sex hormones. Ismail (1967), studying two subjects, showed that their urinary levels of testosterone were lower during periods of abstinence than during periods of their lives when they were sexually active; and Lee (1974) was unable to demonstrate any changes in sex hormones related to coital activity. Details follow, but for a broad review and a variety of perspectives, see Gooren (1987), Ismail (1967), Knussmann (1986), Lange (1980), Rowland (1987), Tsitouris (1984).

Box 4-1

Biological activity and the appropriate hormones have been recently elucidated by Nankin (1986).

Figure 4-3

Redrawn from Davidson (1983).

Sexual Behavior in Men Is More Complex

For an example of the sophisticated research being conducted in this area, see Bozman (1991).

Penile Sensitivity Declines with Age

The third 1989 study mentioned was first authored by Rowland (1989).

The work of Lange (1980) reviewed the speed of detumescence and level of testosterone.

Erection—When Does It Originate in the Mind...?

Three relevant studies include Bancroft (1983) and Bancroft (1988).

Biologically inclined readers may want to look at a classic paper describing how mental affects can be studied at a physiological level by referring to McEwen (1976).

For a study showing the power of visual stimulation in producing an erection, see, for example, Julien (1988).

The Timing of the Cyclic Variation in Male Sexual Behavior

See Reinberg (1978).

As men age, they slow down; see Mulligan (1991), also Bretschneider in Chapter 2.

Less Can Well Mean More to Him

See Pfeiffer (1972), Davidson (1983), Knussmann (1986).

Sexual Behavior and Sex Hormones—Putting It All Together

See also Fox (1972), Lee (1974), Knussmann (1986).

For the Baltimore Longitudinal Study reference, see Tsitouris (1984).

The Effect of Alcohol on Men's Sexual Behavior

See Tsitouris (1984).

Studies on women have shown that with increasing alcohol intake there is an increased perception of arousal and a decreased physical arousal as measured with vaginal tests. See Wilson (1976), Wilson (1978).

When Sex Becomes a Problem

See, for example, Bozman (1991), Montague (1979), O'Carroll (1984), Slag (1983), Tiefer (1987).

Will Testosterone Improve a Healthy Man's Sexual Performance?

See O'Carroll (1984).

Normal Variations In Mood Not Predictive of Sexual Behavior or Hormones

See Davidson (1979).

The Effect of...Prenatal Hormones

Dr. Reinisch's studies and those of Dr. Sanders, her colleague, are listed in the references to the previous chapter. See above. These conclusions have not been universally accepted, however; see Meyer-Bahlburg (1977).

Estrogen Treatment Used for Men with Cancer or Sex Change

See Meyer-Bahlburg (1977).

Certain Widely Prescribed Drugs

See Rosen (1988), Lundberg (1990).

THE REFERENCES

Albeaux, F. M.; Bohler, C.; Karpas, A. (1978) Testicular function in the aging male. In *Geriatric Endocrinology*. Ed. by R. B. Greenblatt. (New York: Raven Press).

Aono, T.; Kurachi, K.; Mizutani, S.; Hamanaka, Y.; Uozumi, T.; Nakasima, A.; Koshiyama, K.; Matsumoto, K. (1972) A decrease in plasma T levels accompanied by increased coitus in men undergoing major thoracic or gastric surgery. *J. Clin. Endrocinol. Metab.* 35:535–42.

Bagatell, C. J., and Bremer, W. J. (1990) Sperm counts and reproductive hormones in male marathoners and lean controls. *Fertil. & Steril.* 53:4:688–92.

Bain, J.; Langevin, R.; Costa, M.; Sanders, R. M.; Hucker, S. (1983) Serum pituitary and steroid hormone levels in the adult male: One value is as good as the mean of three. *Fertil. & Steril.* 49:1:123–26.

Baker, H. W. G.; Burger, H. G.; deKretser, D. M.; Hudson, B.; O'Connor, S.; Wang, C.; Mirovics, A.; Court, J.; Dunlop, M.; Rennie, G. C. (1976) Changes in the pituitary-testicular system with age. *Clin. Endocrinol.* 5:349–72.

Bancroft, J. (1988) Reproductive hormones and male sexual function. In *Handbook of Sexology*, vol. 6: *The Pharmacology and Endocrinology of Sexual Function.*

Bancroft, J. (1988) Sexual desire and the brain. *Sexual and Marital Therapy* 3:1:11–27.

Bancroft, J.; Tennent, G.; Loucas, K.; Cass, J. (1974) The control of deviant sexual behavior by drugs: I. Behavioural changes following oestrogens and anti-androgens. *Brit. J. Psychiat.* 125:310–15.

Bancroft, J., and Wu, F. C. W. (1983) Changes in erectile responsiveness during androgen replacement therapy. *Archives of Sexual Behavior* 12:1:59–66.

Beck-Friis, J.; von Rosen, D.; Kjellman, B. F.; Ljungren, J. G.; Wetterbert,

L. (1984) Melatonin in relation to body measures, sex, age, season and the use of drugs in patients with major affective disorders and healthy subjects. *Psychoneuroendocrinology* 9:261–77.

Berthold, A. A. (1849) Transplantation der Hoden. *Archives of Anatomy and Physiology* 1:42–46.

Bozman, A. W., and Beck, J. B. (1991) Covariation of sexual desire and sexual arousal: The effects of anger and anxiety. *Archives of Sexual Behavior* 20:47–60.

Carani, C.; Bancroft, J.; DelRio, G.; Granata, A. R. M.; Facchinetti, F.; Marrama, P. (1990) The endocrine effects of visual erotic stimuli in normal men. *Psychoneuroendocrinology* 15:207–16.

Christiansen, K.; Knussmann, R.; Couwenbergs, C. (1985) Sex hormones and stress in the human male. *Hormones & Behavior* 19:426–40.

Couwenbergs, C.; Knussmann, R.; Christiansen, K. (1986) Comparisons of the intra and inter-individual variability in sex hormone levels of men. *Ann. Hum. Biol.* 13:301–10.

Cowgill, U. (1966) Season of birth in man. Contemporary situation with special reference to Europe and the Southern Hemisphere. *Ecology* 47:4:614–23.

Davidson, J. M.; Camargo, C. A.; Smith, E. R. (1979) Effects of androgen on sexual behavior in hypogonadal men. *J. Clin. Endocrinol. Metab.* 48:955–58.

Davidson, J. M.; Chen, J. J.; Crapo, L.; Gray, G. D.; Greenleaf, W. J.; Catania, J. A. (1983) Hormonal changes and sexual function in aging men. *J. Clin. Endocrinol. Metab.* 83:41–93.

Davidson, J. M., and Myers, L. S. (1988) Endocrine factors in sexual psychophysiology. In *Patterns of Sexual Arousal,* pp. 158–174. Ed. by R. C. Rosen and J. G. Beck (New York: Guilford Press).

Davidson, J. M.; Smith, E. R.; Levine, S. (1978) Testosterone. In *Coping Men —A Study of Human Psychobiology,* pp. 57–62. Ed. by H. Ursin, E. Baade, and S. Levine (New York: Academic Press).

Davidson, J. M., and Trupin, S. (1973) Neural mediation of steroid-induced sexual behavior in rats. In *Sexual Behavior Pharmacology and Biochemistry,* pp. 13–20. Ed. by M. Sandler, and G. L. Gessa, (New York: Raven Press).

Deslypere, J. P., and Vermeulen, A. (1984) Influence of age and sex on steroid concentration in different tissues in humans. *Abstracts, 7th International Congress of Endocrinology.* Abstract 572. Quebec City.

Deslypere, J. P., and Vermeulen, A. (1984) Leydig cell function in normal men: Effect of age, life-style, residence, diet and activity. *J. Clin. Endocrinol. Metab.* 59:955.

Doering, C.; Kraemer, H.; Brodie, K.; Hamburg, D. (1975) A cycle of plasma testosterone in the human male. *J. Clin. Endocrinol. Metab.* 40:497.

Edwards, A. E., and Husted, J. R. (1976) Penile sensitivity, age and sexual behavior. *J. Clin. Psychol.* 32:697.

Ehrenkranz, J. (1983) Seasonal breeding in humans: Birth records of the Labrador Eskimo. *Fertil. & Steril.* 40:4:485–89.

Feder, H. H.; Storey, A.; Goodwin, D.; Reboulleau, C.; Silver, R. (1977) Testosterone and "5a-dihydrotesterone" levels in peripheral plasma of male and female ring doves *(Streptopellia risoria)* during the reproductive cycle. *Biol. Reprod.* 16:666–77.

Fox, C. A.; Ismail, A. A. A.; Love, D. N.; Kirkham, K. E.; Loraine, J. (1972) Studies on the relationship between plasma testosterone levels and human sexual activity. *J. Endocrinol.* 52:51–58.

Fruhstorfer, B.; Fruhstorfer, H.; Grass, P.; Milerski, H. G. (1985) Daytime noise stress and subsequent night sleep: Interference with sleep pattern, endocrine and neurocrine functions. *Intern. J. Neuroscience* 26:301.

Giblin, P.; Poland, M.; Moghissi, K.; Ager, J.; Olson, J. (1988) Effects of stress and characteristic adaptability on semen quality in healthy men. *Fertil. & Steril.* 49:1:127–32.

Gooren, L. J. G. (1987) Androgen levels and sex functions in testosterone-treated hypogonadal men. *Archives of Sexual Behavior* 16:6:463–73.

Gray, G. D.; Smith, E. R.; Damassa, D. A.; Ehrenkranz, J. R. L.; Davidson, J. M. (1979) Neuroendocrine mechanisms mediating the suppression of circulating testosterone levels associated with chronic stress in rats. Prepublication manuscript.

Harman, S. M., and Tsitouras, P. D. (1980) Reproductive hormones in aging men. I. Measurement of sex steroids, basal luteinizing hormone and Leydig cell response to human chorionic gonadotropin. *J. Clin. Endocrinol. Metab.* 51:35–40.

Hegler, S., and Mortensen, M. (1978) Sexuality and ageing. *Brit. J. Sex. Med.* 5:16–25.

Huss-Ashmore, R. (1988) Seasonal patterns of birth and conception in rural highland Lesotho. *Human Biology* 48:3:493–506.

Ismail, A. A. A., and Harkness, R. A. (1967) Urinary testosterone excretion in men in normal and pathological conditions. *Acta Endocrinol.* (Copenhagen) 56:469–80.

Julien, E.; Over, R. (1988) Male sexual arousal across five modes of erotic stimulation. *Archives of Sexual Behavior* 17:2:131–43.

Kandeel, F. R.; Swerdloff, R. S. (1988) Role of temperature in regulation of spermatogenesis and the use of heating as a method for contraception. *Fertil. & Steril.* 49:1:1–23.

Kennaway, D. J., and Royles, P. (1986) Circadian rhythms of 6-sulphatoxy melatonin, cortisol and electrolyte excretion at the summer and winter solstices in normal men and women. *Acta Endocrinol.* (Copenhagen) 113:450–56.

Kley, H. K., Solbach, H. G.; McKinnan, J. G.; Kroskemper, H. L. (1979) Obesity and estrogen. *Acta Endocrinol.* (Copenhagen) 91:553.

Knussmann, R.; Christiansen, K.; Couwenbergs, C. (1986) Relations between sex hormone levels and sexual behavior in men. *Archives of Sexual Behavior* 15:5:429–45.

Knussmann, R.; Couwenbergs, C.; Christiansen, K.; Fischer-Bruegge, U. (1985) Relations between the intra-individual variability in sex hormone levels of men and several other factors. *Coll. Antropol.* 9:2:189–99.

Kwan, M.; Greenleaf, W. J.; Mann, J.; Crapo, L.; Davidson, J. M. (1983) The nature of androgen action on male sexuality: A combined laboratory–self report study on hypogonadal men. *JCEM* 57:557–62.

Lange, J. D.; Brown, W. A.; Wincze, J. P.; Swick, W. (1980) Serum testosterone concentration and penile tumescence changes in men. *Hormones and Behavior* 14:267–70.

Lee, P. A.; Jaffe, R. B.; Midgley, A. R., Jr. (1974) Lack of alteration of serum gonadotropins in men and women following sexual intercourse. *Am. J. Obstet. Gynecol.* 120:985–87.

Levine, R.; Bordson, B.; Mathew, R.; Brown, M.; Stanley, J.; Starr, T. (1988) Deterioration of semen quality during summer in New Orleans. *Fertil. & Steril.* 49:5:900–07.

Lewy, A. J.; Wehr, T. A.; Goodwin, F. K.; Newsome, D. A.; Markey, S. P. (1980) Light suppresses melatonin secretion in humans. *Science* 210:1267–69.

Lundberg, P. (1989) "Drugs which affect male sexual function." Presentation to the International Academy of Sex Research, Princeton, 1989.

Lynch, H. J.; Jimerson, D. C.; Ozaki, Y.; Post, R. M.; Bunney, W. E.; Wurtman, R. J. (1978) Entrainment of rhythmic melatonin secretion from the human pineal to a 12-hour phase shift in the light dark cycle. *Life Sci.* 23:1557–64.

Lynch, H. J.; Wurtman, R. J.; Moskowitz, M. A.; Archer, M. C.; Ho, M. N. (1975) Daily rhythm in human urinary melatonin. *Science* 187:169–71.

Martin, C. E. (1977) Sexual activity in the aging male. In *Handbook of Sexology,* p. 813. Ed. by J. Money and H. Musaph (Amsterdam: Elsevier/North Holland).

McEwen, B. S. (1976) Endocrine effects on the brain and their relationship to behavior. In *Basic Neurochemistry,* 2nd ed., pp. 737–64. Ed. by Sigel, Albers, Katzman, and Agranoff (Boston: Little, Brown).

Meyer-Bahlburg, H. F. L. (1977) Behavioral effects of estrogen treatment in males. Supplement, Conference on Estrogen Treatment of the Young. Santa Ynez, Calif., Dec. 6–9. 1171–77.

Montague, D. K., et al. (1979) Diagnostic evaluation, classification, and treatment of men with sexual dysfunction. *Urology* 14:545–48.

Monti, P. M.; Brown, W. A.; Corriveau, D. P. (1977) Testosterone and comparison of aggressive and sexual behavior in man. *Am. J. Psychiat.* 134:692–94.

Mulligan, T.; Moss, C. R. (1991) Sexuality and aging in male veterans: A cross-sectional study of interest, ability, and activity. *Archives of Sexual Behavior* 20:17–25.

Nankin, H. R., and Calkins, J. H. (1986) Decreased bioavailable testosterone in aging normal and impotent men. *J. Clin. Endocrinol. Metab.* 63:6:1418–20.

Newman, H. F. (1970) Vibratory sensitivity of the penis. *Fertil. & Steril.* 21:971–73.

O'Carroll, R., and Bancroft, J. (1984) Testosterone therapy for low sexual interest and erectile dysfunction in men: A controlled study. *Brit. J. Psychiatry* 145:146–51.

Parker, L.; Eugene, J.; Farber, D.; Lifrak, E.; Lai, M.; Juler, G. (1985) Dissociation of adrenal androgen and cortisol levels in acute stress. *Horm. Metab. Res.* 17:209.

Persky, H.; Lief, H.; O'Brien, C.; Strauss, D. (1977) Reproductive hormone levels and sexual behaviors of young couples during the menstrual cycle. In *Progress in Sexology,* pp. 293–310. Ed. by R. Gemme and C. C. Wheeler (New York: Plenum Press).

Persky, H.; Lief, H.; Strauss, D.; Miller, W.; O'Brien, C. (1978) Plasma testosterone level and sexual behavior of couples. *Archives of Sexual Behavior* 7:157–73.

Pfeiffer, E., and Verwoerdt, A. (1972) Determinants of sexual behavior in middle and old age. *J. Am. Geriatr. Soc.* 20:151.

Politoff, L.; Birkhauser, M.; Almendral, A.; Zorn, A. (1989) New data confirming a circannual rhythm in spermatogenesis. *Fertil. & Steril.* 52:486–89.

Reading, A. E. (1983) A comparison of the accuracy and reactivity of methods of monitoring male sexual behavior. *J. of Behavioral Assessment* 5:1:11–23.

Reinberg, A., and Lagoguey, M. (1978) Circadian and circannual rhythms in sexual activity and plasma hormones (FSH, LH, testosterone) of five human males. *Archives of Sexual Behavior* 7:13–30.

Reinberg, A.; Smolensky, M. H.; Hallek, M. (1988) Annual variation in semen characteristics and plasma hormone levels in men undergoing vasectomy. *Fertil. & Steril.* 49:2:309–15.

Resko, J. A. (1967) Plasma androgen levels of the rhesus monkey: Effects of age and season. *Endocrinology* 81:1203–11.

Rosen, R. C.; Kostis, J. B.; Jekelis, A. W. (1988) Beta-blocker effects on sexual function in normal males. *Archives of Sexual Behavior* 17:3:241–55.

Rowland, D. L.; Greenleaf, W. J.; Mas, M.; Davidson, J. M. (1989) Penile and finger sensory thresholds in aging and diabetes. *Archives of Sexual Behavior* 1:1–12.

Rowland, D. L.; Heiman, J. R., Gladue, B. A.; Hatch, J. P.; Doering, C. H.; Weiler, S. J. (1987) Endocrine, psychological and genital response to sexual arousal in men. *Psychoneuroendocrinology* 12:2:149–58.

Rubin, R.; Govin, P.; Lubin, A.; Poland, R.; Pirke, K. (1975) Nocturnal increase of plasma testosterone in men: Relation to gonadotropins and prolactin. *J. Clin. Endocrinol. Metab.* 40:1027–33.

Schiavi, R. C.; Schreiner-Engel, P.; White, D.; Mandeli, J. (1988) Pituitary gonadal function during sleep in men with hypoactive sexual desire. *Psychosomatic Med.* 50:304–18.

Schlegel, P.; Chang, T.; Marshall, F. (1991) Antibiotics: potential hazards to male fertility. *Fertil. & Steril.* 55:2:235–42.

Slag, M. F.; Morley, J. E.; Elson, M. K.; Trence, D. L.; Nelson, C. J.; Nelson, A. E.; Kinlaw, W. B.; Beyer, H. S.; Muttal, F. Q.; Shafer, R. B. (1983) Impotence in medical clinic outpatients. *JAMA* 249:1736–40.

Smals, A. G. H.; Kloppenborg, P. W. C.; Benraad, T. J. (1976) Circannual cycle in plasma testosterone levels in man. *J. Clin. Endocrinol. Metab.* 42:979–82.

Stephenson, N. (1990) In Finland's North. *Phila. Inquirer* 12/29 (via Reuters).

Tiefer, L., and Melman, A. (1987) Adherence to recommendations and improvement over time in men with erectile dysfunction. *Archives of Sexual Behavior* 16:4:301–10.

Touitou, Y.; Lagoguey, M.; Bogdan, A.; Reinbert, A.; Beck, H. (1983) Seasonal rhythms of plasma gonadotropins: Their persistence in elderly men and women. *J. Endocrinol.* 96:15.

Tsitouras, P. D., and Hagen T. C. (1984) Testosterone, HL, FSH, prolactin and sperm in healthy aging men. *Abstracts, 7th Int'l Congress on Endocrinology.* Abstract 1951. Quebec City.

Tsitouras, P. D.; Martin, C. E.; Hartman, S. M. (1984) Sexual activity in healthy elderly men. *J. Gerontology* 37:288–93.

Udry, J. R.; Billy, J. O. G.; Morris, N. M.; Groff, T. R.; Raj, M. H. (1985) Serum androgenic hormones motivate sexual behavior in adolescent boys. *Fertil. & Steril.* 45:90–94.

Van Kemenade, J. F. L. M.; Cohen-Kettenis, P. T.; Cohen, L.; Gooren, L. J. G. (1989) Effects of the pure antiandrogen RU 23.903 (Anandron) on sexuality, aggression, and mood in male-to-female transsexuals. *Archives of Sexual Behavior* 18:3:217–28.

Vermeulen, A., and Deslypere, J. P. (1985) Testicular endocrine function in the aging male. *Maturitas* 7:273–79.

Vermeulen, A.; Rubens, R.; Verdonck, L. (1972) Testosterone secretion and metabolism in male senescence. *JCEM* 34:730–35.

Wilson, G. T., and Lawson, D. M. (1976) Effects of alcohol on sexual arousal in women. *J. Abnorm. Psychol.* 85:489–97.

Wilson, G. T., and Lawson, D. M. (1978) Expectancies, alcohol and sexual arousal in women. *J. Abnorm. Psychol.* 87:358–67.

World Health Organization Task Force on Psychosocial Research in Family Planning, Special Programme of Research, Development and Research Training in Human Reproduction (1982). Hormonal contraception for men: Acceptability and effects on sexuality. *Studies in Family Planning* 13:11:328–42.

CHAPTER 5: FERTILITY CYCLES OF WOMEN AND MEN

The Infertility Epidemic

The 28 percent rate was given by Page (1989). See also Greenhall (1990), who conclude that 24 percent of all women attempting to conceive experience an episode of subfertility at some stage of their reproductive life.

Recently attention has begun to be paid to this problem (of the suffering of infertility patients). For example, see Wright (1991), which reviews some of the salient issues that infertility patients face.

Figure 5-1

This figure was redrawn from Medical Research International (1991).

Sex Makes the Cycles

For couples trying to conceive, the need for sexual behavior at the time of ovulation has long been understood, and was discussed in the medical-practice approach to an infertility patient. What has rarely been considered previously is the critical role of the timing of nonovulatory sexual connections. By 1990 that had begun to change. And in 1991 a paper comparing four different standard infertility treatments to a prescription for timed intercourse showed that timed intercourse was as good as the treatments in producing the desired outcome in certain cases. See Martinez (1991).

This approach is new to the 1990s. Before this the new technologies, still in their first blush of power, were often considered primary. However, at a chemical level the timing of the ovarian response to the natural hormonal changes and the technological intervention has been intensively studied. See, for example, DeVoto (1989), Gonen (1989), Hutchinson-Williams (1989), Kemeter (1975), Punonen (1981), Petraglia (1985) and (1986).

Women suffering from unexplained infertility, when investigated, have been shown to have a high incidence of luteal-phase defects—the very condition that regular luteal-phase sexual activity seems to protect against; see, for example, Vuorento (1990).

The Fat-Fertility Link

Fat cells serve as miniature factories converting circulating androgen into the estrogens. The more fat a woman has, the greater this conversion rate and consequently the higher her levels of estrogen. For details of this research, see the medical textbook I wrote in 1984 with Dr. Garcia for comprehensive discussion, cited above.

The androgens are produced both in the ovaries and in the adrenal glands. From the ovary—in some cases as part of the normal menstrual cycle, in others as part of the aging process—androgens are secreted into the bloodstream.

Stress can also stimulate androgen (and cortisol) secretion by turning on the adrenal glands' hormone production, a well-known response to stress.

Foods for Fertility?

See Pirke (1986), Pirke (1989), Schweiger (1987), Schweiger (1988). For a review of other studies leading to the same conclusion, see the nutrition chapter of *Hysterectomy: Before and After*.

Exercising Your Fertility

See Shangold (1979), showing how marathon training altered the luteal phase when data were prospectively collected and not able to be seen when data were retrospectively collected (Shangold, 1982). See also the studies of Dr. Prior cited below; for example, 1982, 1985, 1987, 1990 (two studies in Chapter 1).

Altitude Affects Fertility

See Abelson (1974), Abelson (1976).

Career Infertility

Regarding the menopausal transition, see Metcalf (1979) and *Menopause: A Guide for Women and the Men Who Love Them,* cited above. See also Treloar (1967), Treloar (1974).

Figure 3-10b shows the FSH changes throughout the years and is discussed on page 81.

Sexual Behavior and Fertility

For representative studies in the small mammal, the rat or vole, and the way scientists analyze these problems, see Davidson (1973), Gray (1974), Gray (1974), Gray (1976), Herbert (1977), McEwen (1973).

Studies of the red deer and monkeys include Guinness (1971), Michael (1968), Michael (1976), and Resko (1974).

Increasing the Odds of Successful Pregnancy with Fertile-Type Sexual Behavior

The finding that more than half of the couples who seek infertility treatments do not engage in fertile-type sex are described in Cutler (1979), Luteal-Phase Defects (cited in Chapter 1).

For a discussion of the studies that show that regular weekly sex does indeed promote fertile-type endocrine cycles, see Chapter 1.

Have Sex

For a recent study demonstrating the difficulties men have in exerting self-control over their erections, see Mahoney (1991).

Sex Promotes Fertility

Table 5-1 is redrawn from Cutler (1985).
Figure 5-2 is redrawn from Campbell (1985).
Tables 5-2 and 5-3 are redrawn from Cutler (1985).

A Graphic Look at Fertility

The cervical mucus provides a fertile chemical environment each fertile month during an approximate twenty-four-hour window; see Nulsen (1987).

The Meaning of the Numbers

Another investigator evaluating menstrual-cycle changes in response to intercourse looked at his data differently yet also concluded that sexual behavior affects the timing of the menstrual cycle; see Stanislaw (1987).

Sex Helps Fertility Technology Work Better

In 1991 Martinez et al. published their study showing supportive results to those of Marconi. Their infertility treatments were no better than timed intercourse in producing pregnancies.

The Stress of Success

See also Voigt (1990).

The Stress Infertility Experiment

See Demyttenaere (1989).

Stress in Nature

In order to consider the possibility of a relationship of power and fertility, a complex equation must be drawn, and it is probably circular.

The experience of infertility can produce so much stress that it does diminish one's sense of personal power. Dunphy et al. (1990) showed that infertile couples who had a long-term experience of infertility before coming to their center responded to the attendance at a "tertiary referral center for infertility" with a significant increase in their fecundity. Without any medical treatment 22 percent of the couples with prolonged infertility conceived. The researchers concluded that this positive outcome was probably related to the relief from the stress of knowing they were infertile and to the sense that at last help was at hand. The subgroup who did conceive became pregnant just as fast as the data show that other couples who have no fertility problems and who stopped using contraceptives did. The study seemed to suggest that for a significant percentage of infertile couples stress is directly related to their problem.

The circularity of the problem is troublesome in that stress may cause infertility, the treatment of which may increase the stress and increase the difficulty of conception.

The 1978 study from Davidson's group at Stanford is listed under Jevning (1978).

The Pill

An enormous worldwide literature has been published reflecting studies of women on the pill.

Several recent ones evaluating the response to low-dose contraception show apparently minor alterations in blood lipids, hormonal changes, and a conception delay after stopping oral contraceptives. For an example of the way these studies are reported, see Alvarez (1990), Bracken (1990), Lussier-Caran (1990).

Although changes do occur in users of oral contraceptives, one must balance the risks against those of the alternative, e.g., the risk of an unwanted pregnancy with the apparently mild ones these studies report. Having done this, one's individual genetic predisposition for disease or relative immunity from it needs to be factored into one's decisions.

The issues are complex. Oral contraceptives are highly effective. Take them and you are likely to avoid pregnancy.

THE REFERENCES

Abelson, A. (1976) Altitude and fertility. *Human Biology* 48:83–92.

Abelson, A. E.; Baker, T. S.; Baker, P. T. (1974) Altitude, migration and fertility in the Andes. *Social Biol.* 21:12–27.

Altmann, J.; Altmann, S.; Hausfater, G. (1978) Primate infants' effects on mother's future reproduction. *Science* 201:1028–30.

Alvarez, M. A., Cropp, C. S.; Smith, B. S.; Burkman, R. T.; Zacur, H. A. (1990) Effect of low-dose oral contraceptive on gonadotropins, androgens and sex hormones binding globulin in nonhirsute women. *Fertil. & Steril.* 53:1:35–39.

Arey, L. B. (1939) The degree of normal menstrual irregularity. *Am. J. Obstet. Gynecol.* 37:12.

Benson, H. (1975) *The Relaxation Response* (New York: William Morrow).

Bernard, R. P.; Bhatt, R. V.; Potts, D. M.; Rao, A. P. (1978) Seasonality of birth in India. *J. Biosoc. Sci.* 10:409–21.

Bracken, M. B.; Hellenbrand, K. G.; Holford, T. R. (1990) Conception delay after oral contraceptive use: The effect of estrogen dose. *Fertil. & Steril.* 53:1:21–27.

Campbell, K. L. (1985) Methods of monitoring ovarian function and predicting ovulation: Summary of a meeting. *PARFR* 3:5:1–16.

Cutler, W. B.; Garcia, C. R.; Krieger, A. M. (1979) Luteal phase defects: A possible relationship between short hyperthermic phase and sporadic sexual behavior in women. *Hormones and Behavior* 13:214–18.

Cutler, W. B.; Preti, G.; Huggins, G. R.; Erickson, B.; Garcia, C. R. (1985) Sexual behavior frequency and biphasic ovulatory type menstrual cycles. *Physiol. Behav.* 34:805–10.

Davidson, J. M.; Smith, E. R.; Bowers, C. Y. (1973) Effects of mating on gonadotropin release in the female rat. *Endocrinology* 93:5:1185.

Demyttenaere, K.; Nijs, P.; Evers-Kiebooms, G.; Koninckx, P. R. (1989) The effect of a specific emotional stressor on prolactin, cortisol and testosterone concentrations in women varies with their trait anxiety. *Fertil. & Steril.* 52:6:942–48.

DeVoto, L.; Vega, M.; Navarro, V.; Sir, T.; Alba F.; Castro, O. (1989) Regulation of steroid hormone synthesis by human corpora lutea: Failure of follicle-stimulating hormone to support steroidogenesis in vivo and in vitro. *Fertil. & Steril.* 51:4:628–33.

Domar, A. D.; Seibel. M. M.; Benson, H. (1990) The mind/body program for infertility: A new behavioral treatment approach for women with infertility. *Fertil. & Steril.* 53:246–53.

Downey, J.; Yingling, S.; McKinney, M.; Husami, N.; Jewelewicz, R.; Maidman, J. (1989) Mood disorders, psychiatric symptoms and distress in women presenting for infertility evaluation. *Fertil. & Steril.* 52:425–32.

Dunphy, B. C.; Kay, R.; Robinson, J. N.; Cooke, I. D. (1990) The placebo response of subfertile couples to attending a tertiary referral centre. *Fertil. & Steril.* 54:1072–75.

Gonen, Y.; Casper, R.; Jacobson, W.; Blankier, J. (1989) Endometrial thickness and growth during ovarian stimulation: A possible predictor of implantation in in vitro fertilization. *Fertil. & Steril.* 52:446–50.

Goodman, A. L.; Descalzi, C. G.; Johnson, D. K.; Hodgen, G. D. (1977) Composite pattern of circulating LH, FSH, estradiol, and progesterone during the menstrual cycle in cynomolgus monkeys. *Proceedings of the Society for Experimental Biology and Medicine* 155:479–81.

Goodman, A. L., and Hodgen, G. D. (1979) Between ovary interaction in the regulation of follicle growth, corpus luteum function, and gonadotropin secretion in the primate ovarian cycle. I. Effects of follicle cautery and hemiovariectomy during the follicular phase in cynomolgus monkeys. *Endocrinology* 104:5:1304–09.

Goodman, A. L., and Hodgen, G. D. (1979) Between ovary interaction in the regulation of follicle growth, corpus luteum function, and gonadotropin secretion in the primate ovarian cycle. II. Effects of luteectomy and hemiovariectomy during the luteal phase in cynomolgus monkeys. *Endocrinology* 104:5:1310–16.

Goodman, A. L., and Hodgen, G. D. (1978) Postpartum patterns of circulating FSH, LH, prolactin, estradiol and progesterone in nonsuckling cynomolgus monkeys. *Steroids* 31:5:731–44.

Goodman, A. L.; Nixon, W. E.; Hodgen, G. D. (1979) Between ovary interaction in the regulation of follicle growth, corpus luteum function, and gonadotropin secretion in the primate ovarian cycle. III. Temporal and spatial dissociation of folliculogenesis and negative feedback regulation of tonic gonadotropin release after luteectomy in rhesus monkeys. *Endocrinology* 105:1:69–73.

Goodman, A. L.; Nixon, W. E.; Johnson, D.; Hodgen, G. D. (1977) Regulation of folliculogenesis in the cycling rhesus monkey: Selection of the dominant follicle. *Endocrinology* 100:155.

Gray, G. D.; Davis, H. N.; et al. (1974) Anoestrus and induced ovulation in montane voles. *J. Reprod. Fertil.* 38:193–96.

Gray, G. D.; Davis, H. N.; et al. (1976) Effect of mating on plasma levels of LH and progesterone in montane voles *(Microtus montanus). J. Reprod. Fertil.* 47:89–91.

Gray, G. D.; Zerylnick, M.; Davis, H. N.; Dewsbury, D. A. (1974) Effects of variations in male copulatory behavior on ovulation and implantation in prairie voles *(Microtuochrogaster). Hormones and Behavior* 5:389–96.

Green, B. B.; Weiss, N. S.; Daling, J. R. (1988) Risk of ovulatory infertility in relation to body weight. *Fertil. & Steril.* 50:5:721–26.

Greenhall, E., and Vessey, M. (1990) The prevalence of subfertility: A review of the current confusion and a report of two new studies. *Fertil. & Steril.* 54:978–83.

Guinness, F.; Lincoln, G.; Short, R. V. (1971) The reproductive cycle of female red deer *(Cervus elaphus l.). J. Reprod. Fertil.* 27:427–38.

Herbert, J. (1977) External factors and ovarian activity in mammals. In *Physiology,* vol. II, pp. 457–505. Ed. by Prof. Lord Zuckerman and Barbara Weir (New York, San Francisco, London: Academic Press).

Hutchinson-Williams, K. A.; Lunefeld, B.; Diamond, M. P.; Lavy, G.; Boyers, S. P.; DeCherney, A. H. (1989) Human chorionic gonadotropin, estradiol and progesterone profiles in conception and nonconception cycles in an in vitro fertilization program. *Fertil. & Steril.* 52:441–46.

Jevning, R.; Wilson, A. F.; Davidson, J. M. (1978) Adrenocortical activity during meditation. *Hormones and Behavior* 10:1:54–60.

Kemeter, P.; Salzer, H.; Breitenecker, G.; Friedrich, F. (1975) Progesterone, oestradiol-17B & testosterone levels in the follicular fluid of tertiary follicles and Graafian follicles of human ovaries. *Acta Endocrinol.* 80:686–704.

Lussier-Caran, A.; Nestruck, A. C.; Arslanian, H.; Xhignesse, M.; Davignon, J.; Kafrissen, M. E.; Chapdelaine, A. (1990) Influence of a triphasic oral contraceptive preparation on plasma lipids and lipoproteins. *Fertil. & Steril.* 53:1:28–34.

Mahoney, J. M., and Strassberg, D. S. (1991) Voluntary control of male sexual arousal. *Archives of Sexual Behavior* 2:1–16.

Marconi, G.; Auge, L.; Oses, R.; Quintana, R.; Raffo, F.; Young, E. (1989) Does sexual intercourse improve pregnancy rates in gamete intrafallopian transfer? *Fertil. & Steril.* 51:2:357–59.

Martinez, A. R.; Bernardus, R. E.; Voorhorst, F. J.; Vermeiden, J. P. W.; Schoemaker, J. (1991) Pregnancy rates after timed intercourse or intrauterine insemination after human menopausal gonadotropin stimulation of normal ovulatory cycles: A controlled study. *Fertil. & Steril.* 55:258–65.

McEwen, B. S., and Pfaff, D. W. (1973) Chemical and physiological approaches to neuroendocrine mechanisms: Attempt at integration. In *Frontiers in Neuroendocrinology,* pp. 267–335. Ed. by W. F. Ganong and L. Martini (New York: Oxford University Press).

Medical Research International. Society for Assisted Reproductive Technology. The American Fertility Society (1991) In vitro fertilization-embryo transfer (IVF-ET) in the U.S.: 1989 results from the IVF-ET registry. *Fertil. & Steril.* 55:14–24.

Metcalf, M. G. (1979) Incidence of ovulatory cycles in women approaching the menopause. *J. Biosoc. Sci.* 11:39–48.

Michael, R. P., and Welegally, J. (1968) Ovarian hormones and the sexual behavior of the female rhesus monkey *(Macaca mulatta)* under laboratory conditions. *J. Endocrinal.* 41:407–420.

Michael, R. P., and Zumpe, D. (1976) Environmental and endocrine factors influencing annual changes in sexual potency in primates. *Psychoneuroendocrinology* 1:303–13.

Newton, C. R.; Hearn, M. T.; Yuzpe, A. A. (1990) Psychological assessment and follow-up after in vitro fertilization. *Fertil. & Steril.* 54:879–86.

Nulsen, J.; Wheeler, C.; Ausmanas, M.; Blasco, L. (1987) Cervical mucus changes in relationship to urinary luteinizing hormone. *Fertil. & Steril.* 48:5:783–86.

Page, H. (1989) Estimation of the prevalence and incidence of infertility in a population: A pilot study. *Fertil. & Steril.* 51:571–77.

Petraglia, F.; DiMeo, G.; DeLeo, V.; Nappi, C.; Facchinetti, F.; Genazzani, A. R. (1986) Plasma B-endorphin levels of anovulatory states: Changes after treatments for the induction of ovulation. *Fertil. & Steril.* 45:2:185–90.

Petraglia, F.; Segre, A.; Facchinetti, F.; Campanini, D.; Kuspa, M.; Genazzani, A. R. (1985) B-endorphin and metenkephalin in peritoneal and ovarian follicular fluids of fertile and postmenopausal women. *Am. Fertility Soc.* 44:5:615–21.

Pirke, K. M.; Schweiger, U.; Laessle, R.; Dickaut, B.; Schweiger, M.; Waechteler, M. (1986) Dieting influences the menstrual cycle: Vegetarian vs. nonvegetarian diet. *Fertil. & Steril.* 46:1083.

Pirke, K. M.; Schweiger, U.; Strowitzki, T.; Tuschl, R.; Laessle, R. G.; Broocks, A.; Huber, B.; Middendorf, R. (1989) Dieting causes menstrual irregularities in normal-weight young women through impairment of episodic luteinizing hormone secretion. *Fertil. & Steril.* 51:2:263–68.

Prior, J. C. (1982) Endocrine "conditioning" with endurance training: A preliminary review. *Can. J. Appl. Sport Sci.* 7:3:148–57.

Prior, J. C. (1985) Luteal phase defects and anovulation: Adaptive alterations occurring with conditioning exercise. *Seminars in Reproductive Endocrinology* 3:1:27–33.

Prior, J. C.; Cameron, K.; Yuen, B. H.; Thomas, J. (1982) Menstrual cycle changes with marathon training: Anovulation and short luteal phase. *Can. J. Appl. Sport Sci.* 7:3:173–77.

Prior, J. C.; Jensen, L.; Ho Yuen, B.; Higgins, H.; Browlie, L. (1982) Prolactin changes with exercise vary with breast motion: Analysis of running versus cycling. *Fertil. & Steril.* 38:2:272.

Prior. J. C., and Vigna, Y. (1985) Gonadal steroids in athletic women: Contraception, complications and performance. *Sports Medicine* 2:287–95.

Prior, J. C.; Vigna, Y.; Sciarretta, D.; Alojado, N.; Schultzer, M. (1987) Conditioning exercise decreases premenstrual symptoms: A prospective, controlled 6-month trial. *Fertil. & Steril.* 47:402–08.

Punonen, R., and Lukola, A. (1981) Binding of estrogen and progestin in the human fallopian tube. *Fertil. & Steril.* 38:610–14.

Resko, J. A.; Normal, R. L.; Niswender, D. G.; Spies, H. G. (1974) The relationship between progestins and gonadotropins during the late luteal phase of the menstrual cycle in rhesus monkeys. *Endocrinology* 94:128.

Rivier, C.; Rivier, J.; Vale, W. (1986) Stress-induced inhibition of reproductive functions: Role of endogenous corticotropin-releasing factor. *Science* 231: 607–9.

Sapolsky, R. M. (1982) The endocrine stress response and social status in the wild baboon. *Hormones & Behavior* 16:279–92.

Sapolsky, R. M., (1983) Endocrine aspects of social instability in the olive baboon. *Am. J. Primatology* 5:365–79.

Sapolsky, R. M. (1985) Stress-induced suppression of testicular function in the wild baboon: Role of glucocorticoids. *The Endocrine Society.* 116:6:1–6.

Sapolsky, R. M. (1985) Stress-induced suppression of testicular function in the wild baboon: Role of glucocorticoids. *Endocrinology* 116:2273.

Sapolsky, R. M. (1986) Stress-induced elevation of testosterone concentrations in high-ranking baboons: Role of catecholamines. *Endocrinology* 118:4:1–6.

Sapolsky, R. M. (1986) Stress, social status and reproductive physiology in free-living baboons. In *Psychobiology of Reproduction: An Evolutionary Perspective,* pp. 291–322. Ed. by D. Crews (Englewood Cliffs, N.J.: Prentice-Hall).

Sarrel, P. M., and DeCherney, A. H. (1985) Psychotherapeutic intervention for treatment of couples with secondary infertility. *Fertil. & Steril.* 43:897.

Saxena, B. N.; Poshyachinda, V.; Dusitin, N. (1976) A study of the use of intermittent serum luteinizing hormone, progesterone and oestradiol measurements for the detection of ovulation. *Brit. J. Obstet. Gynecol.* 83:660–64.

Schiavi, R. C.; Schreiner-Engel, P.; White, D.; Mandeli, J. (1988) Pituitary gonadal function during sleep in men with hypoactive sexual desire. *Psychosomatic Med.* 50:304–18.

Schweiger, U.; Laessle, R.; Pfister, H.; Hoehl, C. (1987) Diet-induced menstrual irregularities: Effects of age and weight loss. *Fertil. & Steril.* 48:5:746–51.

Schweiger, U.; Herrmann, F.; Laessle, R.; Riedel, W.; Schweiger, M.; Pirke, K.-M. (1988) Caloric intake, stress and menstrual function in athletes. *Fertil. & Steril.* 49:3:447–50.

Shangold, M. M.; Freeman, R.; Thysen, B.; Gatz, M. (1979) The relationship between long-distance running, plasma progesterone, and luteal phase length. *Fertil. & Steril.* 31:2:130–33.

Shangold, M. M., and Levine, H. S. (1982) The effect of marathon training upon menstrual function. *Amer. J. of Obstetrics & Gynecology* 143:8:862–69.

Smals, A. G. H.; Kloppenborg, P. W. C.; Benraad, T. J. (1976) Circannual cycle in plasma testosterone levels in man. *J. Clin. Endocrinol. Metab.* 42:979–82.

Stanislaw, H., and Rice, F. J. (1987) Acceleration of the menstrual cycle by intercourse. *Psychophysiology* 24:6:714–18.

Syrop, C. H., and Hammond, M. G. (1987) Diurnal variations in midluteal serum progesterone measurements. *Fertil. & Steril.* 47:1:67–70.

Tervila, L. (1958) The weight of the ovaries after stress ending in death. *Ann. Chir. Gynaecol. Fenn.* 47:232–44.

Treloar, A. E. (1974) Menarche, menopause and intervening fecundability. *Human Biology* 16:89–107.

Treloar, A. E. (1981) Menstrual cyclicity and menopause. *Maturitas* 3:249–64.

Veith, J. L.; Anderson, J.; Slade, S. A.; Thompson, P.; Laugel, G. R.; Getzlaf, S. (1984) Plasma B-endorphin, pain thresholds and anxiety levels across the human menstrual cycle. *Physiol. Behav.* 32:31–34.

Voigt, K.; Ziegler, M.; Grunert-Fuchs, M.; Bickel, U.; Fehm-Wolfsdorf, G. (1990) Hormonal responses to exhausting physical exercise: The role of predictability and controllability of the situation. *Psychoneuroendocrinology* 15:173–84.

Vuorento, T.; Hovatta, O.; Kurunmaki, H.; Ratsula, K.; Huhtaniemi, I. (1990) Measurements of salivary progesterone throughout the menstrual cycle in women suffering from unexplained infertility reveal high frequency of luteal phase defects. *Fertil. & Steril.* 54:211–16.

Wright, J.; Duchesne, C.; Sabourin, S.; Bissonnette, F.; Benoit, J.; Girard, Y. (1991) Psychosocial distress and infertility: Men and women respond differently. *Fertil. & Steril.* 55:1:100–08.

Zavos, P. M., and Goodpasture, J. C. (1989) Clinical improvements of specific seminal deficiencies via intercourse with a seminal collection device versus masturbation. *Fertil. & Steril.* 51:1:190–91.

CHAPTER 6: PHEROMONES: MALE AND FEMALE ESSENCES

Box 6-1: Martha McClintock's Discovery of Menstrual Synchrony

For examples of replications, see Graham (1980), Jarett (1984), Quadagno (1981).

There has been dissenting opinion by those who have failed to find it and also, perhaps more significantly, a paper that argued that it was the environmental influence, not the women in close proximity, that accounted for the synchrony. See Little (1989). However, as I will show in the next chapter, I believe that both environmental and pheromonal influences are at work.

Pheromones in Nature

For examples of animal studies, see Brown (1987), Brown (1989), McClintock (1983), McClintock (1984), Singh (1990).

Pheromones in Humans

For an example of an elegant paper commenting on pheromones as human primers, see Burger (1985).

The Monell Chemical Senses Center

Dr. Richard Doty was working at the Monell Chemical Senses Center when I was a graduate student, initially in psychology and later in biology, at the University of Pennsylvania. I met him there and had several interesting exchanges looking at his enormous data base of references on menstrual cycles. Subsequently Dr. Doty moved from Monell into the university to head up a chemical senses and smell center. Colleagues within biology, psychology, the newly developed center by Dr. Doty, and Monell were all working on different aspects of pheromonal research.

The Male Essence

See also Deslypere (1983), who showed that in September through December menopausal women secrete about twice the concentration of androgen as in May.

The Pheromone Findings

See Preti (1986).

The Power of Male Essence

For these studies and discussions of their biological meaning, see Cutler (1986), Cutler (1989), Cutler (1989) again.

A critic, Clyde Wilson (1988), presented a dissenting opinion on our conclusions. My response followed his in the journal. See Cutler (1988).

The Chemical Basis of Intimacy

The follicular developmental-length requirements are described in Gougeon (1982).

Putting Male and Female Essence Together

For another way of looking at the influence of sleeping with a man on menstrual cycles, see Veith (1983).

Japanese Business Encounters Feminine Science

These cultural differences between men and women have been quite eloquently described in two books on the culture. For one by a male, see Morita (1986).

And for one by a female who analyzed what it was like to be a woman in the Japan of the 1980s, see Condon (1986).

The Active Ingredient

For further chemical analysis, see also Nixon (1988).

THE REFERENCES

Brown, R.; Roser, B.; Singh, P. (1989) Class I and Class II regions of the major histocompatibility complex both contribute to individual odors in congenic inbred strains of rats. *Behav. Gen.* 19:659–74.

Brown, R.; Singh, P.; Roser, B. (1987) The major histocompatibility complex and the chemosensory recognition of individuality in rats. *Physiol. Behav.* 40:65–73.

Buckley, T. (1982) Menstruation and the power of Yurok women: Methods in cultural reconstruction. *Am. Ethnologist* 9:47–60.

Burger, J., and Gochfeld, M. (1985) A hypothesis on the role of pheromones on age of menarche. *Medical Hypotheses* 17:39–46.

Condon, Jane (1986) *A Half Step Behind: Japanese Women of the 80's* (New York; Dodd, Mead).

Cutler, W. B. (1988) Reply to Wilson. *Hormones and Behavior* 22:272–77.

Cutler, W. B. (1989) Interpersonal influences on female reproductive endocrinology. *Clin. Practice in Sexuality* 3:24–28.

Cutler, W. B. (1989) Regular sex: Endocrine benefit for women. *Medical Aspects of Human Sexuality* 23:8:52.

Cutler, W. B.; Preti, G.; Krieger, A.; Huggins, G. R.; Garcia, C. R.; Lawley, H. J. (1986) Human axillary secretions influence women's menstrual cycles: The role of donor extract from men. *Hormones and Behavior* 20:463–73.

Deslypere, J. P.; de Biscop, G.; Vermeulen, A. (1983) Seasonal variation of plasma dehydroepiandrosterone sulphate and urinary androgen excretion in postmenopausal women. *Clin. Endocrinol.* (Oxford) 18:25–30.

Goegeon, A. (1982) Rate of follicular growth in the human ovary. *Follicular Maturation and Ovulation,* pp. 155–63. Ed. by R. Rolland, E. van Hall, S. G. Hillier, K. P. McNatty, J. Schoemaker (Princeton, N.J.: Excerpta Medica).

Graham, C. A.; McGrew, W. C. (1980) Menstrual synchrony in female undergraduates living on a coeducational campus. *Psychoneuroendocrinology* 5:245–52.

Jarett, L. R. (1984) Psychosocial and biological influences on menstruation: Synchrony, cycle length and regularity. *Psychoneuroendocrinology* 9:1:21–28.

Little, B. B.; Guzick, D. S.; Malina, R. M.; Rocha, F. (1989) Environmental influences cause menstrual synchrony, not pheromones. *Am. J. Human Biology* 1:53–57.

McClintock, M. K. (1971) Menstrual synchrony and suppression. *Nature* 229:244–45.

McClintock, M. K. (1983) Pheromonal regulation of the ovarian cycle: Enhancement, suppression, and synchrony. In *Pheromones and Reproduction in Mammals,* pp. 113–49. Ed. by J. G. Vandenbergh (New York: Academic Press).

McClintock, M. K. (1984) Estrous synchrony: Modulation of ovarian cycle length by female pheromones. *Physiol. Behav.* 32:701–705.

Morita, A. (1986) *Made in Japan* (New York: Dutton).

Nixon, A.; Mallet, A. I.; Gower, D. B. (1988) Simultaneous quantification of five odorous steroids (16-androstenes) in the axillary hair of men. *J. Steroid Biochem.* 29:5:501–10.

Preti, G.; Cutler, W. B.; Christensen, C. M.; Lawley, H.; Huggins, G. R.; Garcia, C. R. (1987) Human axillary extracts: Analysis of compounds from samples which influence menstrual timing. *J. Chemical Ecology* 13:4:717–31.

Preti, G.; Cutler, W. B.; Garcia, C. R.; Huggins, G. R.; Lawley, H. J. Human axillary secretions influence women's menstrual cycles: The role of donor extract of females. *Hormones and Behavior* 20:474–82.

Quadagno, D. M.; Shubeita, H.; Deck, J.; Francoeur, D. (1981) Influence of male social contacts, exercise and all female living conditions on the menstrual cycle. *Psychoneuroendocrinology.* 6:3:239–44.

Russell, M. J.; Switz, G. M.; Thompson, K. (1980) Olfactory influences on the human menstrual cycle. *Pharmac. Biochem. Behav.* 13:737–38.

Sanders, S.; Ziemba-Davis, M. M.; Reinisch, J. M. (1987) Menstrual cycle patterns in lesbian couples. Seventh Conference of the Society for Menstrual Cycle Research, June 4–6, 1987, Ann Arbor, Mich., School of Nursing, University of Michigan.

Singh, P.; Herbert, J.; Roser, B.; Arnott, L.; Tucker, D.; Brown, R. (1990) Rearing rats in a germ-free environment eliminates their odors of individuality. *J. Chem. Ecol.* 16:1667–82.

Veith, J.; Buck, M.; Getzlaf, S.; Van Dalfsen, P.; Slade, S. (1983) Exposure to men influences the occurrence of ovulation in women. *Physiology and Behavior* 31:313–15.

Wilson, C. (1988) Letter to Editor, *Hormones and Behavior* 221:270–72.

CHAPTER 7: SENSUALITY CYCLES

Your Internal Sensual Wiring

See Allen (1978), Allen (1981).

The Life Cycle of Libido

See Smith (1985).

The Dawn of Desire

See also Alzate (1989), Clement (1989), Kinsey (1953), McConaghy (1987).

The Urge to Merge

See Jones (1990), Weizman (1987).

Hormonally Driven Desire

See Adamopoulos (1988), Bancroft (1979), Bancroft (1983), Greenblatt (1987), Hedricks (1987), Morris (1987), Persky (1982), Schreiner-Engel (1982), Schreiner-Engel (1989), Sherwin (1985) (cited in Chapter 3), Shively (1990), Stanislaw (1988), Udry (1986), Warner (1988).

To show the cycle of testosterone, see Goebelsmann (1974); but also see Leiblum (1988), who has addressed the interpersonal dynamic and its influence.

Times of day as well as weekends influence coital rate also; other sensual measures, such as olfactory sensitivity, also vary with the menstrual-cycle hormonal changes. See Doty (1981).

Likewise in the different trimesters of pregnancy, changes in sexual desire have been reported; see Bogren (1991).

A mental tonic effect of estrogen has been noted, which may help account for hormonally driven changes in sexual desire (although this is conjecture); see Utian (1972), Sherwin (1989).

Desire Can Diminish When Satiety Is Reached

For the Stanford Menopause Study 1987 reference, see Cutler (1987). Perimenopausal sexuality cited in Chapter 2, above.

The Swedish study described is that of Hallstrom (1990). The absence of a relationship is shown in Stuart (1987) and Schreiner-Engel (1989); and see also Bretschneider (1988), cited in Chapter 2 above.

Equivalent declines for men are shown in Jones (1990), White (1990), Weizman (1987).

The beneficial effects of exercise on desire in men are shown in White (1990).

Women May Be Different

Studies have shown that the physiological machinery in men and women is different. Growth-hormone responses to exercises show differences in men and women, as do the watts per kilogram of body weight during vigorous exercise and the maximum uptake of oxygen in milligrams per kilogram. See Voigt (1990), cited in Chapter 5, above.

When Libido Is Lacking

This issue is also reviewed in Chapter 4, above. See Schiavi (1988), cited in Chapter 4, above, as well as Schreiner-Engel (1986), Spector (1990).

Diseases also affect desire. For example, Prather (1988) estimates that half of diabetic women lose all capacity for orgasm. And arousal-phase dysfunctions are common after pelvic or genital cancers. See Schover (1987).

The Impact of Pelvic Surgery

The data continue to accumulate; see in Chapter 3 above, Bellerose (1989).

Experiments in monkeys showed divergent responses depending on the species. Rhesus monkeys stopped all sexual behavior after ovariectomy; see Michael (1978).

But another species of monkey, the stump-tail macaque, continued to be sexual after ovariectomy; see Baum (1978).

For an extensive review of this literature, see *Hysterectomy: Before and After*.

The Willingness to Respond

For other references, see also Townsend (1987), Swan (1981).

Figure 7-2 is redrawn from the data supplied by Rowe (1989).

The Colombian University students were studied by Alzate (1989).

Women Are Becoming More Willing

See Catania (1990), Clement (1989), Garde (1980).

The ABCs of Arousal

See, for example, Wincze (1988).

Sound Body, Sound Mind

The Italian investigator is Pancheri (1986).

A Woman's Physical Capacity to Be Aroused...

For a continuation of this work, see also Tsai (1987), Hoon (1976), Hoon (1982).

The rabbit study was authored by Batra (1985).

Physiological Arousal Is Different from Sensation and Perception

See Morrell (1984), Masters (1959).

The Arousal Pattern in Men Mirrors That of Women

See Cutler (1987) on perimenopausal sexuality, cited in Chapter 2; Davidson (1985), McCoy (1985).

Instructing Your Partner...

Another investigator provided support for the idea that men who ejaculate prematurely are more sensitive to sensory stimulation, in other words, a lower threshold of sensitivity. This suggests that they have to work harder to maintain self-control over their erections. See Strassberg (1990).

Subtle Signs of Sexual Arousal

See Gregoire (1960), who quantifies fifteen different amino acids in vaginal fluid; Masters (1966), Sullivan (1986).

The idea of a female ejaculate propelled through the urethra has undergone a tremendous amount of controversy in the sexological literature. For an interesting, balanced review of the topic, see Alzate (1986) and also Weisberg (1984).

The Simple Facts About Orgasm

Variation in intensity is reported by Garde (1980).

A man's ability to control his erections by delaying orgasm or promoting erection is limited, according to recent studies by Mahoney (1991), Chapter 5; see also other references in Chapter 5.

Once a woman manages to learn the experience of real orgasm, she tends to continue that experience; see Milan (1988).

This probably accounts for the data shown by Garde (1980) that 96 percent are capable of orgasm by the time they are forty.

What Really Is Known

Regarding the vibration principles, see Gillan (1979).

Dr. Per Lundberg, chief of Neurology at Upsala University, explained to the convening scientists at the International Academy of Sex Research the neurology that underlies this vibration principle in support of these concepts.

The Changing Clitoris

Clitoral size is related to androgen levels and tends to increase with increasing age; see Tagatz (1979) and Cutler, *The Medical Management of Menopause,* cited above.

Vaginal stimulation was studied not only by Perry but also in Colombia by Alzate, whose series of papers provides support for the idea that it is the rhythm of the stimulation rather than the exact location of it that may most profoundly influence the capacity for orgasm; see Alzate (1990), (1985), (1985), (1985), (1988), (1987), (1986), (1984).

The Thrust of the Matter

According to Fox (1971), the sperm, even with its motility to swim, requires the assistance of the uterine contractions to help propel it up into the tubes. Fox confirms that the human uterus contracts and relaxes during coitus as in other mammals and that external abdominal pressure, as when a man puts his hand on the abdomen with pressure or during the male-superior position during coitus, increases the intrauterine pressure and facilitates fertility. Krantz (1959) demonstrated the innervation with nerves of the human uterus.

"Keeping her still" is pervasive in the mammalian sexual coital relationship. One cannot fail to see, on viewing any of the mammals in copulating, that the male moves and the female is essentially still.

Fit for Orgasm

The conclusion that the greater the strength of the PC muscle, the easier it is to achieve orgasm has been supported by some in the literature and not others. Those who say yes include Graber (1979), Maly (1980), Perry (1982).

Those who were unable to find this phenomenon in their studies include Chambliss (1982), Freese (1984).

Chambliss did her study by placing women's feet in stirrups as in a gynecological exam. In this position one might not find muscle strength comparable to that of a coital position that allowed the knees to come closer together. Studies at the Athena Institute for Women's Wellness are also under way on this question, but sufficient data had not yet been gathered as of this writing.

The literature on urinary incontinence and muscle retraining is reviewed in both *Hysterectomy: Before and After* and *Menopause: A Guide for Women and the Men Who Love Them.*

Data describing the discovery of a G-spot region are described in Perry (1981), Weisberg (1984), Whipple (1985).

The role of the endogenous opiates in vaginal stimulation has peripheral and maybe transferrable support from other studies. Endorphins have been shown to increase during successful treatment for migraine-headache sufferers, suggesting that endorphins do help relieve pain; see Helm-Hylkema (1990).

Endorphins appear to be manufactured in the ovaries as well as in the central nervous system; see Comitine (1989).

Endorphins also increase when monkeys pet each other in their characteristic grooming behavior that appears to give them so much comfort; see Keverne (1989).

The uterine fluid, particularly the endometrium, appears rich in these endorphins; see Petraglia (1986).

Recently the vagina has been tested for which areas are most electrically sensitive, and the twelve-o'clock-high position has been shown to be the most sensitive; see Schultz (1989). This would support the Perry and Whipple demonstration of the location of the area with the greatest sensitivity being at the twelve-o'clock-high position, perhaps one or two inches internal from the opening into the vagina.

Does Penis Size Matter?

Regarding the *Kama Sutra,* see Burton (1962).

For Chapter 5, "The Intrinsic Innervation . . . ," see Rodin (1973), cited in Chapter 3, above.

For Chapter 3, "Comparative Anatomy . . . ," see Hafez (1973).

Figure 7-5 was redrawn from the original provided in Hafez (1973).

Working with What You've Got

Again, the studies of Fox (1971) support the abdominal-pressure value for internal pelvic pressure. Either by the weight of the man if he is lying across the woman's abdomen or with pressure from his hand placed there, the simultaneous increase in pelvic pressure that is achieved by penile and surface pressing should enhance arousal physiologically.

THE REFERENCES

Adamopoulos, D. A.; Georgiacodis, F.; Abrahamian-Michalakis, A. (1988) Effects of antiandrogen-estrogen treatment on sexual and endocrine parameters in hirsute women. *Archives of Sexual Behavior* 17:5:421–29.

Allen, T., and Adler, N. T. (1978) Localized uptake of (14) deoxyglucose by the preoptic area of female rats in response to vaginocervical stimulation. *Neuroscience Abstracts* 4.

Allen, T.; Adler, N. T.; Greenberg, J. H.; Reivich, M. (1981) Vaginal cervical stimulation selectively increases metabolic activity in the rat brain. *Science* 211:1070–72.

Alzate, H. (1985) Letter to the editor: Clarification to Perry. *J. Sex & Marital Therapy* 11:1:67–68.

Alzate, H. (1985) Vaginal eroticism and female orgasm: A current appraisal. *J. Sex & Marital Therapy* 11:4:271–84.

Alzate, H. (1985) Vaginal eroticism: A replication study. *Archives of Sexual Behavior* 14:6:529–37.

Alzate, H. (1989) Sexual behavior of unmarried Colombian University students: A follow-up. *Archives of Sexual Behavior* 18:3:239–50.

Alzate, H. (1990) Vaginal erogeneity, the "G-spot," and "female ejaculation." *J. Sex Education and Therapy* 16:2:137–40.

Alzate, H., and Hoch, Z. (1986) The "G-spot" and "female ejaculation": A current appraisal. *J. Sex & Marital Therapy* 12:3:211–20.

Alzate, H., and Hoch, Z. (1988) Letter to the editor: Reply to Zaviacic. *J. Sex & Marital Therapy* 14:4:299–301.

Alzate, H., and Londono, M. L. (1984) Vaginal erotic sensitivity. *J. Sex & Marital Therapy* 10:1:49–56.

Alzate, H., and Londono, M. L. (1987) Subjects' reactions to a sexual experimental situation. *J. of Sexual Research* 23:3:362–400.

Bancroft, J.; Davidson, D. W.; Warner, P.; Tyrer, G. (1979) Androgens and sexual behavior in women using oral contraceptives. *Clin. Endocrinol.* 12:327–40.

Bancroft, J.; Sanders, D.; Davidson, D. W.; Warner, P. (1983) Mood, sexuality hormones and the menstrual cycle. III. Sexuality and the role of androgens. *Psychosomatic Medicine* 45:509–516.

Batra, S.; Bjellin, L.; Losif, S.; Martensson, L.; Sjorgren, C. (1985) Effect of oestrogen and progesterone on the blood flowing the lower urinary tract of the rabbit. *Acta Physiol. Scand.* 123:2:191–94.

Baum, M. J.; Slob, A. K.; De Jong, F. H.; Westbroek, D. L. (1978) Persistence of sexual behavior in ovariectomized stumptail macaques following dexamethasone treatment or adrenalectomy. *Hormones and Behavior* 11:323–47.

Bogren, L. F. (1991) Changes in sexuality in women during pregnancy. *Archives of Sexual Behavior* 20:35–45.

Burgio, K. L.; Robinson, J. C.; Engel, B. T. (1986) The role of biofeedback in Kegel exercise training for stress urinary incontinence. *Am. J. Obstet. Gynecol.* 154:58–64.

Burton, Sir Richard F. (1962) *The Kama Sutra: The Classic Hindu Treatise on Love and Sexual Conduct* (New York: Dutton).

Catania, J.; Pollack, L.; McDermott, L.; Qualls, S.; Cole, L. (1990) Helpseeking behaviors of people with sexual problems. *Archives of Sexual Behavior* 19:235–49.

Chambless, D.; Stein, T.; Sultan, F.; Williams, A.; Goldstein, A.; Hazzard-Lineberger, M.; Lifshitz, P.; Kelly, L. (1982) The pubococcygeus and female orgasm: A correlational study with normal subjects. *Archives of Sexual Behavior* 11:479–90.

Clark, R., III, and Hatfield, E. (1989) Gender differences in receptivity to sexual offers. *J. Psych. & Human Sex.* 2:1:39–55.

Clement, U. (1989) Profile analysis as a method of comparing intergenerational differences in sexual behavior. *Archives of Sexual Behavior* 18:3:229–37.

Comitine, G.; Petraglia, F.; Facchinetti, F.; Monaco, M.; Volpe, A.; Genazzani, A. R. (1989) Effect of oral contraceptives or dexamethasone on plasma b-endorphin during the menstrual cycle. *Fertil. & Steril.* 51:1:46–50.

Davidson, J. M. (1985) Sexual behavior and its relationship to ovarian hormones in the menopause. *Maturitas* 7:193–201.

Doty, R. L.; Huggins, G. R.; Snyder, P. J.; Lowery, L. D. (1981) Endocrine, cardiovascular, and psychological correlates of olfactory sensitivity changes during the human menstrual cycle. *J. Comp. Physiol. Psychol.* 95:45–60.

Fox, C. A., and Fox, B. (1971) Comparative study of coital physiology. *J. Reprod. Fertil.* 24:319–36.

Freese, M. P., and Levitt, E. E. (1984) Relationships among intravaginal pressure, orgasmic function, parity factors, and urinary leakage. *Archives of Sexual Behavior* 13:3:261–68.

Garde, K., and Lunde, I. (1990) Female sexual behavior: A study in a random sample of 40-year-old women. *Maturitas* 2:225–40.

Gillian, P. and Brindley, G. S. (1979) Vaginal and pelvic floor response to sexual stimulation. *Psychophysiology* 16:471–81.

Goebelsmann, U.; Arce, H.; Thorneycroft, I. H.; Mishell, D. R. (1974) Serum testosterone concentrations in women throughout the menstrual cycle and following HCG administration. *Am. J. Obstet. Gynecol.* 19:445–52.

Graber, T., and Kline-Graber, G. (1979) Female orgasm: Role of the pubococcygeus muscle. *J. Clin. Psychol.* 40:348–51.

Greenblatt, R. B. (1987) The use of androgens in the menopause and other gynecic disorders. *Obstet. Gynecol. Clinic N.A.* 14:251–68.

Gregoire, A. T.; Lang, W. R.; Ward, K. (1960) The qualitative identification of free amino acids in human vaginal fluid. *Annals of the N.Y. Acad. Sci.* 83:185–88.

Hafez, E. S. E. (1973) The comparative anatomy of the mammalian cervix. In *The Biology of the Cervix.* Ed. by R. J. Blandau and K. Moghissi (Chicago: University of Chicago Press).

Hallstrom, T., and Samuelsson, S. (1990) Changes in women's sexual desire in middle life: The longitudinal study of women in Gothenburg. *Archives of Sexual Behavior* 19:259–67.

Hedricks, C.; Piccinino, L. J.; Udry, J. R.; Chimbira, T. H. K. (1987) Peak coital rate coincides with onset of luteinizing hormone surge. *Fertil. & Steril.* 48:2:234–38.

Heiman, J. R.; and Rowland, D. L. (1983) Affective and physiological sexual response patterns: The effects of instructions on sexually functional and dysfunctional men. *J. Psychosomatic Research* 27:2:105–16.

Helm-Hylkema, H.; Orlebeke, J.; Enting, L.; Thussen, J.; Van Ree, J. (1990) Effects of behavior therapy on migraine and plasma-endorphin in young migraine patients. *Psychoneuroendocrinology* 15:39–45.

Hoon, P. W.; Bruce, K.; Kinchloe, B. (1982) Does the menstrual cycle play a role in sexual arousal? *Psychophysiology* 19:21–26.

Hoon, P. W.; Wincze, J. P.; Hoon, E. F. (1976) Physiological assessment of sexual arousal in women. *Psychophysiology* 13:196–204.

Jones, J., and Barlow, D. (1990) Self-reported frequency of sexual urges, fantasies and masturbatory fantasies in heterosexual males and females. *Archives of Sexual Behavior* 19:269–79.

Kegel, A. H. (1951) Physiologic therapy for urinary stress incontinence. *JAMA* 146:915–17.

Kegel, A. H. (1956) Stress incontinence of urine in women: Physiologic treatment. *J. Int. Coll. of Surgeons* 25:487–99.

Keverne, E. B.; Martensz, N. D.; Tuite, B. (1989) Beta-endorphin concentrations in cerebrospinal fluid of monkeys are influenced by grooming relationships. *Psychoneuroendocrinology* 14:155–61.

Kinsey, A. C.; Pomeroy, W. B.; Martin, C. E. (1948) *Sexual Behavior in the Human Male* (Philadelphia: W. B. Saunders).

Kinsey, A. C.; Pomeroy, W. B.; Martin, C. E.; Gebhard, P. H. (1953) *Sexual Behavior in the Human Female* (Philadelphia: W. B. Saunders).

Krantz, K. E. (1960) The gross and microscopic anatomy of the human vagina. *Annals of the N.Y. Acad. Sci.* 83:89–100.

Krantz, K. E. (1959) Innervation of the human uterus. *Annals of the N.Y. Acad. Sci.* 75:770–84.

Leiblum, S.; Bachmann, G.; Kemmann, E.; Colburn, D.; Swartzman, L. (1983) Vaginal atrophy in the postmenopausal woman: The importance of sexual activity and hormones. *JAMA* 249:2195–98.

Leiblum, S. R., and Rosen, R. C. (1988) Chapter 1. Introduction: Changing perspectives on sexual desire. In *Sexual Desire Disorders,* pp. 1–17. Ed. by Leiblum and Rosen (New York: Guilford Press).

Maly, B. J. (1980) Rehabilitation principles in the care of gynecologic and obstetric patients. *Arch. Phys. Med. Rehab.* 61:78–81.

Masters, W. H. (1959) The sexual response cycle of the human female: Vaginal lubrication. *Annals of the N.Y. Acad. Sci.* 83:301–17.

Masters, W. H., and Johnson, V. (1966) *Human Sexual Response.* (Boston: Little, Brown).

McConaghy, N. (1987) Heterosexuality/homosexuality: Dichotomy or continuum. *Archives of Sexual Behavior* 16:5:411–24.

McCoy, N.; Cutler, W. B.,; Davidson, J. M. (1985) Relationships among sexual behavior, hot flashes and hormone levels in perimenopausal women. *Archives of Sexual Behavior* 14:385–94.

Michael, R. P.; Richter, M. C.; Cain, J. A.; Bonsall, R. W. (1978) Artificial menstrual cycles, behavior and the role of androgens in female rhesus monkeys. *Nature* 275:439–44.

Milan, R. J.; Kilmann, R. P.; Boland, J. P. (1988) Treatment outcome of secondary orgasmic dysfunction: A two- to six-year follow-up. *Archives of Sexual Behavior* 17:6:463–80.

Morrell, M. J.; Dixen, J. M.; Carter, C. S.; Davidson, J. M. (1984) The influence of age and cycling status on sexual arousability in women. *Am. J. Obstet. Gynecol.* 148:66–71.

Morris, N. M.; Udry, J. R.; Khan-Danwood, M. Y. (1987) Marital sex frequency and midcycle female testosterone. *Archives of Sexual Behavior* 16:1: 27–37.

Palmer, J. D.; Udry, J. R.; Morris, N. M. (1982) Diurnal and weekly but no lunar rhythms in human copulation. *Human Biology* 54:111–21.

Pancheri, P.; Falaschi, P.,; Maione Marchini, A.; Rocco, A.; Zibellini, M.; DiCesare, G. (1986) Sexually arousing stimuli, coping mechanisms and hormonal response. *New Trends in Exp. & Clin. Psychiatry* 2:221–40.

Perry, J. D., and Whipple, B. (1981) Pelvic muscle strength of female ejaculators: Evidence in support of a new theory of orgasm. *J. Sex. Res.* 17:22–39.

Perry, J. D., and Whipple, B. (1982) Vaginal myography. In *Circum-vaginal Musculature and Sexual Function,* pp. 61–73. Ed. by B. Graber (New York: S. Karger).

Persky, H., et al. (1982) The relation of plasma androgen levels to sexual behaviors and attitudes of women. *Psychosomatic Med.* 44:305–19.

Persky, H.; Lief, H.; O'Brien, C., Strauss, D. (1977) Reproductive hormone levels and sexual behaviors of young couples during the menstrual cycle. In *Progress in Sexology,* pp. 299–310. Ed. by R. Gemme and C. C. Wheeler (New York: Plenum Press).

Persky, H.; Lief, H.; Strauss, D.; Miller, W.; O'Brien, C. (1978) Plasma testosterone level and sexual behavior of couples. *Archives of Sexual Behavior* 7:157–73.

Petraglia, F.; Facchinetti, F.; M'Futa, K.; Ruspa, M.; Bonavera, J. J.; Gandolfi, F.; Genazzani, A. R. (1986) Endogenous opioid peptides in uterine fluid. *Fertil. & Steril.* 46:2:247–51.

Prather, R. C. (1988) Sexual dysfunction in the diabetic female: A review. *Archives of Sexual Behavior* 17:3:277–84.

Rowe, D.; Rodgers, J. L.; Meseck-Bushey, S. (1989) An "epidemic" model of sexual intercourse prevalences for black and white adolescents. *Social Biology* 36:127–44.

Sanders, G.; Freilicher, J.; Lightman, S. (1990) Psychological stress of exposure to uncontrollable noise increases plasma oxytocin in high emotionality women. *Psychoneuroendocrinology* 15:47–58.

Schover, L. R.; Evans, R. B.; von Eschenbach, A. D. (1987) Sexual rehabilitation in a cancer center: Diagnosis and outcome in 384 consultations. *Archives of Sexual Behavior* 16:6:445–61.

Schreiner-Engel, P., and Schiavi, R. C. (1986) Lifetime psychopathology in individuals with low sexual desire. *J. Nervous & Mental Diseases* 174:646–51.

Schreiner-Engel, P.; Schiavi, R. C.; Smith, H.; White, D. (1982) Plasma testosterone and female sexual behavior. *Proceedings of 5th World Congress of Sexology.* Ed. by Z. Hoch and H. I. Lief (Amsterdam: Excerpta Medica).

Schreiner-Engel, P.; Schiavi, R.; White, D.; Ghizzani, A. (1989) Low sexual desire in women: The role of reproductive hormones. *Hormones and Behavior* 23:221–34.

Schultz, W. C. M. W.; van de Wiel, H. B. M.; Klatter, J. A.; Sturm, B. E.;

Nauta, J. (1989) Vaginal sensitivity to electric stimuli: Theoretical and practical implications. *Archives of Sexual Behavior* 18:2:87–95.

Semmens, J. P. (1983) Letter to the editor. Reply to Wulf Utian. *JAMA* 249:195.

Semmens, J. P., and Wagner, G. (1982) Estrogen deprivation and vaginal function in postmenopausal women. *JAMA* 248:445–48.

Sherwin, B., and Gelfand, M. (1989) A prospective one-year study of estrogen and progestin in postmenopausal women: Effects on clinical symptoms and lipoprotein lipids. *Obstet. & Gynecol.* 73:759–66.

Shively, C.; Manuck, S.; Kaplan, J.; Koritnik, D. (1990) Oral contraceptive administration, interfemale relationships and sexual behavior in *Macaca fascicularis*. *Archives of Sexual Behavior* 19:101–17.

Smith, E. A.; Udry, J. R.; Morris, N. M. (1985) Pubertal development and friends: A biosocial explanation of adolescent sexual behavior. *J. Health & Soc. Beh.* 26:183–92.

Spector, I., and Carey, M. P. (1990) Incidence and prevalence of the sexual dysfunctions: A critical review of the empirical literature. *Archives of Sexual Behavior* 19:389–407.

Stanislaw, H., and Rice, F. J. (1988) Correlation between sexual desire and menstrual cycle characteristics. *Archives of Sexual Behavior* 17:6:499–508.

Strassberg, D.; Mahoney, J.; Schaugaard, M.; Hale, V. (1990) The role of anxiety in premature ejaculation: A psychophysiological model. *Archives of Sexual Behavior* 19:251–57.

Stuart, F. M.; Hammond, D. C.; Pett, M. A. (1987) Inhibited sexual desire in women. *Archives of Sexual Behavior* 16:2:91–106.

Sullivan, M. J. L., and Brender, W. (1986) Facial electromyography: A measure of affective processes during sexual arousal. *Psychophysiology* 23:182–88.

Swan, S., and Brown, W. (1981) Oral contraceptive use, sexual activity, and cervical carcinoma. *Am. J. Obstet. Gynecol.* 139:52–57.

Tagatz, G. E.; Kopher, R. A.; Nagel, T. C.; Okagaki, T. (1979) A bioassay of androgenic stimulation. *Obstet. Gynecol.* 54:562–64.

Townsend, J. M. (1987) Sex differences in sexuality among medical students: Effects of increasing socioeconomic status. *Archives of Sexual Behavior* 16:5:425–43.

Tsai, C. C., et al. (1987) Vaginal physiology in postmenopausal women: pH value, transvaginal electropotential difference and estimated blood flow. *South. Med. J.* 80:8:987–90.

Udry, J. R. (1979) Age at menarche, at first intercourse and at first pregnancy. *J. Bios. Sci.* 11:433–41.

Udry, J. R. and Morris, N. M. (1968) Distribution of coitus in the menstrual cycle. *Nature* 220:593–96.

Udry, J. R., and Morris, N. M. (1968) Effect of contraceptive pills on the distribution of sexual activity in the menstrual cycle. *Nature* 227:502.

Udry, J. R.; Talbert, L. M.; Morris, N. M. (1986) Biosocial foundations for adolescent female sexuality. *Demography* 23:217–29.

Utian, W. H. (1972) The mental tonic effect of oestrogens administered to oophorectomized females. *Afr. Med. J.* 46:1079–82.

Warner, P., and Bancroft, J. (1988) Mood, sexuality, oral contraceptives and the menstrual cycle. *J. Psychosomatic Research* 32:4–5:417–27.

Weisberg, M. (1984) Physiology of female sexual function. *Clin. Obstet. & Gynecol.* 27:697–705.

Weizman, R., and Hart, J. (1987) Sexual behavior in healthy married elderly men. *Archives of Sexual Behavior* 16:1:39–44.

Whipple, B., and Komisaruk, B. R. (1985) Elevation of pain threshold by vaginal stimulation by women. *Pain* 21:357–67.

White, J.; Case, D.; McWhirter, D.; Mattison, A. M. (1990) Enhanced sexual behavior in exercising men. *Archives of Sexual Behavior* 19:193–209.

Wincze, J. P.; Malhotra, C.; Susset, J. G.; Bansal, S.; Balko, A.; Malamud, M. (1988) A comparison of nocturnal penile tumescence during waking states in comprehensively diagnosed groups of males experiencing erectile difficulties. *Archives of Sexual Behavior* 17:4:333–48.

Wyatt, G. E.; Peters, S. D.; Guthrie, D. (1988) Kinsey revisited, part I: Comparisons of the sexual socialization and sexual behavior of white women over 33 years. *Archives of Sexual Behavior* 17:3:201–39.

Wyatt, G. E.; Peters, S. D.; Guthrie, D. (1988) Kinsey revisited, part II: Comparisons of the sexual socialization and sexual behavior of black women over 33 years. *Archives of Sexual Behavior* 17:289–332.

CHAPTER 8: COSMIC CYCLES—LUNAR EVENTS

A cyclic variation in many kinds of trouble, in time of birthing, in aldosterone and creatinine secretion, in lupus, and in self-poisonings have been described by a number of scientists; see Oswald (1981), Paaby (1988), Paaby (1989), Steinberg (1988). Even in young women before the onset of puberty, monthly cycles of gonadotropins have been reported in about 30 percent of those studied; see Hanson (1975).

The Magic of 29.5-Day Cycles

See Arey (1939), Gunn (1937), Treloar (1967) (cited in Chapter 2, above), Vollman (1977) (cited in Chapter 1, above).

Earth Time Via Moon Time

The lunar cycle as it is reflected in animal life and their reproductive rhythms has been studied in a number of animals; see, for example. Brown (1967), Brown (1953), Brown (1960), Brown (1975), Bunning (1964), Cloudsley-Thompson (1961), Franke (1985), Havenschild (1960), Naylor (1958), Naylor (1960), Robertson (1978).

Studies showing circadian (around the twenty-four-hour day) influence on the sexual dance of animals have also been studied and were principally found in lower mammals, such as hamsters. For examples of these studies, see Johnson (1986), Moore-Ede (1977), Moore-Ede (1977), Morin (1977), Reiter (1973), Reiter (1975), Reiter (1980), Ruzak (1975), Sulzman (1977), Turek (1988).

To get an idea of the ways in which the anatomical analysis has been performed, see the references cited just above regarding melatonin.

Looking for the State of the Art

I did find numerous published citations on the influence of moon cycles on nonhuman life on earth. These were cited at the beginning of this chapter.

The human studies that have consistently shown an absence of the moon-menses relationship were Abell (1979), Arrhenius (1898), Gunn (1937), Pochobradsky (1974).

Although monthly cycles of births were suggested by Menaker (1959) and Osley (1973), and were shown in Eskimo records by Ehrenkranz (1983), they were not found by Abell (1979). However, breeding monkeys at the equator were reported to show a monthly cycle in Reiter (1972). There was an early report of a monthly cycle of human serum melatonin, but it did not appear to be supported by subsequent researchers; see Wetterberg (1976).

Setting Up for Successful Revelation

If you look at the circular lunar graph, you can see that there are empty spaces on occasion where one too many sectors was present in any quarter for that particular month's need of days between quarters. I spaced out the days between the quarters and allowed the empty space to fall where it did.

The Results—Charting the Moon/Menses Relationship

For the published paper that resulted from this discovery, see Cutler (1980).

Replication in Brooklyn

One rather extraordinary study showed that the amount of light per day did not affect annual primate fertility cycles—in Atlanta, Georgia; see Plant (1974). Primates showed a testosterone cycle across the year with a September peak (just as has been described in Chapter 4) when they were kept indoors in a laboratory and never saw sunlight. The monkeys lived in a well-controlled environment, where the length of the day was defined by when the lights went on and when the lights were turned off. In spite of constant conditions thorughout the year, the animals still showed the annual cycle that had previously been attributed to the seasonal change in the length of the day.

Studies of rodents show that the length of the lights-on period controls the pattern of the cycle; see Moore-Ede (all), Sulzman (1977), Ruzak (1975).

To see how light deprivation affects breeding capacity, see Reiter (1973), Reiter (1975).

Enhancing the Analysis: Moon Orchestrating Menses

The 1987 paper is cited below as Cutler (1987). Regarding Figure 8-10, this one graph was not published in the 1987 paper cited above because the *American Journal of Human Biology* publishes only studies of natural phenomena. This one graph was an unnatural phenomenon, a result based on an external pheromone rather than what would occur in an unmanipulated cycle. This graph is shown for the first time in this book.

Synchronizing the Rhythms

For some of the studies showing the body rhythms that occur in an approximate twenty-four-hour period in humans, see Beck-Friis (1984), Brzezinski (1988), Kennaway (1986), Lewy (1980).

Disrupting the Synchrony

See Kennaway (1986), Lynch (1975), Lynch (1978).

For the disruption between cortisol and growth hormone, see Born (1988). Other disruptions are shown in Beck-Friis (1984), Lynch (1978).

The Daily Sun Cycle Influences Biorhythms

See Brzezinski (1988), Kauppila (1987), Kennaway (1986), Kivela (1988).

Cosmic Control: What Does the Sun Have to Do with It?

See Jacobsen (1987), Lewy (1987).

Cosmic Control: The Extraordinary Effect of Electromagnetic Radiation

See Wever (1974), Persinger (1974).

Even microwave radiation can affect animal behavior; see, for example, Thomas (1979).

The Internal Human Network

The system is very complex and not completely understood. For example, removing the pineal gland does not affect the daily rhythm of the pineal hormone, melatonin.

The Hormone Involved in Interpreting Cosmic Time

For the studies described in Box 8-1 see Arendt (1977) Beck-Friis (1984), Brzezinski (1987), Brzezinski (1988), Brzezinski (1989), Czeisler (1990), Czeisler (1986), Kaupilla (1987), Kennaway (1986), Lewy (1980), Lynch (1978), Ronnberg (1990).

Sun or Moon: Which Is Dominant?

The revelation of the influence of the sun and moon on sexual cycles is shown in many ways; see, for example, Pol (1989), Rodgers (1988), Sundararaj (1978), Udry (1967). For examples of the studies that show cycles at the level of the cell, see Cugini (1985), Hughes (1976), Ochiai (1980), Reppert (1988).

THE REFERENCES

Abell, G. O. (1979) The moon and the maternity ward. *The Skeptical Inquirer,* summer issue.

Arendt, J.; Wirz-Justice, A.; Bradtke, J. (1977) Annual rhythm of serum melatonin in man. *Neuroscience Letters* 7:327–30.

Arrhenius, S. (1898) The effect of constant influences upon physiological relationships. *Skandiana Arch. Phys.* 8:367.

Beck-Friis, J.; von Rosen, D.; Kjellman, B. F.; Ljungren, J. G.; Wetterbert, L. (1984) Melatonin in relation to body measures, sex, age, season and the use of drugs in patients with major affective disorders and healthy subjects. *Psychoneuroendocrinology* 9:261–77.

Born, J.; Muth, S.; Fehm, H. L. (1988) The significance of sleep onset and slow wave sleep for nocturnal release of growth hormone (GH) and cortisol. *Psychoneuroendocrinology* 13:3:233–43.

Brown, F. A., Jr. (1960) Response to pervasive geophysical factors and the biological clock problem. *Cold Spring Harbor Symp. Quant. Biol.* 125:57–71.

Brown, F. A., Jr., Fingerman, M.; Web, H. M.; Sandeen, M. I. (1953) Persistent diurnal and tidal rhythms of color change in the fiddler crab *Uca pugnax. J. Exp. Zool.* 123:29–52.

Brown, F. A., Jr., and Park, Y. H. (1967) Synodic monthly modulation of the diurnal rhythm of hamsters. *Proceedings of the Society for Experimental Biology and Medicine* 125:712.

Brown, F. A., Jr., and Park, Y. H. (1975) A persistent monthly variation in responses of planarians to lights, and its annual modulation. *I. J. of Chronobiology* 3:57–62.

Bünning, Erwin (1964) *The Physiological Clock* (Berlin, Germany: Springer Verlag OHG).

Brzezinski, A.; Seibel, M. M.; Lynch, H. J.; Deng, M. H.; Wurtman, R. J. (1987) Melatonin in human preovulatory follicular fluid. *J. Clin. Endocrinol. Metab.* 64:865–67.

Brzezinski, A., and Wurtman, R. J. (1988) The pineal gland: Its possible roles in human reproduction. *Obstet. Gynecol. Survey* 43:197–207.

Cloudsley-Thompson, J. L. (1961) *Rhythmic Activity in Animal Physiology and Behavior* (New York: Academic Press).

Cugini, P.; Lucia, P.; Tomassini, R.; Letizia, C.; Murano, G.; Scavo, D.; Tamburrano, G.; Maldonato, A.; Halberg, F.; Schramm, A.; Pusch, H. J.; Franke, H.; Shoffling, K.; Althoff, P.; Rosak, C. (1985) Temporal correlation of some endocrine circadian rhythms in elderly subjects. *Maturitas* 7:175–86.

Cutler, W. B. (1980) Lunar and menstrual phase locking. *Am. J. Obstet. Gynecol.* 137:834–39.

Cutler, W. B.; Schleidt, W. M.; Friedmann, E.; Preti, G.; Stine, R. (1987) Lunar influences on the reproductive cycle in women. *Human Biology* 59:6:959–72.

Czeisler, C. A.; Allan, J. S.; Strogatz, S. H.; Ronda, J. M.; Sanchez, R.; Rios, D.; Freitage, W. O.; Richardson, G. S.; Kronauer, R. E. (1986) Bright light resets the human circadian pacemaker independent of the timing of the sleep-wake cycle. *Science* 233:667–71.

Czeisler, C. A.; Johnson, M. P.; Duffy, J.; Brown, E. N.; Ronda, J. M.; Kronauer, R. E. (1990) Exposure to bright light and darkness to treat physiologic maladaptation to night work. *NEJM* 322:1253–1308.

Ehrenkranz, J. (1983) Seasonal breeding in humans: Birth records of the Labrador Eskimo. *Fertil. & Steril.* 40:4:485–89.

Franke, H. D. (1985) On a clocklike mechanism timing lunar rhythmic reproduction in *Typosyllis prolifera* (Polychaeta). *J. Comp. Physiol. A* 156:4:533–61.

Friedmann, E. (1981) Menstrual and lunar cycles. *Am. J. Obstet. Gynecol.* 140:350.

Gunn, D. L.; Jenkin, P. M.; Gunn, A. L. (1937) Menstrual periodicity: Statistical observations on a large sample of normal cases. *J. Obstet. Gynecol. Brit. Emp.* 44:839.

Havenschild (1960) Lunar periodicity. *Cold Spring Harbor Symp. Quant. Biol.* 25:491–97.

Hughes, A., et al. (1976) Ovarian independent fluctuations of estradiol receptor levels in mammalian tissues. *Mol. Cell. Endocrinol.* 5:379–88.

Jacobsen, F. M.; Wehr, T. A.; Sack, D. A.; James, S. P.; Rosenthal, N. E. (1987) Seasonal affective disorder: A review of the syndrome and its public health implications. *Am. J. Public Health* 77:55–60.

Johnson, C. H., and Hastings, J. W. (1986) The elusive mechanism of the circadian clock. *American Scientist* 74:29–36.

Kauppila, A.; Kivela, A.; Pakarinen, A.; Vakkuri, O. (1987) Inverse seasonal relationship between melatonin and ovarian activity in humans in a region with a strong seasonal contrast in luminosity. *J. Clin. Endocrinol. Metab.* 65:823–28.

Kennaway, D. J., and Royles, P. (1986) Circadian rhythms of 6-sulphatoxy melatonin, cortisol and electrolyte excretion at the summer and winter solstices in normal men and women. *Acta Endocrinol.* 113:450–56.

Kivela, A.; Kauppila, A.; Ylostalo, P.; Vakkuri, O.; Leppaluoto, J. (1988) Seasonal, menstrual and circadian secretions of melatonin, gonadotropins and prolactin in women. *Acta Physiol. Scand.* 132:321–27.

Lewy, A. J.; Sack, R. L.; Miller, S.; Hoban, T. M. (1987) Antidepressant and circadian phase-shifting effects of light. *Science* 235:352–54.

Lewy, A. J.; Wehr, T. A.; Goodwin, F. K.; Newsome, D. A.; Markey, S. P. (1980) Light suppresses melatonin secretion in humans. *Science* 210:1267–69.

Little, B. B.; Guzick, D. S.; Malina, R. M.; Rocha, F. (1989) Environmental influences cause menstrual synchrony, not pheromones. *Am. J. Human Biology* 1:53–57.

Lynch, H. J.; Jimerson, D. C.; Ozaki, Y.; Post, R. M.; Bunney, W. E.; Wurtman, R. J. (1978) Entrainment of rhythmic melatonin secretion from the human pineal to a 12-hour phase shift in the light dark cycle. *Life Sci.* 23:1557–64.

Lynch, H. J.; Wurtman, R. J.; Moskowitz, M. A.; Archer, M. C.; Ho, M. N. (1975) Daily rhythm in human urinary melatonin. *Science* 187:169–71.

Menaker, W., and Menaker, A. (1959) Lunar periodicity in human reproduction: A likely unit of biological time. *Am. J. Obstet. Gynecol.* 77:905.

Moore-Ede, M. C., and Sulzman, F. M. (1977) The physiological basis of circadian timekeeping in primates. *The Physiologist* 20:17.

Moore-Ede, M. C., and Herd, J. A. (1977) Renal electrolyte circadian rhythms: Independence from feeding and activity patterns. *Amer. J. Physiol.* 232:2:128.

Morgan, P. J.; Williams, I. M. (1989) Central melatonin receptors: Implications for a mode of action. *Experientia* 45:955–65.

Morin, L. P.; Fitzgerald, K. M.; Rusak, B., Zucker, I. (1977) Circadian organization and neural mediation of hamster reproductive rhythms. *Psychoneuroendocrinology* 2:73.

Naylor, E. (1958) Tidal and diurnal rhythms of locomotory activity in *Carcinus maenas* (L). *J. Exp. Biol.* 35:602–10.

Naylor, E. (1960) Locomotory rhythms in *Carcinus maenas* (L) from non-tidal conditions. *J. Exp. Biol.* 37:481–88.

Ochiai, K. (1980) Cyclic variation and distribution in the concentration of cytosol estrogen and progesterone receptors in the normal human uterus and myoma. *Acta Obstet. Gynaec. Jpn.* 32:945–52.

Osley, M.; Summerville, D.; Borst, L. B. (1973) Natality and the moon. *Am. J. Obstet. Gynecol.* 117:413.

Oswald, I.; Golland, I. M.; Adam, K. (1981) Poisonings beneath the Scottish moon. *Brit. Med. J.* 282: Letter to Editor.

Paaby, P.; Nielsen, A.; Moller-Petersen, J,; Raffn, K. (1988) Cyclical changes in endogenous overnight creatinine clearance during the third trimester of pregnancy. *Acta Med. Scan.* 223:459–68.

Paaby, P.; Nielsen, A.; Raffn, K. (1989) A monthly cycle during third trimester pregnancy in the serum concentrations of progesterone and aldosterone and the urinary excretion rate of potassium. *Acta endocrinol.* (Copenhagen) 120:636–43.

Persinger, M. A. (1974) *ELF and VLF Electromagnetic Effects* (New York: Plenum Press).

Plant, T. M.; Zumpe, D.; Sauls, M.; Michael, R. P. (1974) An annual rhythm in the plasma testosterone of adult male rhesus monkeys maintained in the laboratory. *Endocrinology* 62:403–404.

Pochobradsky, J. (1974) Independence of human menstruation on lunar phases and days of the week. *Am. J. Obstet. Gynecol.* 118:8:1136–38.

Pol, P. S.; Beuscart, R.; Leroy-Martin, B.; Hermand, E.; Jablonski, W. (1989) Circannual rhythms of sperm parameters of fertile men. *Fertil. & Steril.* 51:6:1030–33.

Reiter, R. J. (1972) The role of the pineal in reproduction. In *Reproductive Biology*. Ed. by H. Balin and S. Glass (New York: Excerpta Medica).

Reiter, R. J. (1973) Comparative physiology: Pineal gland. *Ann. Rev. Physiol.* 35:305.

Reiter, R. J.; Richardson, B. A.; Johnson, L. Y.; Ferguson, B. N. (1980) Pineal melatonin rhythm: Reduction in aging Syrian hamsters. *Science* 210:1372–73.

Reiter, R. J.; Vaughan, M. K.; Rudeen, P. K.; Vaughan, G. M.; Waring, P. (1975) Melatonin-pineal relationships in female golden hamsters. *Proc. Soc. Exp. Biol. Med.* 149:290–93.

Reppert, S. M.; Weaver, D. R.; Rivkees, S. A.; Stopa, E. G. (1988) Putative melatonin receptors in a human biological clock. *Science* 242:78–81.

Robertson, D. R. (1978) The light-dark cycle and nonlinear analysis of lunar perturbations and barometric pressure associated with the annual locomotor activity of the frog *Rana pipiens*. *Biol. Bull.* 154:302–21.

Rodgers, J. L., and Udry, J. R. (1988) The season of birth paradox. *Social Biology* 35:3–4:171–85.

Ronnberg, L.; Kauppila, A.; Leppaluoto, J.; Martikainen, H.; Vakkuri, O. (1990) Circadian and seasonal variation in human preovulatory follicular fluid melatonin concentration. *J. Clin. Endocrinol. Metab.* 71:493–96.

Ruzak, B., and Zucker, I. (1975) Biological rhythms and behavior. *Ann. Rev. Psychol.* 137–71.

Sulzman, F. M.; Fuller, C. A.; Moore-Ede, M. C. (1977) Feeding time synchronizes primate circadian rhythms. *Physiol. and Behavior* 18:775.

Sundararaj, N.; Chern, M.; Gatewood, L.; Hickman, L.; McHugh, R. (1978) Seasonal behavior of human menstrual cycles: A biometric investigation. *Human Biology* 50:15–31.

Thomas, J. R.; Burch, L. S.; Yeandle, S. S. (1979) Microwave radiation and chlordiazepoxide: Synergistic effects on fixed-interval behavior. *Science* 203:1357–58.

Turek, F. W. (1988) Manipulation of a central circadian clock regulating behavioral and endocrine rhythms with a short acting benzodiazepine used in the treatment of insomnia. *Psychoneuroendocrinology* 13:3:217–32.

Udry, J. R., and Morris, N. M. (1967) Seasonality of coitus and seasonality of birth. *Demography* 4:673–79.

Waldron, I. (1968) The mechanism of coupling of the locust flight oscillator to oscillatory inputs. *Zeitschrift für vergleichende Physiologie* 57:331–47.

Wetterberg, L.; Arendt, J.; Paunier, L.; Sizonenko, P.; van Donselaar, W.; Heyden, T. (1976) Human serum melatonin changes during the menstrual cycle. *J. Clin. Endocrinol. Metab.* 42:185–88.

Wever, R. (1974) ELF-effects on human circadian rhythms. In *ELF and VLF Electromagnetic Field Effects,* pp. 101–4. Ed. by M. A. Persinger (New York: Plenum Press).

CHAPTER 9: MONOGAMY AND RESTRAINT

Philosphy Meets Biology

Provided the estrogen source is natural, cholesterol and cardiovascular health appear to be optimized when levels of estrogen are higher. This natural estrogen may be manufactured either by the body or by drugs. However, the synthetic estrogens applied in some prescription drugs do not carry this same benefit. For a review citing twenty to thirty of the studies see *Menopause: A Guide for Women and the Men Who Love Them,* listed in Foreword above.

The relation between the luteal phase length and breast disease has been described by Sitruk-Ware (1977), see Chapter 1, Mauvais-Jarvis (1985), Chapter 1.

For the relationship of hormones to breast cancer, see Cutler (1990), (1990).

For a review of the relationship between hormones, and neurophysiology and immunology see Saphier (1989).

For a review of the uterine fibroid tumor literature, see *Hysterectomy Before and After,* cited in Foreword above.

Sexual Intercourse During Menses May Abuse the Body

For a description of bleeding changes at the menopause, see *Menopause: A Guide for Women and the Men Who Love Them.*

For a description of the impact of heavy bleeding on hysterectomy, see *Hysterectomy Before and After.*

Sex Can Be Dangerous to Your Well-Being

For the problems men have with controlling their own arousal, see studies of Mahoney (1991), cited above in Chapter 5. See also Lovelace (1980)

THE REFERENCES

Cutler, W. B. (1990) Hormone replacement therapy and the risk of breast cancer. *Infec. Surg.* 9:10–24.

Cutler, W. B., and Genovese, E. (1990) Hormones and breast cancer. In response to Berkvist et al. *NEJM* 322:3:202–03.

Fadem, B. H.; Barfield, R. J.; Whalen, R. E. (1979) Dose-response and time-response relationships between progesterone and the display of patterns of receptive and proceptive behavior in the female rat. *Hormones and Behavior* 13:40–48.

Fadem, B. H., and Barfield, R. J. (1982) Differences between male and female rats in the temporal patterning of copulation. *Behavioral and Neural Biology* 36:411–15.

Hatfield, E., and Rapson, R. (1990) Emotions: A trinity. In *Emotions and the Family,* Ed. by E. A. Bleckman (Hillsdale, N.J.: Lawrence Erlbaum Assocs., Inc.).

Lovelace, L., and McGrady, M. (1980) *Ordeal* (Secaucus, N.J.: Citadel Press).

Pittman, F. (1989) *Private Lies* (New York: Norton).

Prior (1990) See Chapter 1 for reference.

Saphier, D. (1989) Neurophysiological and endocrine consequences of immune activity. *Psychoneuroendocrinology* 14:1, 2:64–87.

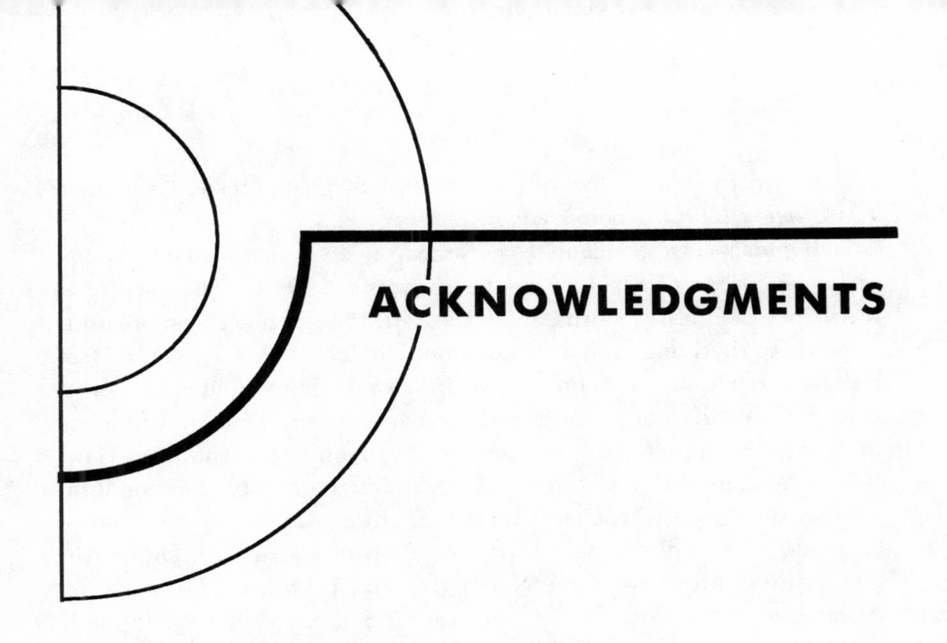

ACKNOWLEDGMENTS

I would like to acknowledge and thank individuals who directly contributed to *Love Cycles*. First, Celso Ramón Garcia, M.D., Professor of Gynecology at the University of Pennsylvania School of Medicine, for his many years as able mentor and critic. He read the entire manuscript critically. I thank him for intellectually interacting with uncompromising rigor to my early ideas and contributing those of his own.

Eliot Stellar, Ph.D., Professor of Anatomy at the University of Pennsylvania, has my gratitude for his scholarly guidance from 1974 to 1979 and his strenuous encouragement, which directed my focus on these studies.

Thanks also to Alan Epstein, M.D., Professor of Biology at the University of Pennsylvania, for his influential guidance on the psychoneuroendocrinology perspective.

The idea for the book had been germinating over ten years. I offer posthumous gratitude to Peter Livingston, my literary agent, who helped me define the direction of the work. Loretta Barrett suggested a separate chapter on fertility cycles. Her enthusiasm and focus were significant.

Stephanie Young, the Health and Fitness Editor of *Glamour* magazine, contributed by editing five of the chapters. Stephanie made *Love Cycles: The Science of Intimacy* more readable.

Individuals correspondents have earned my appreciation. Heli Alzate, M.D., Professor of Sexology at the Universidad De Caldas, Faculty of Medicine, Colombia, South America, for his correspondence on orgasmic response to rhythmic digital stimulation; Dr. John Bancroft, at the Medical Research Council Reproductive Biology Unit in Edinburgh, Scotland, for his correspondence on cyclic variation in sexual interest in mood in women on and off oral contraception; Dr. John F. Cattanach, in Hawthorn, Australia, for his correspondence on the post–tubal ligation difficulties that many women experience; Evan Cutler, Fort Collins, Colorado, for feedback on Chapter 4; Barbara Fadem, Ph.D., at the University of Medicine in the New Jersey Medical School, for her correspondence on her work on the "rape cage"; Suzanne Galloway, formerly Administrative Director of Athena Institute, for her input—as a newly graduated college student she encountered many of these ideas for the first time; Elizabeth Genovese, M.D., at Graduate Hospital, Philadelphia, for her reading as a specialist in internal medicine and comments on the infertility chapter; Julia Heiman, Ph.D., at Harborview Community Mental Health Center in Seattle, Washington, for supplying some of her original studies that I was missing and that are included in sequence in the book; Maryl Hitchings in Flourtown, Pennsylvania, for her readings of the early versions of the first several chapters and critical comments thereon; Susan Kerschensteiner in Wayne, Pennsylvania, for her very valuable feedback on Chapter 5; Dr. Per Lundberg, M.D., Department of Neurology, Uppsala University, Sweden, for educating me about the effects of drugs on libido and potency as well as our discussions on uterine sensual response as mediated by neurophysiology; Lyla McCandless, Stockton, New Jersey, for her very helpful comments on love and permanence as she read a number of early versions of the chapters; Naomi Morris, M.D., M.P.H., at the University of Illinois Community Health Sciences School of Public Health, for supplying me with the works she and Richard Udry have been publishing on teenage sexuality and the cyclic variation in female sexuality; Jerilyn Prior, M.D., Department of Medicine, University of British Columbia, for several years of correspondence about her work in endocrine response to fitness training in women; Dr. David Rowland at Erasmus University in Rotterdam for correspondence on his work on erection response and sensitivities; Dr. June Reinisch, Director of the Kinsey Institute in Bloomington, Indiana, for correspondence relating to prenatal hormone effects; Dr. Raul Schiavi, Director of the Human Sexuality Program at the Mt. Sinai Medical

Center in New York, for critically reviewing the chapter on male sexuality; Dr. Mona Shangold in Philadelphia for supplying her studies on sports and endocrine fitness in women athletes; Louise Smith Shepherd at SmithKline Beecham for exposing me to a study on light-dark-phased artificial lights for depression in human biological rhythms; Sue Silverton, M.D., Ph.D., at the University of Pennsylvania in Philadelphia, for critically reading the entire manuscript and offering numerous subtle and powerful suggestions; Donald Strassberg, Ph.D., at the University of Utah, Department of Psychology, for his correspondence on penile sensitivity experiments (with Mahoney) described in Chapter 5; and Peter Whybrow, M.D., Chairman of the Department of Psychiatry at the University of Pennsylvania, for his critical reading of Chapter 5 and his suggestion that I explore the literature on melatonin in order to address its role in seasonal sexual rhythms. All of these individuals reacted to my ideas and suggested their own, substantially expanding the scope herein.

I am especially grateful to Pamela Sinkler Todd for the beautiful art and her perspicacious renditions of anatomy graphs and tabulated data that appear as the visuals for *Love Cycles.*

Grateful appreciation to Thomas Quay, Esq., for his ongoing support, critical reading of every chapter, and helpful feedback on the entire work.

At Villard/Random House those who worked on this manuscript. Particularly Diane Reverand, the publisher of Villard, for her vision in selecting my proposal and her exquisite work editing the manuscript; Daia Gerson for her fine work in copy editing the manuscript; and the others on staff who though unnamed have contributed. Thanks also to the staff at Athena Institute: Kate Felmet, Kate Paffett, Jennifer Almquist, and Paul Dubbeling for their assistance in putting the manuscript together.

INDEX

Numerals in *italics* indicate illustrations.